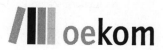

The focal topic "From Knowledge to Action – New Paths towards Sustainable Consumption"
has been funded through the "Social-ecological Research Programme" (SÖF)
by the German Federal Ministry of Education and Research (BMBF).

SPONSORED BY THE

This book was produced in a climate-neutral way. Preventing, reducing and
compensating for CO_2 emissions is a fundamental principle of oekom.
The publisher compensates for unavoidable emissions by investing
in a Gold Standard project. For more information, see www.oekom.de

Bibliographic information concerning the German National Library:
The German National Library has recorded this publication
in the German National Bibliography:
Detailed bibliographic data are available online at http://dnb.d-nb.de.

The original German version was published in 2011 under the title
"Wesen und Wege nachhaltigen Konsums. Ergebnisse aus dem Themenschwerpunkt
'Vom Wissen zum Handeln – Neue Wege zum nachhaltigen Konsum'"
by oekom verlag, München

© 2012 oekom verlag, München
Gesellschaft für ökologische Kommunikation mbH
Waltherstraße 29, 80337 München, Germany

Cover design: oekom verlag
Cover illustration: Giuseppe Porzani/fotolia.com
Layout and Typeset by: Reihs Satzstudio, Lohmar

Printing: Digital Print Group, Nürnberg

This book was printed on FSC®-certified recycling paper.

All rights reserved.
ISBN 978-3-86581-302-2

Rico Defila, Antonietta Di Giulio,
Ruth Kaufmann-Hayoz (eds.)

The Nature of Sustainable Consumption and How to Achieve it

Results from the Focal Topic
»From Knowledge to Action – New Paths towards Sustainable Consumption«

*Translated from German by
Corina Holzherr, John Williams and Trevor Pettit*

Contents

Angelika Zahrnt
Preface …… 9

Rico Defila, Antonietta Di Giulio, Ruth Kaufmann-Hayoz
Introduction …… 11

Part 1 – The synthesis framework

Rico Defila, Antonietta Di Giulio, Ruth Kaufmann-Hayoz, Markus Winkelmann
1 A landscape of research around sustainability and consumption …… 23

Antonietta Di Giulio, Bettina Brohmann, Jens Clausen, Rico Defila, Doris Fuchs, Ruth Kaufmann-Hayoz, Andreas Koch
2 Needs and consumption – a conceptual system and its meaning in the context of sustainability …… 45

Daniel Fischer, Gerd Michelsen, Birgit Blättel-Mink, Antonietta Di Giulio
3 Sustainable consumption: how to evaluate sustainability in consumption acts …… 67

Ruth Kaufmann-Hayoz, Sebastian Bamberg, Rico Defila, Christian Dehmel, Antonietta Di Giulio, Melanie Jaeger-Erben, Ellen Matthies, Georg Sunderer, Stefan Zundel
4 Theoretical perspectives on consumer behaviour – attempt at establishing an order to the theories …… 81

Ruth Kaufmann-Hayoz, Bettina Brohmann, Rico Defila, Antonietta Di Giulio, Elisa Dunkelberg, Lorenz Erdmann, Doris Fuchs, Sebastian Gölz, Andreas Homburg, Ellen Matthies, Malte Nachreiner, Kerstin Tews, Julika Weiß
5 Societal steering of consumption towards sustainability …… 113

Melanie Jaeger-Erben, Martina Schäfer, Dirk Dalichau, Christian Dehmel, Konrad Götz, Daniel Fischer, Andreas Homburg, Marlen Schulz, Stefan Zundel
6 Using 'mixed methods' in sustainable consumption research: approaches, challenges and added value …… 143

Part 2 – Findings from the project groups

A Status of sustainability in investment decisions

Julika Weiß, Immanuel Stieß, Stefan Zundel
1 Motives for and barriers to energy-efficient refurbishment of residential dwellings ... 165

Joachim Schleich, Bradford F. Mills
2 Determinants and distributional implications in the purchase of energy-efficient household appliances ... 181

B Changing everyday consumption patterns

Martina Schäfer, Melanie Jaeger-Erben
3 Life events as windows of opportunity for changing towards sustainable consumption patterns? The change in everyday routines in life-course transitions ... 195

Ellen Matthies, Dirk Thomas
4 Sustainability-related routines in the workplace – prerequisites for successful change ... 211

C Social embedding of consumer behaviour

Matthias Barth, Daniel Fischer, Gerd Michelsen, Horst Rode
5 Schools and their 'culture of consumption': a context for consumer learning ... 229

Konrad Götz, Wolfgang Glatzer, Sebastian Gölz
6 Household production and electricity consumption – possibilities for energy savings in private households ... 245

Melanie Jaeger-Erben, Ursula Offenberger, Julia Nentwich,
Martina Schäfer, Ines Weller

7 Gender in the focal topic "From Knowledge to Action – New Paths towards Sustainable Consumption": findings and perspectives ... 263

Sophia Alcántara, Sandra Wassermann, Marlen Schulz

8 Is "eco-stress" associated with sustainable heat consumption? ... 277

Ursula Offenberger, Julia Nentwich

9 Socio-cultural meanings around heat energy consumption in private households ... 291

D Consumers in new roles

Cordula Kropp, Gerald Beck

10 How open is open innovation?
User roles and barriers to implementation ... 309

Birgit Blättel-Mink, Jens Clausen, Dirk Dalichau

11 Changing consumer roles and opportunities for sustainable consumption in online second-hand trading: the case of eBay ... 323

Ulf Schrader, Frank-Martin Belz

12 Involving users in sustainability innovations ... 335

E Design and efficacy of societal steering

Andreas Koch, Daniel Zech

13 Impact analysis of heat consumption – user behaviour and the consumption of heat energy ... 353

Georg Sunderer, Konrad Götz, Sebastian Gölz

14 The evaluation of feedback instruments in the context of electricity consumption ... 367

Andreas Klesse, Joachim Müller, Ralf-Dieter Person
15 Achieving and measuring energy savings through behavioural changes: the challenge of measurability in the actual operation of university buildings ... 383

Bettina Brohmann, Veit Bürger, Christian Dehmel, Doris Fuchs, Ulrich Hamenstädt, Dörthe Krömker, Volker Schneider, Kerstin Tews
16 Sustainable electricity consumption in German households – framework conditions for political interventions ... 399

Bettina Brohmann, Christian Dehmel, Doris Fuchs, Wilma Mert, Anna Schreuer, Kerstin Tews
17 Bonus schemes and progressive electricity tariffs as instruments to promote sustainable electricity consumption in private households ... 411

Appendix

Profiles of the project groups ... 423
Authors ... 447

Angelika Zahrnt

Preface

In industrialised countries consumption is the engine of economic growth. Its ecological consequences, however, are waste, polluting emissions, resource depletion and the destruction of the natural environment. Having already exceeded the limits of our planet's ecological resilience, economic growth in the developing and emerging countries to match that of industrialised nations would greatly accelerate climate change and species extinction. Furthermore, its negative social and cultural consequences such as exploitation, injustice, substandard working conditions and the loss of cultural diversity are unacceptable. In 1992, at the Earth Summit in Rio, these insights led to the demand that Western industrialised countries develop consumption and production models that would be compatible with sustainable development, i.e. they would have to contribute towards a world where, today and in the future, all humans could satisfy their (basic) needs. This commitment was reiterated at the UN Conference in Johannesburg in 2002, where nations were asked to draw up action plans for sustainable consumption and production. Twenty years after Rio, most countries, including Germany, have not met this requirement.

Nonetheless, some progress has been made in the area of sustainable consumption. New energy-efficient products have been developed, labels for organic and fair-trade foods have been introduced; new ideas for replacing products with services have matured into successful business projects, and old ideas such as multiple and shared use of commodities have been resurrected.

However laudable each of these individual developments may be, so far, neither environmental impact nor other undesirable consequences of prevailing consumption patterns have noticeably decreased – and this in spite of an ever-increasing awareness, growing concern and a pressing urgency for action.

The focal topic of the SÖF (Social-ecological Research Programme) "From Knowledge to Action – New Paths towards Sustainable Consumption" attempts to address this discrepancy between knowledge and action, i.e. why awareness and concern for the situation at hand does not translate into everyday routines. Whether it is an investment decision in the refurbishment of our home, turning the computer on in our office or deciding what we buy on our weekly shopping trip – we are constantly faced with questions about sustainability, and nearly every act of consumption entails some use of energy. At the same time, climate change is now our most pressing social challenge, and the discrepancy between awareness of the problem with its potential solutions and a certain reluctance to act remains palpable. Therefore, the SÖF's focal topic has particularly concentrated on how this discrepancy between thought and action could be reduced, i.e. what conditions favour widespread and fast adoption of sustainable consumption and what circumstances act as barriers.

The upshot of these investigations was a profusion of interesting results relating to a range of issues and projects. The findings should prove especially useful for teachers, lecturers and researchers who are concerned with sustainable consumption, as well as for people at the interface between research and practice who make it their mission to promote sustainable consumption. These findings are also interesting because, in a complex field such as consumption, where research spans the natural and the social sciences, interdisciplinarity can prove invaluable. A transdisciplinary approach, bringing together research and practice, can help track the complex interaction between the many players in this field.

As well as presenting the findings of each research group, this volume also contains the results of the common synthesis development. Here, general and more abstract issues within sustainable consumption have been addressed. What is sustainable consumption and how can it be evaluated? What are the options for society to control consumption and how effective are they? The answers to these questions are presented in condensed form and have been complemented by reflections on methodological issues.

I would like this volume to lead to the dissemination of the results and insights of the research projects and the common synthesis development (moderated by the accompanying research project), and to help boost sustainable consumption in our society. I also hope that consumer research will widen its remit to include not only questions of how consumption can contribute to sustainable development, but that attention will be paid to conscious avoidance of consumption. A further challenge for the research community is the conflict between, on the one hand, economic policy-making, where consumption is chiefly an engine of growth, and, on the other hand, the demand implicit in sustainability that consumption patterns in industrialised countries should be compatible with global well-being.

Rico Defila, Antonietta Di Giulio, Ruth Kaufmann-Hayoz

Introduction

SÖF and the focal topic of sustainable consumption

This book is a product of the focal topic "From Knowledge to Action – New Paths towards Sustainable Consumption", funded by the German Federal Ministry of Education and Research (BMBF) as part of the Social-ecological Research Programme (SÖF).[1] Social-ecological research endeavours to develop strategic knowledge for implementing sustainable development and for dealing with specific sustainability issues. The choice of the topic *consumption* by the BMBF corresponds to a central field of societal action within sustainable development. As early as 1992, at the United Nations Conference on Environment and Development in Rio de Janeiro, Agenda 21 called for a targeted examination of non-sustainable production methods and consumption patterns. This issue was also of central importance at the World Summit for Sustainable Development in Johannesburg in 2002, where a ten-year framework of programmes for sustainable consumption and production was adopted (Marrakech Process).

Consumption is an extremely complex social phenomenon. Not only is it crucial to national and international economic development, but it has socio-cultural as well as ethical/moral dimensions; it also affects non-human nature. As people engage in consumer transactions and make consumption decisions on a daily basis, the demand is that their actions should be compatible with sustainable development. Yet, the economic, environmental and social aspects involved in individual consumer behaviour are so diverse and complex that insight into the necessity for change is not readily followed by corresponding individual or collective action.

This is where the focal topic "From Knowledge to Action – New Paths towards Sustainable Consumption" comes in. Research has been conducted for some time into conditions that hinder or favour (an increase in) sustainable consumer behaviour. However, the researchers in this area come from very diverse disciplines, professional communities and networks. Thus, their publications are quite widely scattered, ren-

---

1 http://www.sozial-oekologische-forschung.org/en/947.php [05.02.2012]

dering an integration of the available knowledge more difficult. Additionally, certain fundamental and normative aspects have so far only been rudimentarily addressed – among these, a more precise (and scientifically supported) definition of what sustainable consumption can mean in concrete terms and how individual consumption can be evaluated in terms of sustainability. It is against this background that the focal topic, first announced in 2006, aims to generate new knowledge; knowledge that is relevant for both action and guidance (*Orientierungs- und Handlungswissen*) and which different groups can use to stimulate and support individual sustainable consumer behaviour.

It is characteristic of "sustainability science" to investigate societal issues in order to generate 'useful' knowledge that can help shape sustainable development. Sustainability science has for some years become increasingly articulated in the international discourse. The Social-ecological Research Programme can be seen as part of sustainability science and is one of the earliest and most substantial funding initiatives of this new kind of research. Such research requires collaboration between the natural sciences, engineering sciences and the social sciences, as well as the involvement of social actors, e.g. consumer organisations, other non-governmental organisations, local authorities and businesses in the research process. In other words, such research is of an inter- and transdisciplinary nature. This also holds for the focal topic of sustainable consumption. Since 2008, ten project groups with a total of 28 sub-projects have been funded (refer to the profiles of the individual project groups in the appendix for information on participating partners in research and in practice, and for their research questions, aims, key results and publications).

In terms of content, a wide range of consumer actions is covered by the ten project groups: considered decisions as well as everyday routines; analysis and reconstruction of the social meaning of consumer behaviours as well as concrete interventions capable of bringing about change; investigations into the design and impact of policy instruments; questions on how to raise awareness of and encourage competence in sustainable consumer behaviour. Several project groups within the focal topic are engaged with those aspects of private consumption that involve the use of energy (Change, ENEF-Haus, Intelliekon, Seco@home, Transpose, Heat Energy). They focus on everyday routines, as well as the decisions involved in the purchase of appliances and the refurbishment of homes. Here, the following issues are being explored from psychological, sociological, economic and political science perspectives: conditions that either facilitate or constrain sustainability actions; different ways of exerting influence within these areas of consumption; effects and potentials of these interventions (in collaboration with the natural and engineering sciences). A second key area in the focal topic is "social innovations" (Consumer/Prosumer, User Integration, BINK,

LifeEvents). Topics include online second-hand trading, user involvement in product development, innovations in educational institutions, and the endeavour to better align communication activities with the realities of the target audiences' everyday life. Again, these are being investigated from very diverse disciplinary perspectives and in close collaboration with field partners. Overall, it has been clearly shown that it is unhelpful for research into sustainable consumption to narrow investigation to the purchase of certain products. Instead, it makes more sense to view consumer behaviour in terms of acts of selection, acquisition, usage (or consumption), as well as disposal or passing on of consumer goods (products, services, infrastructures).

What is sustainable consumption?

The topic of sustainable consumption is to be understood in the context of the discussion around sustainable development. Sustainable development as defined by the United Nations means that the global, regional and national development of human society has to be guided by the overarching principle that current and future (basic) needs of all human beings should be satisfied and that, concurrently, all human beings should be in a position to lead a good life. Thus, 'sustainable consumption' means that goods are acquired, used and disposed of in such a fashion that all humans, now and in the future, are able to satisfy their (basic) needs and that their desire for a good life can be fulfilled.

However, this general description of what sustainability means in terms of consumption requires further elaboration and contextualisation. Otherwise, it will neither lend itself as a suitable object of research, nor will it be capable of guiding practical and political action. Therefore, each project group had to decide what sustainable consumption meant for their specific field of action. Their choices were guided by proposals from the academic and/or political discourse related to their particular research topic. Some of the energy projects related to political imperatives, i.e. concrete, negotiated sustainability targets for Germany for the next few years. Others were guided by what is 'generally' regarded as 'sustainable' (i.e. published guidelines, labels etc.). Finally, some project groups saw it as crucial that their definition of sustainable consumption was linked to how their practice partners understood the term. None of the project groups saw it as their central aim to draw up a scientific definition of sustainable consumption.

Nonetheless, as part of the common synthesis development across the project groups, a joint effort was made towards a closer definition of sustainability in the context of consumption. This process and the results reached so far will be explained below.

Accompanying research and synthesis process

A broad and heterogeneous research programme with inter- and transdisciplinary dimensions such as the focal topic of sustainable consumption calls for a very specific support programme. Expected and desired outcomes were as follows (as stated in the research programme): an overall synthesis; the linking of the project activities; an exchange of experience at both national and international level; a transfer of the project outcomes to major fields of societal action and policy making. Given these expectations, the BMBF decided to complement the ten research projects with a further accompanying research project. Its remit was threefold. (1) It should generate content-related findings across the project groups. In particular, it should make available guidance and practical knowledge for managing the transition towards sustainable consumption patterns. It should also generate new knowledge relating to inter- and transdisciplinary research processes. (2) It should support the thematic project groups in their tasks, particularly in terms of synergies and maximising practical relevance of their results and products. (3) It should accompany and support the dissemination to practice of the focal topic's results (see profile of the accompanying research project in the appendix).

In line with these tasks the accompanying research project has several aims, one of which is the development of a *synthesis*: the results of the individual projects are integrated into a suitable synthesis framework and communicated to the scientific community. Against this background the practical relevance of the results, together with key attempts at solutions, are identified and put at the disposal of corresponding groups of actors, in a suitable form.

In terms of the synthesis activities at the level of the focal topic, the accompanying research project adopts a "content-rich moderation" strategy, i.e. it moderates the synthesis procedures and additionally takes care of certain steps in the integration process.[2] This cooperation between accompanying research project and project groups gives rise to a range of synthesis products aimed at different target audiences. One such product is this volume, which is intended for an interdisciplinary research community concerned with sustainable consumption. The overall concept of this publication has been developed by all the project groups. The joint work and the major decision processes necessary for this publication took place, to a large extent, at several events specifically hosted for researchers across the projects. The events were planned, moderated

---

2 For the term content-rich moderation, refer to Defila R., Di Giulio A., Scheuermann M. (2006): Forschungsverbundmanagement. Handbuch zur Gestaltung inter- und transdisziplinärer Projekte. Zürich: vdf Hochschulverlag AG, p. 126.

and followed up by the accompanying research project. The discussion on the possible content of a synthesis within the focal topic was initiated at a workshop in November 2009. The basis for this initial discussion was the *'Landscape of Sustainable Consumption Research'*, which had previously been prepared by the accompanying research group and was further developed at the event itself. The project groups started off discussing what they saw as the most topical issues in this area of research – both within disciplines and on an inter- and transdisciplinary level. The project groups also attempted to situate their work within this 'research landscape'. Secondly, the discussion focused on the function and possible elements of the *synthesis framework*.

The synthesis framework represents a common frame of reference. All project groups can refer to it and their insights and experiences can flow into it. It does not represent a theory which all participants have to agree upon, and should thus not be regarded as 'the lowest common denominator'. Instead, it is to be understood as a content framework where the different approaches and results of the project groups can be located and where they can complement each other. The framework is formed by topics, terms and approaches that play a central role in the research area of sustainable consumption and which are of relevance, or could be of relevance, to several or all project groups. It also contains overarching perspectives (e.g. terminologies or meta-theories). A joint development of these different elements of the content gives the participants the opportunity to agree on shared approaches and to identify and explicate divergences.

Further development of the synthesis framework was continued at two subsequent meetings (May and November 2010). In parallel, the project groups were given the opportunity to comment on a proposal from the accompanying research project regarding the content of this volume and to come forward with suggestions as to how they could individually contribute. Proposals included the co-authoring of Part 1 (the synthesis framework) and suggestions for the presentation of the results reached by the project groups (Part 2).[3]

---

3 The synthesis process also involved a joint discussion of all texts contained in the volume as part of extensive quality assurance procedures within the focal topic. The results of the internal and external review were brought together at a meeting of the authors in June 2011.

Figure 1: Networking method "Postillon d'Amour"
(seminar on linking project activities 2009)[4]

The most frequently used procedures in the synthesis process correspond to the "project management" and "group" types of procedure as introduced by Rossini and Porter and further developed by Krott.[5] What is meant by "project management" is that one single participant or a small group took the lead in developing concepts and draft texts; these texts then served as a basis for the ensuing discussion in a wider setting. The "group" type of procedure was used in the synthesis seminars, where participants collectively developed classification systems and wordings for the common texts.

---

4 For method refer to Defila et al. 2006, pp. 102 ff.

5 For these types of synthesis development procedures see Defila et al. 2006, pp. 124 ff.

The structure of the book

The synthesis development that extended over more than one year is reflected in the structure of the book. *Part 1* contains the elements of the synthesis framework that were jointly identified and considered in several rounds of discussion. The first contribution is the *Landscape of Sustainable Consumption Research* (Defila et al.). It proposes a structure for ordering the issues that are currently being studied in a heterogeneous field of widely scattered scientific communities, and locates within that structure the main questions treated in the focal topic. The contributions of Di Giulio et al. (2) and Fischer et al. (3) represent fundamental *conceptual and normative analyses and clarifications*. Here, key questions include: which concept of need is compatible with the notion of sustainability on the one hand and perceptions of the good life on the other? What do we understand by 'sustainable consumption'? How can individual consumer behaviour be evaluated in terms of sustainability? The contributions by Kaufmann-Hayoz/Bamberg et al. (4) and Kaufmann-Hayoz/Brohmann et al. (5) deal with *theoretical backgrounds and approaches* that play an important role in research on sustainable consumption. Contribution 4 focuses on individual consumption as action; it classifies phenomena of individual consumer behaviour and its socio-cultural context; it attempts to match currently widespread theories of action with the various consumer behaviour phenomena; it shows which phenomena are explained by existing theories and identifies where there might still be gaps in the theory. Contribution 5 discusses questions and challenges around the range of instruments that can be used for steering individual consumption towards greater sustainability; it also evaluates the effectiveness of these instruments. Finally, Jaeger-Erben/Schäfer et al. (6) discuss the options and the specific benefits of combining quantitative and qualitative social-scientific methods in research on sustainable consumption. This discussion takes place against the backdrop of the experiences gained within the focal topic.

The results of the synthesis development at the level of the focal topic are not intended as conclusive answers to questions such as what should be understood by sustainable consumption, or which instrument is best suited to directing consumer behaviour. Instead, classification systems and structures have been developed that help recognise and interrelate different perspectives amid an abundance of scholarly approaches. They can also help locate and contextualise relevant knowledge.

In contrast, at the level of the project groups, knowledge was gained that was relevant to specific fields of action. Some of this allowed researchers to make specific recommendations. *Part 2* of the book presents a selection of results from the project groups; they offer an insight into the variety of subjects and perspectives characterising the focal topic. This part of the book consists of several sections.

The two contributions in *Section A* provide the reader with a better understanding of the reasons and motives behind sustainable or non-sustainable investment decisions. They focus on 'reflected consumer actions', in line with the classification of phenomena by Kaufmann-Hayoz/Bamberg et al. in *Part 1*. Weiß et al. (1) deal with decisions relating to the refurbishment of private homes, and Schleich and Mills (2) look at the purchase of energy-efficient household appliances.

In contrast to the 'reflected consumer actions' of *Section A*, *Section B* deals with habitual consumption behaviour that forms part of our daily routines. The contributions shed light on the role of regular habits in private and professional life and identify the potentials as well as the limits of measures that are designed to change our habits. Schaefer and Jaeger-Erben (3) look into the question whether major changes in the course of people's lives represent opportunities to modify their consumer behaviour towards greater sustainability. Matthies and Thomas (4) report on the potentials and the impacts of psychologically well-founded interventions designed to curb energy consumption in the workplace.

Section C deals with various aspects of the socio-cultural embeddedness of consumer behaviour and with the role of social identities. The articles contribute to a better understanding of the interplay between social structures and individual consumer behaviour; they endeavour to paint a larger picture, where consumer behaviour is embedded in organisational and socio-cultural contexts. These contexts should be taken into account when it comes to interventions aimed at encouraging sustainable consumption. Ignoring them could reduce the effectiveness of such interventions and/or lead to undesirable side-effects. Jaeger-Erben/Offenberger et al. (7), Alcántara et al. (8) and Offenberger and Nentwich (9) focus on gender issues. Barth et al. (5) shed light on organisational consumer culture in educational organisations and Götz et al. (6) consider the role of household production.

Section D is dedicated to the more recent blurring of the strict division between consumption and production, which is associated with the changing roles of consumers. These phenomena are discussed in the light of sustainability. Blättel-Mink et al. (11) examine if and how online trading in second-hand products can change individual consumer behaviour and what sustainability potential is linked to this activity. Kropp and Beck (10) and Schrader and Belz (12) cast a critical eye on consumer participation in corporate product development. They ask if such practices can or could promote the development of 'sustainable' products and services.

Section E looks at the impacts of changes in consumer behaviour, as well as at the efficacy of measures designed to direct consumer behaviour towards sustainability. The articles show how the challenges of causality and measurability can be met. Koch and Zech (13) show how user behaviour can influence the heat energy demand of buildings and call for increased consideration of this aspect in the setting of standards and norms. Klesse et al. (15) focus on the challenges of reliable measurements of electricity consumption in buildings that are shared by many different users, and argue that reliable measurement is a prerequisite for assessing the efficacy of strategies aimed at changing behaviour. Sunderer et al. (14) report on how feedback via smart metering can reduce household electricity consumption. Finally, Brohmann/Bürger et al. (16) and Brohmann/Dehmel et al. (17) present a number of international comparative studies on the effectiveness of economic instruments for regulating the electricity consumption of private households.

Acknowledgements

We would like to express our thanks – on behalf not only of ourselves, but of all the leaders and participants of the project groups. We are grateful to all individuals involved in the focal topic *sustainable consumption* – the teams at the BMBF and at the German Aviation and Aerospace Centre (PT DLR; project management agency of BMBF) – for their support with the synthesis and networking activities. We would particularly like to thank Mrs. Bärbel Kahn-Neetix (BMBF) for her generous and sympathetic support and her openness towards some unusual ideas and unorthodox approaches; Mr. Martin Schmied (PT DLR) for knowledgeable and ongoing support, and his open and trusting collaboration; Mr. Ralph Wilhelm for his constructive and uncomplicated approach as a series editor. We are also greatly indebted to the team of more than 30 external reviewers who, through careful reading and constructive feedback, decisively contributed to the quality of the texts. Special thanks go to Mrs. Susanne Darabas, Mrs. Sarah Schneider and Mrs. Silvia Stammen of oekom publishers for their far-sighted and patient support in the face of an ambitious time schedule. This is equally true for Mr. Clemens Herrmann, Mr. Matthias Reihs and Mrs. Claudia Mantel-Rehbach, who took care of the English edition on behalf of oekom publishers. Thanks also go to Thomas Brückmann, Peter Kobel, Arthur Mohr, Andrea Mordasini and Markus Winkelmann. Their committed and meticulous work in the sometimes hectic preparations for the cross-project workshops and in the taking of the minutes amid animated discussions during the synthesis seminars, has significantly contributed to the overall success of the project. Arthur Mohr, who painstakingly and patiently reviewed many hundreds of pages several times over, has contributed substantially to the formal coherence of both the German and the English edition of the volume, and Rhea Belfanti and Lukas Oechslin were a big help in coordinating and overseeing the translation. Finally, this book could not have been produced without Peter Kobel, who was the hub of this project, coordinating the efforts of the hundred or so people involved in writing, reviewing, correcting and type-setting. For the English edition we are very grateful to Corina Holzherr and her colleagues John Williams and Trevor Pettit for their competent translation and their never-ending patience in dealing with authors' questions and feedback.

Part 1

The synthesis framework

Rico Defila, Antonietta Di Giulio, Ruth Kaufmann-Hayoz, Markus Winkelmann[1]

1 A landscape of research around sustainability and consumption

This contribution presents the research landscape on sustainability and consumption (hereafter simply referred to as 'sustainable consumption'), produced as part of the common synthesis development within the focal topic. The aim of this overview is to describe both settled and ongoing research issues around sustainability and consumption, and thus to place in context the work produced by the project groups. The "Landscape of Sustainable Consumption Research, version 2.1" is intended to acknowledge the breadth of current research on sustainable consumption, both within and outside the focal topic – it does not, however, attempt to answer the question of what it means to apply sustainability principles to consumption.

The research landscape is the fruit of several rounds of discussion within the focal topic over a period of some 18 months – discussions that were prepared, moderated and followed up by the accompanying research project. The first section of this article outlines the purpose of the research landscape. This is followed by a description of its development (1.2), its form (1.3) and its thematic focus (1.4); sections 1.5 and 1.6 introduce the research landscape itself.

---

[1] Discussion participants were Tanja Albrecht (ENEF-Haus), Marlen Arnold (User Integration), Sebastian Bamberg (LifeEvents), Matthias Barth (BINK), Siegfried Behrendt (Consumer/Prosumer), Barbara Birzle-Harder (Intelliekon), Birgit Blättel-Mink (Consumer/Prosumer), Bettina Brohmann (Transpose/Seco@home), Jens Clausen (Consumer/Prosumer), Henriette Cornet (User Integration), Dirk Dalichau (Consumer/Prosumer), Jutta Deffner (Intelliekon/ENEF-Haus), Christian Dehmel (Transpose), Benjamin Diehl (User Integration), Daniel Fischer (BINK), Doris Fuchs (Transpose), Jürgen Gabriel (Heat Energy), Sebastian Gölz (Intelliekon), Ulrich Hamenstädt (Transpose), Andreas Homburg (BINK), Melanie Jaeger-Erben (LifeEvents), Andreas Klesse (Change), Andreas Koch (Heat Energy), Pia Laborgne (Heat Energy), Ellen Matthies (Change), Gerd Michelsen (BINK), Harald Mieg (BINK), Joachim Müller (Change), Ralf-Dieter Person (Change), Klaus Rennings (Seco@home), Kerstin Tews (Transpose), Claus Tully (BINK), Victoria van der Land (ENEF-Haus), Sandra Wassermann (Heat Energy), Julika Weiß (ENEF-Haus), Daniel Zech (Heat Energy), Stefan Zundel (ENEF-Haus).

1.1 Purpose of a research landscape

Numerous research projects focus on "sustainable consumption" – some remain within one discipline, others are of an inter- and transdisciplinary nature. The researchers belong to diverse scholarly communities and networks and thus their publications are scattered among different journals, publishers etc. It is therefore almost impossible to gain an overview of what has been researched in relation to sustainable consumption and the debates that have taken place on discipline-internal as well as trans- and interdisciplinary levels. This state of affairs makes it difficult to refer to established results and can give rise to duplication of research projects. Taken together, these circumstances slow down the common research effort and limit its potential to find practicable pathways towards sustainable consumption. A necessary, though not sufficient step towards linking the different areas and establishing a larger picture is a description of past and present research issues around the topic of sustainable consumption.

Coming up with such an overview of research issues was one of the common aims of the project groups and the accompanying research project within the focal topic "Sustainable Consumption – from Knowledge to Action". The result of this synthesising process was a structured landscape of the research field of "sustainable consumption" and allows past and present research issues to be located in the overall field. This research landscape was first of all used internally to situate the project groups within the focal topic and to devise common products (in terms of content and structure). The research landscape does, however, also contribute to the wider academic discourse around sustainability and is thus presented here for consideration and discussion: what has been developed here can serve as a resource for planning future research around sustainability and consumption, on both a trans- and an interdisciplinary level.

"Landscape of Sustainable Consumption Research, version 2.1" is meant as a snapshot (in time) – sketchy and transient in nature – with no claim to completeness. It is our hope that this current presentation, as indicated by the adjunct 'version 2.1', will serve as a springboard for further development. Below we will outline our approach in the mapping of the research landscape, present the results, and indicate where the research conducted by the project groups fits into this landscape.

1.2 Steps in the development of the research landscape

"Landscape of Sustainable Consumption Research, version 2.1" is the product of an inductive, dialogue-based methodology. A conscious effort was made to steer clear of ready-made interpretations of sustainable consumption and to avoid judging between useful and less useful research or evaluating the quality of research.

The first task of the accompanying research project team was the development of version 1.0 of "Landscape of Sustainable Consumption Research". Here, a two-pronged approach was adopted: first, publications were gathered across a range of disciplines, containing the term *sustainable consumption* either in the title or the keyword list; second, the sowiport database[2] was used to analyse some 300 abstracts of research projects (mostly from the social sciences) between 2004 and 2009. Version 1.0 (a diagrammatic compilation of questions in different thematic areas, with explanatory annotations) was subsequently further developed in the course of a synthesis seminar, which was organised for the 39 participants of all project groups and the team of the accompanying research project (see Figure 1). Participants were invited to bring to the seminar three pertinent research questions that were not dealt with in the focal topic. With the help of a framework of several prepared questions, the content of the research landscape was discussed within the individual project groups. The results of these group discussions were brought together in a plenary, where contentious issues were further clarified. The plenary discussions yielded a number of concrete amendments, and in cases where no satisfactory solution was found, the accompanying research project team was given clear instructions to re-examine unresolved issues. As a further aid to discussion, the accompanying research project team collated all the written comments and proposals from the project groups and made them available to all participants.

On the basis of these discussions and further literature searches, the accompanying research project team revised and presented a new version of the research landscape (2.0) to all participating researchers. A separate document, addressing individual suggestions from the synthesis seminar, was put at the disposal of all participants. It contained a description of either how suggestions had been implemented or a justification for the exclusion of those suggestions that had been deemed unsuitable to be taken further.

As a next step, one telephone interview was conducted with each project group with the aim of gathering additional comments relating to "Landscape of Sustainable

---

2 The sowiport database (www.gesis.org/sowiport; May 22 2011) contains 13 separate sub-databases. The keywords *sustainable consumption* or *sustainability in consumption* were searched for, both in English and in German. They had to appear in the title and/or in the abstract and no attention was paid to how exactly sustainability was interpreted by the project or publication in question.

Figure 1: Synthesis seminar participants discussing the research landscape (November 2009)

Consumption Research, version 2.0". The interview questions had been sent to the project groups in advance with the request to reach a consensus within the groups on a joint response prior to the scheduled interview. The interview questions firstly concerned the description of the research landscape (intelligibility and comprehensiveness of the comments; terminology; gaps; need for changes); secondly, they referred to the research landscape per se (terminology and intelligibility; research questions that could not easily be placed; questions lacking in precision or specificity; overly detailed questions; unnecessary questions; need for changes); thirdly, each project group was invited to locate their research questions within the research landscape.

Thus, "Landscape of Sustainable Consumption Research, version 2.1" is the product of the above procedures. Sections 1.3 to 1.6 provide a detailed account of the research landscape, and section 1.7 discusses possible applications and further developments. In an ultimate attempt to consolidate version 2.1 within the focal topic, it was employed as a framework for a conference organised by the team of the focal topic and this article was externally and internally reviewed.[3]

1.3 Scope and form of the research landscape

"Landscape of Sustainable Consumption Research, version 2.1" does not show what research outcomes on sustainable consumption have been reached (current state of research), nor does it comment on which issues should be researched (research needs). It merely indicates where research is actually taking place (or has taken place over recent years). Given these objectives and the inductive methodology chosen, the research landscape presented here is inevitably retrospective in nature and does not attempt to indicate fields of research with a high innovative potential in the future.

It does, however, give an account of which questions are currently considered pertinent (or have been pertinent over recent years) from a research perspective, where gaps in knowledge have been identified and where external funds have been, or could theoretically be, secured. Furthermore, it can serve as a basis for identifying blind spots within sustainable consumption research and indicating areas for further research, e.g. in the form of new research programmes. (One might, for instance, look at the research landscape against the background of socio-political issues.) Finally, it offers a structuring principle for exploiting, reviewing and representing the current state of research at a disciplinary, interdisciplinary and transdisciplinary level. Having identified current and recent research endeavours, this landscape can be used in future years for tracing developments in the field (i.e. for identifying which issues have been added and which ones might have become less relevant etc.). In terms of the common work within the focal topic, this research landscape ought to prove instrumental in structuring the synthesis outcomes, in identifying synergy potentials between the project groups and in developing a common language.

Having employed an inductive and dialogue-based methodology (drawing together existing research at disciplinary, inter- and transdisciplinary levels), the research landscape does not have an underlying theoretical framework; neither has it been tested according to logical principles. It is primarily a representation of existing phenomena,

---

3   The steps in the development of the research landscape draw on the types of synthesis development identified and described by Rossini and Porter and presented by Krott, i.e. "project management" (draft and development by accompanying research project team), "group" (discussion and decision-making in plenary) and "negotiation" (discussion within the project groups). (See Defila et al. 2006, pp. 124 ff. for the four types.)

rather than a reconstruction in terms of discourse-analytical principles. References to literature and projects have deliberately been avoided. The selection of literature that could be cited within the limited scope of this contribution would inevitably remain arbitrary. Moreover, occasional reference to the literature and specific projects used for generating (by means of content analysis and abstraction) the questions of the research landscape would not increase the comprehensibility of the results.

The current attempt at capturing and structuring the complex field of sustainability and consumption is neither the first of its kind, nor is it unique. Exemplary in this context – although different in their objectives – are the works of Scherhorn et al. (1997) and Warde (2010). Starting out from a specific interpretation of sustainable consumption, Scherhorn et al. (1997) attempt to identify corresponding needs for research. Warde's (2010) volume brings together various social-science contributions to consumption research, spanning several decades and covering numerous aspects. It does not, however, explicitly focus on sustainability within consumption, nor does it come up with an overarching classification of the topic.

We chose to represent this research landscape by using questions which are representative of relevant research that has been conducted and papers that have been published. We deliberately chose questions rather than statements, because they come closest to expressing the purpose of the research landscape, i.e. it is not meant as an overview of research outcomes, but as an outline of research topics. The questions of the research landscape should, however, not be mistaken for research questions, or issues, as posed in concrete research projects. Instead, they have emerged from a process of abstracting from and summarising project descriptions and publications, and are thus the result of interpretation. The extent of interpretation depends on how explicitly the original project descriptions and publications stated their research issues. The questions constituting our research landscape can be traced to the actual research projects, where the issues are contextualised and combined in various ways. It was not our intention to phrase the questions for our research landscape in such a way as to reflect all the aspects of a project or a publication. Rather, an attempt was made to work out from the real and complex questions asked in projects or publications those aspects that could be phrased as single questions of the research landscape.

This research landscape does not indicate the relative importance and frequency of occurrence of specific topics. However, it does illustrate some of the concrete issues that were investigated: an attempt was made to summarise inductively into one question the various 'distilled' questions (as described above). Where projects and publications dealt with very concrete cases, these were added in brackets as examples. (No claim is made to listing all concrete cases mentioned in research projects.) Since some questions could not be summarised into one overarching question without a substantive loss of content, the landscape contains more general as well as more specific questions.

The intended, somewhat 'generic' character of the questions contained in the research landscape meant that, whenever possible and appropriate, time and place indications were avoided. Naturally, accounts of actual research projects and publications tend to include such contextualisations. This research landscape does not attempt to ignore them, but neither does it make a point of explicitly referring to them. Notable exceptions are cases where an issue is historical in character (e.g. the development of consumption) or where the time dimension is particularly important in the context of sustainability (future generations). Similarly, exceptions have been made where a spatial location is essential to the nature of the question (e.g. when non-local people's actions have an impact on a specific locality). The same goes for cases focusing on specific actors, gender differences, domains of consumption etc., where examples are added in brackets; examples are only included in the question if they substantially add to its propositional content.

In the wording of the research questions, terminology proved particularly challenging. Generally, the purpose of this research landscape is not to close off theoretical options, but rather to exhibit a certain pluralism; i.e. it is intended to be compatible with a maximum number of theoretical and conceptual frameworks. With this in mind, and given that terminology is always linked to theory, a 'narrowing down' of theoretical terminology has deliberately been avoided. The result is a somewhat terminological fuzziness in the wording of the questions; whenever possible, the questions contain non-specialised (generally intelligible) vocabulary instead of theoretical and discipline-specific jargon. For instance, terms like 'system' or 'discourse' are not to be understood in terms of a very specific system or discourse theory. In the same vein, 'determinants of consumer behaviour' here simply refers to all the factors that influence human behaviour; it does not suggest that human behaviour can be determined according to deterministic principles. Whilst the terminology used in the research landscape needs to be compatible with the readers' specific theoretical and terminological backgrounds, it does not represent them and should thus not be compared with them.

1.4 Thematic focus and structure of the research landscape

Our research landscape focuses on individual and collective consumer behaviour in the context of sustainable development. Consumer behaviour here refers to processes of preference formation, decision-making for or against specific consumer goods (products, services, infrastructures), the actual acquisition of consumer goods, plus the use or consumption and the disposal or transfer of them. Consumer behaviour as we interpret it also refers to associated processes such as the evaluation of behaviour, norm building and communication. We do not restrict ourselves to research on individual behaviour, but include research on the behaviour of collective actors ('collective' in a broad sense, i.e. the behaviour of organisations or social groups, aggregations of individual behaviour and similar phenomena). Additionally, we included research questions not directly concerned with human behaviour, but with mechanisms, interactions and discourses considered as fundamental to human behaviour. Finally, questions concerned with the sustainability-related impacts of consumer behaviour were also included.

While consumption and production are complementary, they also interrelate with and merge into one another. It is often not possible to draw a sharp distinction between consumption and production. Whether a phenomenon belongs to production or consumption sometimes depends merely on the perspective taken. Nevertheless, in an attempt to separate the two spheres, the research landscape was limited to projects concerned with consumption. Hence, it does not include projects that directly and exclusively focus on the processes and techniques involved in the production of consumer goods. Research on the social and economic effects of production conditions or the consumption of natural resources during production has been included only to the extent to which it was relevant in a sustainability assessment of products. Further questions relating to the production of consumer goods have been included insofar as they include consumers or are situated at the interface between production and consumption (e.g. open innovation research). Excluded are research questions relating to the processes and techniques of waste management, i.e. the disposal of goods that have been thrown away.

The questions and issues that make up the research field of "sustainable consumption" as described above (section 1.3) have been grouped thematically. Here too, an inductive method was used. The emergent research landscape can thus be divided into seven thematic areas (see Figure 2). The questions that shape and structure each area are always indicative of which issues are currently being researched. They do not indicate which area of knowledge a particular project refers to. The questions in the thematic area of "norms/criteria", for instance, do not indicate which norms or criteria should or should not be considered in a project. Instead, they focus on the issues that are being investigated through research in this thematic area. For each thematic area, key questions have been devised that indicate particularly well what types of issues are considered in the area. This (inductive) distribution of issues into seven areas does, of course, not represent the only way of structuring such a research landscape.

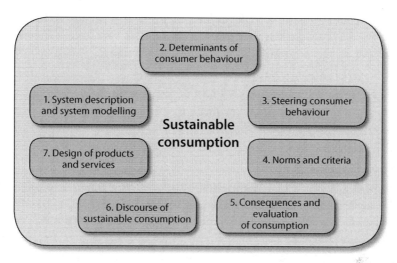

Figure 2: The seven thematic areas of "Landscape of Sustainable Consumption Research, version 2.1"

A thematic area for 'meta research', i.e. 'research into research' on sustainable consumption, was deliberately left out. Accordingly, questions concerned with adequate methods and theories for researching sustainable consumption are not mentioned, the rationale being that, ultimately, such issues are specific to each thematic area and cannot be addressed across all areas of the research landscape (e.g. specific issues relating to the effects and evaluation of consumption or the discourse of sustainable consumption). Thus, the issue of which methods and approaches are appropriate cannot usefully be addressed without considering the other research questions within the research landscape. If it was addressed in isolation, it would lead to abstractions and generalisations removed from the context of sustainable consumption.

1.5 Description of the individual thematic areas

Each thematic area is briefly described below; Figures 3 to 9 and section 1.6 present the questions pertaining to the individual areas and the research conducted in the project groups. The description of each area simultaneously demarcates it from the other areas. However, such a demarcation is only possible up to a certain point. The thematic areas of the research landscape represent analytical categories that have been inductively arrived at. Describing different perspectives rather than discrete phenomena, they all relate to the same 'pre-scientific' phenomenon: sustainability in the context of human consumer behaviour. Each area contains questions that are similar in perspective, preoccupations and in the scope of their theoretical approaches. The areas are not necessarily found in a pure form within specific research projects. They mirror 'pre-scientific' phenomena (e.g. individual values, group norms and rules of sustainable consumption) that interact with each other in real life but may be treated separately for research purposes. Furthermore – in order to come up with academically rigorous proposals for achieving sustainability in consumption – insights from several areas need to be related to each other. The grouping of the questions into the seven areas of the research landscape is therefore a good option, but not the only possible option for making sense of the whole field.

In each of the seven areas, the generation of systems, target and transformation knowledge is weighted somewhat differently. Therefore, each area comes with an indication of the type of knowledge that usually played the central role in the generation of results. The three types of knowledge referred to here were – as far as we know – first introduced into academic discourse by the Conference of the Swiss Scientific Academies (CASS 1997, p. 15):

- Systems knowledge ('knowledge concerning the current situation'): knowledge of structures, processes, statistical variabilities, modes and mechanisms of functioning;
- Target knowledge ('knowledge concerning the target situation'): evaluation of current situations, prognoses, scenarios; knowledge of critical levels/criteria/ethical boundary conditions;
- Transformation knowledge ('knowledge concerning the transition from the current to the target situation'): knowledge of how to achieve target situations; knowledge of what can promote and hinder transformation; knowledge of how to shape and implement the transition.

Area 1 – System description and system modelling

System description and system modelling	This area consists of issues around the description or the modelling of the consumption system as a whole or of its parts (relations between production, trade, consumption, disposal).

This area is concerned with what constitutes consumption as a social phenomenon and how it can be described. Questions are centred around how consumption emerged, what social functions it fulfils, the various elements of consumption as a social phenomenon (actors, processes etc.) as well as associated functional mechanisms (e.g. in the economic arena) and interdependences. Thus, this area contains questions that – under different headings – represent a (somewhat abstract) systemic perspective on the phenomenon of consumption. They are concerned with the functioning of the consumption system as a whole or of its parts (e.g. individual domains of consumption). In this area the generation of systems knowledge is primary.

Research devoted to the issues in area 1 (see Figure 3) provides important factual-methodological knowledge for explaining and steering consumer behaviour. Of primary interest are, however, the functional mechanisms and an appropriate description of the overall system and/or its parts. This area can thus be differentiated from other areas in that it does neither focus on understanding individual or collective consumer behaviour (area 2), nor on steering (area 3) or evaluating it (areas 4 and 5).

Core questions

What are the functional mechanisms (global, international, national, subnational) in the consumption system (relations between production, trade, consumption, disposal)?

What are the functional mechanisms of a specific domain of consumption (e.g. home, food)?

Questions

How can the consumption of/demand for certain resources or ecosystem services (e.g. water, energy) be modelled and how can models adequately represent user behaviour?
Project group Change

What types of consumers can be distinguished (in general or in relation to particular domains of consumption)?
Project group Consumer/Prosumer

Are there any systemic mechanisms that work against a sustainability-oriented consumption system?

How do globalisation processes affect
a) the consumption system
b) specific domains of consumption?

What are the functions of consumption in a given society and how can they be described?

What are the systemic effects of individual and collective consumer behaviour?

What influence do consumers have on the mechanisms of the consumption system (consumer power, ecological citizenship)?

What is the influence of complementary currencies on factors relevant to sustainable consumption?

What are the trends of and assumptions about the development of society (e.g. demographic and technological development) relevant to sustainable consumption?

How are consumer behaviour, consumer expectations and the use of natural resources interrelated (e.g. tourism services)?

How did consumption emerge historically and how has it developed?

How can future generations be appropriately included in a description of the consumption system?

How can the provision of certain resources or ecosystem services be described?

How do market mechanisms change when sustainability considerations are taken into account (e.g. by economic actors)?

How do consumption patterns spread (e.g. in social groups, in emerging economies)?

Figure 3: Questions in area 1 – system description and system modelling

Area 2 – Determinants of consumer behaviour

Determinants of consumer behaviour	This area consists of questions relating to the identification, description and analysis of factors that influence individual and collective consumer behaviour.

The objective of the questions in this area is to understand, explain or predict individual and collective actors' consumer behaviour. Thus, the goal is to investigate what shapes the preparation, enactment and subjective evaluation of consumer behaviour. Questions

Core questions

What are the incentives for and barriers to (e.g. institutional, economic, social) sustainable consumption?
Project groups Consumer/ Prosumer, ENEF-Haus, User Integration, Transpose, Heat Energy

What characteristics of consumers and of actors in the consumer environment and what political, socio-cultural, economic and physical-material conditions have an influence on individual and collective consumer behaviour?
Project groups BINK, ENEF-Haus, Intelliekon, LifeEvents, Seco@home

Questions

What motivates people to embrace sustainable consumption (e.g. motives for purchasing fair trade products, tradeoffs between different attributes of products)?
Project group Seco@home

What factors are responsible for inconsistencies in consumer behaviour (e.g. between different domains of consumption, stages of life)?

What influence does consumer culture (e.g. brand consciousness) have on consumer behaviour (of different social groups)?

What influence do cultural context and social milieu have on consumer behaviour?

What influence do physical-material conditions have on consumer behaviour (e.g. weather, nature, housing, infrastructure)?

What influence do social identities, different stages of life and exceptional events have on consumer behaviour?

What influence do the structures and organisation of everyday life have on consumer behaviour?

What influence do subnational, national and international political discourses/decisions/agreements have on consumer behaviour?

What kind of influence do fundamental values and ideologies have on consumer behaviour?

What kind of influence do extraordinary events (e.g. terrorist attacks, natural disasters, economic crises) have on the behaviour and self-perception of consumers?

How do socio-demographic attributes (age, gender, education, income) influence consumer behaviour and the perception of quality of life?

How do different actors (e.g. retailers, producers, educational institutions) influence consumer behaviour?

How do perceptions of the good life, of quality of life and happiness influence consumer behaviour?

How does market supply influence consumer behaviour (e.g. range of products and services, consumption infrastructure, prices, advertising)?

Figure 4: Questions in area 2 – determinants of consumer behaviour

in this area may be concerned with factors relating to the consumers themselves (e.g. motives, perceptions of quality of life, basic values and control beliefs). Some questions may also relate to external factors that have an effect on actors. These questions may include physical-material, political, economic and socio-cultural factors, as well as an understanding of how consumer behaviour is socially embedded. Overall, the questions in this area cover a wide range of influences on consumer behaviour (from weather and extraordinary events to demographic characteristics, social identities, major life events, social milieux, economic conditions and infrastructure facilities). They all generate systems knowledge relating to the various aspects that influence consumer behaviour.

Research focusing on questions in area 2 (see Figure 4) provides important knowledge for steering consumer behaviour, but the main interest is to first explain consumer behaviour. Area 2 is different from other areas in that it is neither concerned with steering behaviour (area 3) nor with identifying norms in relation to consumer behaviour (area 4). Whilst external factors are considered, no overall systemic perspective (as in area 1) is taken. Finally, this area is not about the consequences and effects of consumer behaviour (as in area 5).

Area 3 – Steering consumer behaviour

Steering consumer behaviour	This area consists of questions concerning the deliberate influencing or steering of individual and collective consumer behaviour in the context of sustainable development.

The questions in this area aim to generate knowledge on how individuals and collective actors can be steered towards consumer behaviour that is in tune with, or can contribute to, sustainable development. Questions in this area are mostly concerned with the functioning and effectiveness of steering instruments (e.g. labels, economic instruments, feedback). The area also includes questions about the role of consumers within this reorientation process, as well as about strategies and policies for achieving sustainable consumption. The questions may be quite general or they may focus on particular domains of consumption or groups of consumers.

In area 3, the emphasis is on the generation of transformation knowledge. Questions (see Figure 5) are concerned with how consumer behaviour can be steered towards sustainability. Research in this area mostly draws on knowledge about the functioning of the consumption system (area 1) and on the determinants of consumer behav-

Core questions

What models of social regulation are particularly relevant in the context of consumer behaviour (e.g. governance approaches)?

What steering instruments, interventions (e.g. regulatory, cooperative, economic, communication instruments) can be used to achieve sustainable consumption; what are their implications and how effective are they?

How can sustainable consumption on the part of individual and collective actors be promoted (e.g. private households, public procurement, universities)?
Project groups BINK, Change, ENEF-Haus, Seco@home, Transpose

Questions

How can different actors (e.g. NGOs, the state, retailers, producers) influence consumer behaviour?
Project group Transpose

How can new roles for actors (e.g. Change Agents, Lead-User, Prosumer) be used productively to achieve sustainable consumption?
Project group User Integration

How can life changes be used to promote sustainable consumption?
Project group LifeEvents

How can the consumer behaviour of specific population groups, consumer types and lifestyle groups be influenced (e.g. children, migrants, yuppies)?
Project groups Intelliekon, LifeEvents

How can sustainable consumption be implemented in particular societal subsystems (e.g. education)?
Project group BINK

How can sustainability be achieved in a particular domain of consumption (e.g. food) or in the context of a particular resource/utility (e.g. energy)?
Project groups Change, Heat Energy

What competencies do individuals need in order to act in accordance with sustainable consumption principles and how can these be taught?

What sustainability strategies can be formulated in the context of consumption and what is the role of sufficiency, efficiency and consistency in this context?

How can products and behaviours be classified in terms of their suitability for different kinds of interventions?

What are the effects of recommendations by expert bodies (e.g. Council for Sustainable Development) on consumer behaviour?

How can different industry sectors contribute towards sustainable consumption (e.g. construction industry, catering industry, energy providers, online trading) and how can their contributions be implemented?

How can innovative technologies be used in the promotion of sustainable consumption?

How can the impacts that products have on sustainability (e.g. virtual water consumption, grey energy, fair trade) be communicated to the public (e.g. via labels)?

How can we enhance the attractiveness of sustainable consumption behaviour?

How can the market power of sustainability-conscious consumers be increased?

How would we describe a consumer policy that is focused on sustainability (e.g. in cross-national comparison, in a specific domain of consumption)?

To what extent can sustainable consumption be attained by means of social marketing (e.g. advocacy, public voluntary commitment, feedback)?

Figure 5: Questions in area 3 – steering consumer behaviour

iour (area 2). Whilst knowledge about the intended goal and the criteria necessary to evaluate behaviour (area 4 and 5) is a prerequisite, research in this area does not focus on producing such knowledge.

Area 4 – Norms and criteria

Norms and criteria	This area consists of questions concerning the justification of norms and criteria, as well as ethical principles relating to sustainable consumption.

The questions in this area relate to norms, rights and obligations. The aim is to determine how sustainability can be applied to the consumption system and to consumer behaviour, both in a broad sense and within specific domains of consumption. A further objective is to provide factual-methodological knowledge of criteria and proce-

---

Core questions

What does sustainability mean in the context of consumption? How can sustainable consumption be defined?

What are the responsibilities of individual actors in the consumption system with respect to sustainability (e.g. consumers, producers, the state, businesses, NGOs)?

How can conflicting goals in the field of sustainable consumption be dealt with (e.g. regional production vs. global availability of foods, energy security vs. protection of the environment, product diversity vs. sustainability standards)?

Questions

What criteria and methods are suitable for evaluating the sustainability of specific products, services and technologies etc.? *Project group Heat Energy*

What criteria and methods can be used to evaluate the sustainability of consumption in a given society (e.g. characteristic values, indicators designed to evaluate sustainable consumption, comparison between countries)?

What are the equity issues in the context of (sustainable) consumption (e.g. inter-generational, intra-generational, gender-specific)?

What negative consequences of consumer behaviour are tolerable from the point of view of sustainability?

What level of consumption can be generalised (to the whole world)?

What are justifiable visions of sustainable lifestyles, alternative forms of consumption, working models etc.?

Who are the holders of rights that have to be considered in the context of sustainable consumption and what are the rights in question?

What are the criteria for and goals of sustainability in a particular domain of consumption (e.g. the home)?

Figure 6: Questions in area 4 – norms and criteria

dures with which to evaluate specific behaviours, products and services on the one hand, and general consumer behaviour on the other. This area also includes questions about the responsibility of individual actors and about theoretically justifiable approaches to a) the hierarchy of goods and values and b) conflicts such as that between ample consumer choice and sustainability; other questions address the ethical bases of consumption and sustainability (e.g. the interests of future generations). Therefore, this area is mainly concerned with generating target knowledge.

Research concerned with questions in area 4 (see Figure 6) provides a theoretically grounded justification for the steering of consumer behaviour (area 3). The focus is not on how steering can/should take place, but on which direction this steering should adopt. Whilst research in this area is not evaluative in itself, it should generate knowledge on how to evaluate the consequences of consumer behaviour (area 5).

Area 5 – Effects and evaluation of consumption

Effects and evaluation of consumption	The questions in this area pertain to the direct and indirect consequences and the evaluation of consumer behaviour – in the context of sustainable development.

This area contains a number of questions aimed at eliciting information about the consequences and impacts of consumer behaviour and the functioning mechanisms of the consumption system – both within particular domains of consumption and more generally. Other questions are concerned with whether a particular product, technology or form of behaviour can be classified as sustainable or not, and with the undesirable effects of modified behaviour (e.g. rebound effects). Finally, this area contains questions that address the need for action in specific domains of consumption. Thus, this area is primarily concerned with generating target knowledge.

Research devoted to the questions in area 5 (see Figure 7) provides knowledge that is relevant to the steering of consumer behaviour. It is not, however, concerned with justifying or challenging the criteria (area 4) that are applied, or with how to steer consumer behaviour (area 3). The specific interest here is to find out how close or how far a society is from achieving the goal of sustainable development.

A landscape of research around sustainability and consumption

Core questions

What social, cultural, economic and ecological consequences (including geographically wide-ranging, long-term or unintended consequences) does individual and collective consumer behaviour have in specific domains of consumption?
Project group Consumer/Prosumer

How sustainable are particular products, product categories, services, working patterns, technologies etc. (e.g. bioenergy, regional products, teleworking)?

Questions

How can a change in consumer behaviour contribute to the sustainable development of a society?
Project group Intelliekon

What rebound effects are to be expected from changes in consumer behaviour towards greater sustainability?

What local consequences for sustainability result from specific nonlocal consumers (e.g. tourists)?

How can the functioning of a particular domain of consumption impact on sustainability?

How should sustainable investments be evaluated (e.g. real estate, equities)?

How does the behaviour of particular groups of consumers affect the consumption of resources (e.g. commuters, university staff)?

What action is needed in a given domain of consumption (e.g. nutrition)?

Figure 7: Questions in area 5 – consequences and evaluation of consumption

Area 6 – Discourse of sustainable consumption

Discourse of sustainable consumption	This area consists of questions concerning the description and analysis of the social discourse in relation to sustainable consumption.

Questions in this area aim to analyse the current discourse on sustainable consumption in society. One objective is to gain knowledge about perceptions, concepts and assumptions that characterise the discourse of particular groups in society (e.g. academia, political parties). A further aim is to understand how the discourse of sustainable consumption has developed over time (e.g. with respect to the 'life cycle' of issues). Some questions focus on knowledge transfer relating to sustainable or unsustainable consumption both within and between groups and networks, or on how certain lifestyles, consumption patterns etc. are marketed as sustainable (e.g. marketing strategies), and on the communicative role of advertising. All questions in this area depend on the analysis of the 'outcomes of the discourse' for their answers, regardless of the medium or form of the discourse (i.e. oral, written, pictures etc.). Here, the focus is always on systems knowledge.

> **Core questions**
>
> What perceptions, concepts and assumptions relating to sustainable consumption can be identified in the discourse of specific groups in society?
>
> How is the term 'sustainable consumption' used in these discourses (e.g. in marketing strategies, political manifestoes)?
>
> **Questions**
>
> What perceptions, concepts and assumptions relating to sustainable consumption can be identified in specific academic disciplines/research areas (e.g. human ecology)?
>
> What lifestyles, consumption patterns etc. are marketed as being 'sustainable', and what perceptions and concepts are they based on?
>
> What perceptions, concepts and assumptions are being propagated through the discourse on sustainable consumption (e.g. gender roles)?
>
> What knowledge about sustainable consumption is held by which groups in society and how is this knowledge propagated across groups?
>
> How does communication about sustainable consumption function within different social groups/networks?
>
> How has the discourse of sustainable consumption developed over time (e.g. with respect to the 'life cycle' of issues)?
>
> How are the problems of unsustainable consumption described in different media discourses (e.g. newspapers, films, novels, blogs, advertisements)?

Figure 8: Questions in area 6 – discourse of sustainable consumption

Research relating to the questions in area 6 (see Figure 8) is primarily concerned with how the discourse of sustainable consumption operates. While such research also deals with the values underlying sustainable consumption, it analyses them in the context of communication and does not set standards for sustainable consumption (area 4). Where evaluations are made, they relate to the discourse itself and not to the consumption patterns or products (area 5).

Area 7 – Design of products and services

Design of products and services	This area consists of questions concerning the interaction between consumption and production, with respect to the design of sustainable products and the establishment of a market presence.

Questions in this area focus on the interface between production and consumption in the context of the design and provision of consumer goods. How can and do consumers influence the development and the design of sustainable products? Essentially, the objective is to identify the potential for sustainable consumption habits in innovation-oriented relations between producers and consumers. This area is thus concerned with transformation knowledge.

Research in area 7 (see Figure 9) is not concerned with the role of the interaction between consumers and producers within the consumption system (area 1), or with the influence of producers on the behaviour of individual and collective consumers (area 2). Neither is it concerned with the (possible) role of producers in the steering of consumer behaviour towards sustainability (area 3), with producers' specific responsibility (area 4), or with their marketing strategies (area 6). Finally, it is not concerned with the assessment of products in terms of their sustainability (area 5). Questions that focus on production processes and the provision of consumer goods in the strict sense of the term are not included in the research landscape. The questions in area 7 highlight the interface between the existing research landscape of sustainable *consumption* with a potential research landscape of sustainable *production*.

> **Core questions**
>
> How do ideas for sustainable products come about and how can the products become established on the market?
>
> **Questions**
>
> | | |
> |---|---|
> | What potential do new innovation regimes hold for sustainable products (e.g. Open Innovation, cooperation networks)? *Project group User Integration* | What role does the co-evolution of technical sustainability innovation and consumer behaviour play in the design of sustainable products/services? |
> | How can new types of relations between actors (e.g. business-client relations) lead to sustainable products? | How can changes in consumer behaviour influence production? |

Figure 9: Questions in area 7 – design of products and services

1.6 The questions belonging to the individual areas

Figures 3–9 contain the questions belonging to the various areas of the research landscape. The questions show the range of considerations covered in each area, but they should not be seen as completely separate (also see section 1.3). Concrete examples from project descriptions or literature are added in brackets. Each area contains at least one 'core question'. Core questions are not intended to indicate the special prominence or relative frequency of an issue. They have simply been selected from the pool of questions within the research landscape in terms of how well they characterise the essence of that particular research area. Whilst the core questions provide the reader with a 'headline' for the research in each area, it is only by looking at all the questions that the wealth of research in the area is revealed. Thus, the core questions neither summarise the other questions nor subsume all of the other questions.

The core questions are placed at the top of the list of questions. They are followed by the additional (i.e. non-core) questions that the project groups particularly focused on (up to three questions per project group).

1.7 Uses of the research landscape and further development

The "Landscape of Sustainable Consumption Research, version 2.1" as described in this chapter represents a discipline-independent, structured overview of sustainable consumption issues within various disciplinary, interdisciplinary and transdisciplinary research contexts. In terms of research planning, it can be seen as a basis for identifying blind spots as well as future research needs within sustainable consumption. From a historical perspective, it can be used in the future as a basis for tracking research developments (e.g., it could be interesting to look at the relationship between trends in research and trends in society). This could even be undertaken today by comparing the research landscape presented here with what was proposed in Scherhorn et al. 1997 – even though the two publications can only be compared to a limited extent.

The research landscape can be used to identify, build on and present work undertaken in different disciplinary, interdisciplinary and transdisciplinary contexts. Possibilities are currently being explored to make available the research landscape with an accompanying online literature database. This could help researchers from one field identify relevant work in areas that are unfamiliar to them.

It was not our intention to narrow down our research landscape in terms of specific geographical research locations (i.e. the material was not analysed in this respect). However, in view of the original database used and the participating researchers, the research questions could be seen as Euro- or even Germano-centric. A review and expansion of the research landscape to include more countries would certainly prove fruitful.

A second, thematic expansion could take the form of developing comparable research landscapes on related topics (e.g. sustainability in the production or the disposal of goods and services). These would represent a natural follow-up to the present undertaking.

References

CASS (1997): Conference of the Swiss Scientific Academies (CASS), Forum for Climate and Global Change (ProClim-): Visions by Swiss Researchers. Research on Sustainability and Global Change – Visions in Science Policy by Swiss Researchers. Bern: ProClim-/SANW. [http://www.proclim.ch/4dcgi/proclim/en/media?1122; 03.05.2012]

Defila R., Di Giulio A., Scheuermann M. (2006): Forschungsverbundmanagement. Handbuch für die Gestaltung inter- und transdisziplinärer Projekte. Zürich: vdf Hochschulverlag an der ETH Zürich.

Scherhorn G., Reisch L., Schrödl S. (1997): Wege zu nachhaltigen Konsummustern. Überblick über den Stand der Forschung und vorrangige Forschungsthemen. Ergebnisbericht über den Expertenworkshop "Wege zu nachhaltigen Konsummustern" des Bundesministeriums für Bildung, Wissenschaft, Forschung und Technologie (BMBF). Marburg: Metropolis-Verlag.

Warde A. (ed.) (2010): Consumption. Los Angeles: Sage. (4 volumes).

Antonietta Di Giulio, Bettina Brohmann, Jens Clausen, Rico Defila, Doris Fuchs, Ruth Kaufmann-Hayoz, Andreas Koch[1]

2 Needs and consumption – a conceptual system and its meaning in the context of sustainability

Given that the concept of need is central to both the idea of sustainability and the notion of consumption, it makes sense to approach the debate on sustainability in consumption by starting with the concept of 'need'. Such a concept should be attuned to both the concept of consumption and the idea of sustainability. The concept of need that has to be developed and the conceptual system that arises from it should be capable of serving as a basis for discussions about norms and for empirical research (on a disciplinary, interdisciplinary and transdisciplinary level). The objective of this article is to propose such a conceptualisation of needs and thereby contribute to the terminological underpinning of the interdisciplinary debate on sustainability in consumption. This can then be used in the generation of proposals on how to evaluate individual consumer behaviour – which accounts for a large part of overall consumption – within the context of sustainability (cf. Fischer et al. in this volume for an example of such a proposal).

2.1 Introduction – scope of the conceptual system

'Consumption' is commonly understood as the use of goods (products and services/ infrastructure) to satisfy individual human needs. Goods fulfil both an instrumental and a symbolic/communicative function, i.e. they help people construct their identity, form groups and (re)produce social structures and boundaries. Far from being passive onlookers, consumers play an active part in the creation and circulation of goods (e.g. Papastefanou 2007; Van Vliet et al. 2005, p. 17; Fichter 2005; Bartiaux 2003, p. 1240;

---

[1] Discussion participants were Barbara Birzle-Harder (Intelliekon), Henriette Cornet (User Integration), Angelika Just (LifeEvents), Andreas Klesse (Change), Peter Kobel (SÖF-Konsum-BF), Joachim Schleich (Seco@home), Immanuel Stieß (ENEF-Haus), Daniel Zech (Heat Energy).

van Vliet 2002, p. 53; Giddens 1984). Thus, the possession and usage of goods is not an end in itself, but instrumentally related to needs. (For an elaboration of the term 'consumption', see also Fischer et al. and Kaufmann-Hayoz/Bamberg et al. in this volume.)

The concept of need plays a central role in the idea of sustainability – in particular when circumscribing the goal of sustainable development (e.g. Rauschmayer et al. 2011; Michaelis 2000). A further key concept in this context is the notion of a good life: there is a general consensus about the (normative) idea of sustainability, namely that the development of human society (at a global, regional and national level) should be oriented towards the superordinate goal of ensuring that, both now and in the future, all human beings should be able to satisfy their needs and lead a good life. Accordingly, what is required of strategies, action plans and programmes of sustainable development is that they should define what exactly constitutes the good life. Contained within the notion of sustainability is the intention that future generations should have as much leeway as possible to determine and satisfy their needs. This implies that each generation has to make assumptions about the needs of future generations. There is general agreement that the nature of the idea of sustainability is regulative, in the sense that its precise meanings and interpretations are subject to negotiation processes (at international, national and subnational level) in each generation anew. It is therefore impossible to determine once and for all what sustainable development looks like in concrete terms. It must be tailored to the specific historical and cultural context, and it depends on the existing knowledge and the (socially) negotiated hierarchy of goods and values. The idea of sustainability is a normative political idea, i.e. states share in the obligation to implement it. The terminology proposed in this article should therefore be compatible with the idea that the state bears part of this responsibility. Ultimately, the idea of sustainability is visionary, as it goes hand in hand with the aspiration to set a positive goal for the development of human society.

The question of how to define a good life, and the question of how to relate the notion of a good life and that of human needs are two central issues in the discourse of sustainable development (e.g. Di Giulio 2008, 2004; Michaelis 2000; Manstetten 1996). The aspects of consumption that are particularly important from a perspective of sustainability are therefore the effects of acquiring, using, disposing of and transferring goods on the opportunities of people to satisfy needs and lead a good life (in the present as well as in the future). Given the anthropocentric character of sustainability, effects on non-human nature are, strictly speaking, not relevant – except insofar as the opportunity for humans to lead a good life is affected by non-human nature.

If sustainability in consumption is the aim, it follows that a concept of need is required that is compatible with the notion of consumption as well as with the idea of sustainability and the notion of a good life (e.g. Soper 2006; Jackson et al. 2004; Michaelis 2000;

Røpke 1999). Such a concept of need should be suitable for formulating criteria with which to evaluate sustainability in consumption. It should serve as a basis not only for developing specific ethical norms (cf. Fischer et al. in this volume for the consumer behaviour of individuals), but also for empirical research (on a disciplinary, interdisciplinary and transdisciplinary level). Ultimately, the concept has to be compatible with the distinctions and definitions used in different theoretical approaches (which does not mean that there need be a one-to-one correspondence).

The aim of this article is to set out such a concept of need and a conceptual system developed out of this concept. The discussion will range from theories of a good life to whether the concept lends itself to empirical research and whether it is compatible with the 'need area approach' ('Bedürfnisfeld-Ansatz'). The latter point is included because the need area approach not only specifically proceeds from the notion of needs, but also already plays a certain role within sustainable development research. A concept of need that is compatible with the need area approach is therefore desirable. This article seeks to provide a basis for an interdisciplinary debate on the conceptual and normative basis of sustainability in consumption. It goes without saying that an exhaustive treatment of the subject or review of the literature would be beyond the scope of the present contribution.

2.2 A concept of needs compatible with the idea of sustainability, the notion of a good life and the notion of consumption

The term 'good life' has been used in philosophy for theoretical descriptions of what constitutes a fulfilled human life. Such descriptions are by definition positive, supra-individual and ahistorical. (Given that any approach is always contingent on its historical context, the word 'ahistorical' should not be understood in an absolute sense, but in the sense that it implies a certain long-term validity.) The philosophical concept of a good life can therefore provide a positive content, neither time-bound nor culturally bound, to the superordinate and very abstract goal of sustainable development. This content in turn can then serve as a basis for the tailoring of sustainable development to different specific contexts (for a more detailed explanation see Di Giulio et al. 2010).

There is a general consensus that the good life in the context of sustainability is best defined by so-called "objective theories". These theories define elements of a good life that are universally valid and therefore independent of subjective wishes and individual inclinations. By contrast, so-called "subjective theories" define a good life exclusively in terms of subjectively experienced well-being. (Objective theories do not negate the importance of subjectively experienced well-being, but they do not regard it as the only decisive factor to define a good life.) Amongst objective theories, it is the anthropological approaches that are most prominent in the discourse of sustainability (e.g. Rauschmayer et al. 2011; Jackson et al. 2004; Michaelis 2000). One such approach – particularly advocated by Martha Nussbaum and Amartya Sen – is the "capability approach". In anthropological approaches, it is argued that certain characteristics and capabilities are universal to human life (e.g. humour or play). Based on this argument it is postulated that living a good life is having the opportunity to develop these characteristics and capabilities according to one's own physical and psychological capabilities, as well as one's personal values and inclinations. Therefore, an anthropological approach neither dictates a certain lifestyle nor pronounces on what a fulfilled life should represent for a specific individual. At the same time, anthropological approaches are not relativistic. Thus, from these approaches, the duty arises to provide all people with these opportunities – regardless of whether they take them or not. The obligation is thus to provide the external conditions necessary to develop characteristics and capabilities that are deemed to be universal for all human beings, enabling them to lead a life that they individually perceive as meaningful. Such an obligation can legitimately be imposed also on states. (The goal of sustainability is therefore to be stated more precisely as follows: not to guarantee a good life as such for all humans, but to guarantee the prerequisites for a good life for all humans.) Basic characteristics and capabilities that are, for instance, defined by the "capability approach" (e.g. Nussbaum 1992) include having a human body (including features such as hunger, thirst, sexual desire, aversion to pain etc.), cognitive capabilities (perceiving, imagining, thinking), relatedness to other species and to nature (knowing about other living creatures, knowing about our dependency on non-human nature etc.), humour and play (laughter, significance of playing etc.), the ability to move from place to place, the capability for developing attachments to things and persons outside oneself, and for living with concern for and in tune with nature.

The most important arguments in favour of an anthropological approach to the notion of a good life in the context of sustainability are the following:

- An anthropological approach does not reduce a good life to the physiological aspects of survival; this would be regarded as cynical. But, for the following reasons, it nevertheless allows one to postulate an obligation on the part of the state to ensure that the external conditions for a good life are in place. A good life is not merely equated with the living out of individual preferences and the experience of subjective well-being. A distinction is made between individual prerequisites and internal states on the one hand, and external conditions of a good life on the other.
- An anthropological approach presupposes universal characteristics and capabilities that are representative of human life. It is not relativistic and it does not bind the

issue of a good life to any particular and therefore any context-specific definition of what constitutes a good life. At the same time, it allows for and even advocates the context-specific definition of a good life because individuals have to define and to live their own version of it.

- An anthropological approach is capable of potentially distinguishing between legitimate and illegitimate concerns. This is essential in the context of sustainability.
- An anthropological approach draws a distinction (at least theoretically) between a good life and the range of material endowments. Such a separation is an indispensable requirement for an informed debate on what material endowments are necessary for any given purpose.

Objective approaches to determining of what constitutes a good life are the subject of some dispute. In sociology and political science, and particularly in the sub-disciplines of international relations and development policy, universal approaches come under periodic crossfire (e.g. as untenable or as 'Western' and thereby not universal). For instance, proponents of post-colonial or post-structuralist approaches contest the possibility of making universally valid (and normative) assertions about the good life, social development or human rights (Chakrabarty 2000; Lee 1995). Martha Nussbaum reacted to such relativistic arguments as early as 1992 by articulating once more the contents and strengths of the "capability approach". Nussbaum's anti-relativistic position is very close to that of Amartya Sen who – together with other economists – succeeded in establishing the underlying principles of the capability approach in international policy debate. It inspired the creation of the UN's Human Development Index (HDI), which was developed by Mahbub ul Haq, Amartya Sen and Maghnad Desai and subsequently adopted by the United Nations' Development Programme (UNDP). The objective of the HDI is to use a small number of indicators (mainly per capita income, income distribution, life expectancy and educational level) to capture the possibilities that are open to people all over the world to develop those capabilities and characteristics required for a good life. In other words, its aim is to capture the external conditions for a good life.

The term 'need' has been given various interpretations, some of which appear to be incompatible (see also e.g. Leßmann 2011; O'Neill 2011; Rauschmayer et al. 2011; Jackson et al. 2004; Michaelis 2000). It can, for instance, stand for freely chosen as well as indispensable (no longer disputable) goals; it can stand for a (inner) guiding force behind an act as well as for the external prerequisites required for achieving a specific goal; it is used to represent a sensation (e.g. hunger) or the object of a sensation (e.g. food). The term can also be used to refer to the purpose that is pursued as the result of a sensation (e.g. survival). The term 'need' can acquire an ethical and political charge in this context only when the various interpretations have been clearly differentiated and a meaningful sustainability-related sense for the term has been established (see also Cruz 2011; Michaelis 2000).

As a starting point on the road to our objective, the following generic distinction of meaning is made: In one meaning 'need' is used in a positive sense to represent people's (physiological and psychological) drivers of wanting or motivations to act. In the other meaning it is used in a 'deficit-oriented' sense, standing for either an objectively determinable physical or psychological state of deficiency (so-called "objective needs"), or a subjective perception of deficiency (so-called "subjective needs"). In this second use of the word, an important distinction is made regarding who is entitled to determine (the legitimacy of) needs, i.e. the individual ("subjective needs") or an instance beyond the individual ("objective needs").

In the context of the idea of sustainability, and with the aim of complementing the concept of the good life with the concept of need, it makes sense to view needs as historical, i.e. time-bound entities (e.g. Soper 2006). In this way, the use of this term becomes attached to existing human subjects who feel a motivation to act or perceive/have a deficiency and can (at least in theory) demand that this deficiency be made good. Needs are thus always situated in time and tied to particular individuals and their sensibilities. Complementary to the (comparatively) ahistorical concept of a good life, a concept of need that relates strongly to a particular 'here and now' can be used to anchor the overarching goal of sustainable development in concrete circumstances, with people in specific life situations, at specific points in time (see e.g. Di Giulio et al. 2010). This presupposes that the concept of need is not defined in a 'deficit-oriented' sense, as this would not be compatible with the positive concept of a good life and the positive and visionary idea of sustainability. Furthermore, in order to be compatible with the idea of sustainability, it is necessary to consider the concept of need in terms of objective rather than subjective needs, because it cannot justifiably be a state's obligation to remove all feelings of deficiency, or to support all subjective volitions (see e.g. Røpke 1999). To summarise, a concept of need that is compatible with the notion of sustainability and meaningfully complements the notion of a good life must be framed in terms of objective needs (even though needs are obviously always subjectively experienced by individuals) and must not be deficit-oriented.

In order to link a concept of need as described above with an anthropological approach to a good life, we draw on Soper (2006) in proposing that "'objective needs' (…) refers to (…) (individual) constructs of wanting that comprehensibly (and thereby convincingly) refer to capabilities and characteristics that are determined as universally valid (and thus objective) elements of a good life by an anthropological approach to a good life" (Di Giulio et al. 2010, pp. 20 f.; translated by C. Holzherr). This definition resembles the concept of need as defined by Jackson et al. (2004), who consider 'need' as a universal motivating force. Similarly, Rauschmayer et al. (2011) define 'need' as a constitutive aspect of human flourishing, which requires no further justification as a reason to act, since it cannot be related to any other purposes – or O'Neill (2011) who proceeds from categorical needs, which he understands as indispensable, irreducible and non-substitutable objectives affecting people detrimentally if they are not met. (For similar approaches see for example Baumeister/Leary 1995, who justify universally valid human needs on the basis that their non-satisfaction would lead to empirically measurable disorders/diseases.)

A definition of needs as constructs of wanting allows to distinguish – at least terminologically – between a need (e.g. to be pain free), the means to satisfy this need (e.g. food, health care), and the sensations that arise from its fulfilment or non-fulfilment (e.g. hunger, satiety, pain), which can, in turn, become motivators for purposeful action (e.g. procurement of food) (see e.g. Jackson et al. 2004). In the context of consumption, this distinction is of particular importance as it permits to relate needs to goods, whilst simultaneously distinguishing between needs and the goods that satisfy these needs. This approach in turn ties in with proposals in the literature to classify goods in terms of their potential for satisfying needs. Jackson et al. (2004), for instance, stress the stability of (universal) needs in contrast to the instability of 'satisfiers' (means for satisfying needs). Drawing on Manfred Max-Neef (e.g. 1991), they propose a typology of 'satisfiers', where the criteria for distinction are, for instance, a) whether they do, indeed, satisfy a need, b) whether they are detrimental to the satisfaction of one or more needs, c) whether they can potentially fulfil more than one need, or d) whether they are indispensable for the satisfaction of a need (for similar proposals see Cruz 2011; Rauschmayer et al. 2011).

The definition of objective needs as outlined in the present article additionally allows for both intra- and intergenerational justice to be taken into account. In a world of finite resources, it is impossible to satisfy without limits subjective constructs of wanting. In the present definition, objective needs are limited to only one part of these constructs of wanting by introducing a precondition: In order to qualify as an objective need, a construct of wanting has to satisfy the specific precondition that it can be traced back to those characteristics and capabilities that are required for a good life. As a result, it is possible to draw conclusions about resource distribution without negating the reality and justification of sensations of wanting as a consequence. (It cannot be ruled out, of course, that existing natural resources may be insufficient for the provision of the external conditions that will enable all people to live a good life.)

The concept of need as it is presented here is thus a useful complement to the concept of a good life, and it can easily be linked to both the idea of sustainability and the notion of consumption. In order for this concept to be viable, however, the list of capabilities and characteristics that needs have to refer to in order to be accepted as objective must be kept within limits. In other words, the list must not include everything that humans undertake or would aim to undertake. Otherwise, objective needs could be justified solely by persuasive argument. In other words, it would simply be a case of arguing iteratively until any given sensation of wanting could be traced back to these capabilities and characteristics. In this light, the list proposed by Martha Nussbaum might be too long (shorter lists are proposed e.g. by Costanza et al. 2007 or Veenhoven 2000). Thus, the "capability approach" is not the only anthropological approach to a good life whose suitability within the context of sustainability is being discussed. Jackson et al. (2004), Michaelis (2000) or McAllister (2005) for instance present and discuss alternative approaches; these cannot be discussed further here. A basic assessment of how suitable the "capability approach" is for the context of sustainability is provided by Leßmann (2011). It is suggested here that these approaches, and the lists of capabilities and characteristics postulated as part of these approaches, be critically examined in the light of the concept of need presented in this article. Furthermore, it is necessary to discuss whether there is a requirement for a list of capabilities and characteristics specifically tailored to consumption. Such an undertaking would, however, have to be the task of an interdisciplinary research project.

2.3 Presentation of the conceptual system

The concept of need outlined above forms the basis of the conceptual system for the sphere of consumption that is proposed below (see also Figure 1). This system will then be evaluated in two respects: in terms of its normative content and in terms of what could be gained if it were exploited in empirical disciplinary, interdisciplinary and transdisciplinary research. It does not claim any validity beyond the sphere of consumption. (It is thus not intended as a conceptual system for sustainability, such as that proposed e.g. by Leßmann 2011.) Where links between terms within the system have been elaborated in the discussion below, the intention was to point up causal relationships between the terms within the system, and not to imply any empirical cause and effect.

The following interrelated terms make up the system (see Figure 1; for an in-depth discussion of the core elements refer to Di Giulio et al. 2010):
- Individual constructs of wanting: objective needs, subjective desires, demands, and ideas about degree and breadth of satisfaction of objective needs and subjective desires,

- Consumer goods: products and services/infrastructures,
- Components of nature: resources, processes and states (incl. requirements in terms of quantity and quality).

Objective needs

Objective needs are individual constructs of wanting of humans that comprehensibly (and therefore convincingly) refer to those capabilities and characteristics proposed and deemed universally valid (and thus objective) elements of a good life by an anthropological approach to a good life (see above). Objective needs are ends in themselves for the satisfaction of which consumer goods are used. (The converse argument, however, that all objective needs can be satisfied entirely by consumer goods, does of course not follow from this.) Objective needs are legitimate needs and cannot be contested on ethical grounds. Individuals and states have an obligation to provide conditions under which people can – now and in the future – satisfy their objective needs, and/or conditions that do not make it impossible to satisfy these needs.

Subjective desires

Subjective desires are individual constructs of wanting of humans that cannot lay claim to the status of objective (and therefore legitimate) needs, because, in contrast to objective needs, they do not comprehensibly (and therefore not convincingly) refer to those capabilities and characteristics proposed and deemed universally valid (and thus objective) elements of a good life by an anthropological approach to a good life. Subjective desires are ends in themselves, for the satisfaction of which consumer goods are used. (Again, this does not imply that subjective desires are satisfied entirely by consumer goods.) Subjective desires are legitimate only as far as their satisfaction does not prevent other humans from satisfying their objective needs. They can thus be contested on ethical grounds. Neither individuals nor states have an obligation to ensure that people can satisfy these sensations of wanting.

Ideas about the degree and breadth of satisfaction of objective needs and subjective desires

Ideas about the degree and breadth to which objective needs and subjective desires should be satisfied are also individual constructs of wanting. These constructs indicate when an individual considers an objective need, or a subjective desire, as being satisfied. They are, on the one hand, linked to the individual conceptions of a fulfilled and meaningful life, and, on the other hand, strongly influenced by the socio-cultural environment of an individual. Generally, it is possible to fulfil the same desire (e.g. to travel to foreign countries) or need (e.g. to move from place to place) by means of different consumer goods, in different proportions. Ideas about the degree and breadth to which objective needs and subjective desires can be satisfied prevalent in an individual's socio-cultural environment have the power to kindle new, previously non-existent subjective desires. Individual ideas about the degree and breadth of satisfaction of needs and desires are legitimate only so far as their fulfilment does not prevent the satisfaction of other people's objective needs. They can thus be contested on ethical grounds. Neither individuals nor states have an obligation to ensure that people can satisfy these sensations of wanting.

Demands
Human beings have certain demands (requirements) of products and services/infrastructures (e.g. with regard to design, performance, quality). Demands that people make are also individual constructs of wanting, but refer to and depend on specific products and services/infrastructures. They are not ends in themselves for the satisfaction of which products and services/infrastructures are used. Demands are based on objective needs, subjective desires and ideas about the degree and breadth to which objective needs and subjective desires should be satisfied. The demands that people make of consumer goods are legitimate so far as their fulfilment does not prevent other people from satisfying their objective needs. Therefore, they can be contested on ethical grounds. Neither individuals nor states have an obligation to ensure that people can fulfil these sensations of wanting.

Consumer goods
Consumer goods are products and services/infrastructures that serve to satisfy objective needs and subjective desires. People make use of them as a means of meeting their objective needs and fulfilling their subjective desires. (It does not follow that people will entirely rely on consumer goods to satisfy these needs and desires.) Changing ideas about the degree and breadth to which objective needs and subjective desires should be satisfied, as well as newly generated subjective desires, can lead to the creation of new consumer goods. Conversely, the availability of consumer goods may influence people's ideas about the degree and breadth to which needs and desires should be satisfied, and can create new subjective desires. By contrast, objective needs as such are not influenced by what is on offer. Consumer goods can be contested on ethical grounds insofar as their availability and their use must not prevent other people from satisfying their objective needs. Insofar as consumer goods are essential for the satisfaction of objective needs, there is an obligation to ensure their availability.

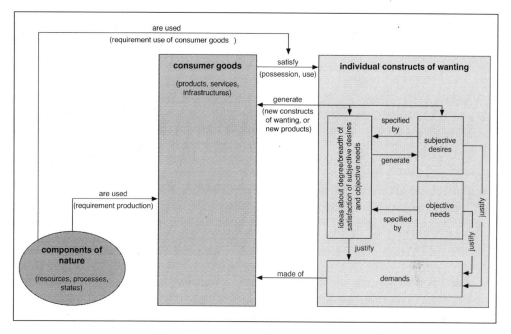

Figure 1: Conceptual system for the sphere of consumption developed on the basis of the concept of need[2]

Components of nature

Components of nature are natural resources, processes and states in non-human nature. In order to satisfy their needs and desires, people make direct or indirect use of these components: The manufacture, distribution, use, as well as the disposal and transfer of nearly all consumer goods rely on, or affect, components of nature. The use of components of nature is thereby not an end in itself, but a means to an end. In other words, the individual's construct of wanting that requires satisfaction is not the use of natural resources, processes or states – these are only instrumental to a purpose (e.g. warmth, relaxation) and, in our context of consumption, they are 'mediated' in the form of consumer goods (e.g. heating systems, public parks). The requirements regarding quality and quantity of natural resources, processes or states in nature depend on the production and use of consumer goods. The extent of the use of components of nature can be contested on ethical grounds, i.e. they should be used only insofar as their use does not prevent other people from satisfying their objective needs. Additionally, the extent to which components of nature are used is dependent on the amount available (now and in the future). This applies specifically to natural resources. In the present discourse,

---

[2] The diagram summarises the relationships between the separate elements of the proposed system.

the term 'ecosystem services' (to include ecosystem goods and services) is employed to relate the components of nature to human well-being (in an anthropocentric way) by describing their welfare-related services. Such 'ecosystem services' include, for instance, peace and quiet, fertile soils or inputs into the production of marketable goods (e.g. Staub et al. 2011). The question of whether and how the ecosystem services approach and our conceptual system could be linked in the context of consumption is outside the scope of this article as this would require an in-depth discussion.

2.4 Bringing the conceptual system 'down to earth' – what could be gained by its empirical application?

In order to use our conceptual system for empirical research and for assessing sustainability in concrete consumer actions, it is essential that the phenomena designated by the individual terms (e.g. objective needs and subjective desires) can be empirically identified and distinguished from each other. Additionally, it is essential that the postulated relationships between, for instance, consumer goods and particular objective needs, or between particular subjective desires and associated components of nature, can be backed up empirically. Thus, the terms in the conceptual system must be categories that do not only have normative value, but can also be established empirically. For such 'empirical grounding' an interdisciplinary approach, spanning the social and natural sciences, is indispensable. It is the task of specific research projects to establish the extent to which the different conceptually identified phenomena can be identified empirically, too. A great deal of promising research has already been undertaken in this area. (For example, lists of consumer goods matching needs have been drawn up, as in Rauschmayer et al. 2011, Costanza et al. 2007 or Jackson et al. 2004, and as part of numerous psychological studies, e.g. Lavers 2008 and Baumeister/Leary 1995, who have attempted to distinguish empirically between subjective desires and objective needs.)

In this section – in order to show how the conceptual system could be applied in research and what benefits this may bring – we will indicate how it could be employed by using one of the projects within the focal topic as an example. Given that our conceptual system had not been developed as a common theoretical approach prior to the research in the project groups, but is the result of a subsequent synthesising process that took place at the level of the entire focal topic, this simulated application is abbreviated and reconstructive. In order to show how the system could be of benefit in the context of development-oriented projects, it will, also in a reconstructive way, be applied to a project that was not part of the focal topic. Finally, we will discuss the compatibility of our system with the need area approach.

2.4.1 Project group "Heat Energy" – applying the conceptual system to the 'consumption of heat energy' (written by Andreas Koch)

The need for warmth in housing is an objective need, which can be related in particular to physical integrity as a universal element of a good life. In concrete terms, this equates to the need for a hygienic room temperature (perceived also as comfortable), and the need for hot water.

Individual ideas about the degree and breadth of satisfaction of objective needs – for instance ideas about adequate room temperature or ventilation – can result in an increased demand for heat energy. Individual demand for heat energy may equally be influenced by subjective desires, such as the desire to wear certain clothes inside the house or the desire for more living space. The resulting overall energy demand can therefore be described as the sum of objective needs and subjective desires (Koch et al. 2008), although these two terms have so far not been systematically differentiated on an empirical basis. An attempt at such a differentiation can be found in calculations where the room temperature that should be provided is determined by the average number of people who rate a change in the state of temperature as uncomfortable (Schramek et al. 2008; see also DIN EN ISO 7730). A systematic differentiation between needs and desires could, for instance, furnish the basis for discussing what room temperature is legitimate and therefore must be provided, or for discussing quantity and temperature of domestic hot water that have to be provided.

By systematically differentiating between needs and desires, one might also be able to link rebound effects to the different constructs of wanting. This way rebound effects could be included in the debate about legitimate and less legitimate concerns. At present, to give an example, energy efficiency often tends to be assessed only by looking at the so-called 'specific' energy demand per square metre. A possible increase in living space, which may 'cancel out' the energy saved by improved insulation, is not taken into account. This, in turn, is unfortunate, as in Germany, the "decisive factor in the increase in direct energy consumption of households" between 1995 and 2005 according to a study of the German Federal Environment Office was in fact "the expansion of living space" (UBA: German Federal Environment Office 2006; translated by C. Holzherr): The 13 percent increase in living space could not be compensated by a 9 percent reduction of the specific energy demand achieved by improvements in efficiency.

With reference to the proposed conceptual system, different energy saving strategies are conceivable: To start with, one could, as part of an efficiency strategy, try to minimise the amount of resources used to satisfy the objective need for warmth. Possible measures include improving insulation and optimising the control of the heating system. Then a consistency strategy could be adopted which would involve the use of renewable energies. The two strategies may, of course, be combined. Although the legitimacy of subjective desires can by definition be contested on ethical grounds, it now can be examined to what extent desires could be fulfilled by these two strategies. The desire for more living space could, for instance, be compensated by improving the insulation of a building. (Of course this raises the issue about the adequate benchmarks, as this strongly influences the extent of achievable compensation (kWh of primary energy, t CO_2).) Alternatively, a sufficiency strategy could be pursued for both desires and ideas about the extent and breadth of the satisfaction of needs. The differentiated conceptual system thus opens up the discussion of different pathways towards sustainable consumption. In the example above, for instance, the legitimate energy demand might be calculated by also taking into consideration the size and composition of the household. Such an approach stands in contrast to current EnEV[3] calculations of, for example, domestic hot water demand, whereby usable surface area is the only criterion.

2.4.2 "Sustainable City Quarters"[4] – applying the conceptual system to construction/housing (written by Bettina Brohmann)

In construction and housing, consumer goods range from residential buildings and open spaces (size, design) to social and cultural facilities. They satisfy a variety of objective needs and subjective desires, and are strongly interdependent. The aim of this section is to establish if, and to what extent, the conceptual system elaborated in this article could be advantageously applied to the development of an urban neighbourhood.

To distinguish objective needs, subjective desires, consumer goods, and ideas about the degree and breadth of the satisfaction of needs and desires in the process of urban development opens up the opportunity to look at each of the single components separately as a first step. These components can then be compared, evaluated and subsequently weighted. In so doing, the consumer goods involved can explicitly be discussed in terms of their functions against the backdrop of needs, desires and ideas. In other words, the constructs of wanting (and thereby the envisaged objectives) are considered first, and the question of how things, e.g. the transport infrastructure, should be designed and what consumer goods are appropriate are discussed as a next step. Furthermore, by distinguishing between objective needs (e.g. for housing space or social cohesion) and subjective desires (e.g. regarding the design of residential buildings), the intentions of stakeholders can be identified and compared in terms of where they coincide and where they diverge. (What objective needs should be considered? What subjective desires and ideas do the future residents hold about (the extent of) satisfac-

---

3   EnEV is the German Energy Savings Ordinance.

4   For a description of the project see www.oeko.de/service/cities.

tion of their needs? What desires and ideas were taken into account by planners and investors of a housing estate in the planning of a particular infrastructure?) It is thus possible to address specific requirements with regard to building materials, design of living space, food supply, cultural facilities, transport infrastructure etc. against the backdrop of objectives that have already been clarified.

In order to weight the subjective desires and the individuals' ideas about the extent of need satisfaction, the decision was made in this project to match the consumer goods (that should or could be provided) with any given needs, desires and ideas. (What is a 'satisfier' for which need, desire or idea?) Additionally, in order to assess the effects on sustainable development of providing and making use of these consumer goods, a number of quantitative (and if required qualitative) indicators were developed (TU Munich 2000; FEST 2000; Ismaier 2000). These indicators have a reference value (e.g. a baseline year or the status quo) as a starting point, on the basis of which a positive or negative development can be tracked. In this way, the participants can use these indicators to assess how far the satisfaction of their desires and ideas would contribute to sustainable development. As a result, they can make informed choices about which consumer goods to opt for. The development and implementation of such a measuring system at a local level are possible if the necessary basic data are available and if the persons concerned, as well as any interested third parties, are actively involved in the development process (Brohmann et al. 2002). It should be noted that the global dimension can only be incorporated (by comparing these local indicators with superordinate targets, such as national sustainability targets) in those cases where the indicators are the same on all levels.

The fact that the conceptual system presented in this article allows for separate, but interrelated explorations of objectives, interests and means renders it helpful for projects focusing on dialogue, participation and development – such as the urban development project described above.

2.4.3 Compatibility with the need area approach

The need area approach (related to the activity-oriented approach) was developed in the 1990s in the context of environmental research programmes (e.g. by Thomas Dyllik, Gertrude Hirsch Hadorn, Jürg Minsch, Uwe Schneidewind, Jochen Reiche, Karl-Otto Henseling; for the approach itself see e.g. Behrens et al. 2005; Mogalle 2001; Klann/Nitsch 1999; Schneidewind 1997). The aim was to provide a framework for the discourse and analysis of environmental impacts that was compatible with social science reasoning and could simultaneously accommodate everyday behaviour. Taking human needs as a starting point for research was seen as the best way to frame actions, actors, structures (generated and reproduced by human actions) and products as an overall system oriented to a specific need. On that basis, problems, requirements for change, scope for action, and research issues should be identified (e.g. as part of transdisciplinary research projects). The main aims of such an integrated approach were

- to delineate a (new) and extended framework, e.g. for material flow analyses, capable of generating a practicable description and analysis of system boundaries;
- to identify impacts and interactions in relation to human behaviour and its changes (scenario building);
- to address actors in an activity-oriented manner and to facilitate a clear identification of desired and feasible transformations in a given area of need;
- to involve actors in the analysis and structuring of the 'area of need'.

The approach has since been further developed and taken up by social-ecological research. So-called 'areas of need' that have been the object of research are nutrition (e.g. www.ernaehrungswende.de), mobility (e.g. www.renewbility.de) and construction/housing (e.g. Buchert et al. 2004). In collaboration with actors at different levels of decision-making, scenarios have been developed in these 'areas of need' that can throw light on the effects of different courses of action. Communication, health and leisure/tourism are further 'areas of need' that have recently been established.

The approach has proved successful for discussing consumption patterns, their effects in terms of sustainability, and changes in consumer behaviour with different actors. It has also been successful for identifying priorities in sustainable policy making. The approach is promising insofar as it is capable of bridging 'the logic of research' in academia with 'the logic of action' pertaining to actors in the real world. Furthermore, it has proved useful to examine whether steering measures (interventions, policy instruments) are properly focused on the appropriate actors, and whether with regard to a given 'area of need' the relevant actors and the relevant problems have been identified (e.g. www.eupopp.net).

The need area approach proceeds from a different concept of need from the one developed here. It is not objective needs in the sense defined here that form the basis of identifying and analysing 'areas of need', but combinations of constructs of wanting (that have not been further specified) and consumer goods linked to these. Nevertheless, the concept of need developed here can be linked to the need area approach: the proposed conceptual framework could be instrumental in a nuanced analysis within 'areas of need', and the 'areas of need' could be related to a classification of objective needs. Whether this could contribute to the further development and implementation of the need area approach remains to be explored by future research projects.

2.5 Normative content of the conceptual system

We all make use of and rely on consumer goods to varying degrees (as individuals, members of households or organisations). Consumption cannot be understood in purely functional terms, but is always socially and symbolically charged (see also Kaufmann-Hayoz/Bamberg et al. in this volume). Consumer behaviour has economic, socio-cultural and ecological repercussions that affect both the present and the future. An act of consumption is thus always an expression of a moral and ethical stance and thereby inevitably subject to ethical scrutiny.

An evaluation of consumption in terms of sustainability represents such an ethical assessment, i.e. the idea of sustainability is a possible vantage point from which to assess consumption. In this context, it is crucial to define clearly the normative content of the concept of need (see e.g. Michaelis 2000). The conceptual system presented above has been conceived in this spirit, i.e. it has been developed with the intention of contributing to the academic and political debate about the normative content of needs in the context of sustainability in consumption. It is therefore time to sketch out some key normative consequences of what has been said thus far. These considerations are complemented by the contribution of Fischer et al. in this volume, which focuses on one sub-area of consumption, i.e. individual consumer behaviour, and suggests how its sustainability could be assessed.

The key normative consequences arising from our conceptual system can be summarised as follows:

- Objective needs cannot be contested on ethical grounds.
- Individuals and states have an obligation to ensure a) the necessary external (i.e. not the internal ones of individual humans) conditions for satisfying objective needs, and b) the provision of those consumer goods that are essential, i.e. indispensable and non-substitutable for the satisfaction of objective needs.
- Contestable on ethical grounds are the following: subjective desires, ideas about the degree and breadth of satisfaction of objective needs and subjective desires, the demands made of consumer goods, the provision of consumer goods in excess of what is required for fulfilling objective needs, and the extent to which components of nature are used.
- In the debate about the permissible recourse to components of nature, it is important to pay attention to the 'physical boundaries' regarding the quantity and quality of these components.

Our conceptual system thus represents a systematisation that allows for a differentiated discussion of sustainability in consumption – at least in theory.

In order to determine and define sustainability in consumption in more detail, we will list below a number of components that emerge from our conceptual system. Similar components have been identified by, among others, O'Neill (2011), Rauschmayer et al. (2011), Jackson et al. (2004), Michaelis (2000), and Røpke (1999):

- *Criteria:* The decisive criteria for sustainability in consumption are the external conditions necessary for the satisfaction of objective needs of present and future generations. The 'physical boundaries' dictated by the components of nature form part of these conditions. While subjective desires, ideas about the degree and breadth of satisfaction of objective needs and subjective desires, the demands made of consumer goods, and the provision of consumer goods that are not strictly necessary are not reprehensible per se and need not be rejected unreservedly, they do not represent criteria for sustainability. Instead, they have to be judged according to the decisive criterion.
- *Communication:* It is possible to discuss sustainable consumption without a) sending the message that sustainability in consumption equates to a diminution in the quality of life, b) condemning subjective desires wholesale, and c) calling for a general (and unrealistic) renunciation of consumption. Instead, it is now possible to frame consumption positively in communication and to associate sustainability in consumption with the idea of a fulfilled life.
- *State interventions:* State interventions affecting the consumer behaviour of individuals are legitimate because of a) the obligation to provide the external conditions allowing people to satisfy their objective needs (including the provision of indispensable consumer goods and the necessary components of nature, in terms of both quality and quantity), and b) the obligation to restrict those actions that would otherwise compromise suitable external conditions for current and future generations.
- *Strategies:* It is possible to formulate differentiated strategies relating to the single elements of our conceptual system to promote sustainability in consumption. Sufficiency and efficiency strategies can be understood as both interlinked and complementary: efficiency strategies could be implemented in order to satisfy subjective desires to the greatest possible extent. Sufficiency strategies could come into effect afterwards and be unmistakeably restricted to those elements of the conceptual system that can be contested on ethical grounds. The same principle applies to the 'clash of discourses', whereby some people require quantifiable standards and others insist on a qualitative description of what is required. In terms of the system presented in this article, it is evident that needs have to be described qualitatively and that quantification is appropriate only in relation to some of the external conditions necessary for a good life.

With a view to promoting sustainability in consumption, the question of whether this can legitimately be steered by the state is of vital importance. Following on from the conceptual system presented above, and in particular the paired terms objective needs and subjective desires, the following principle for steering by the state applies: the 'objectivity' of particular needs gives rise to an entitlement to demand of the state that these needs are met, i.e. that it provides a kind of basic service (see e.g. O'Neill 2011). The satisfaction of constructs of wanting in the sphere of objective needs has thus to be ensured by the state, whereas it is legitimate for the state to intervene when it comes to the satisfaction of subjective desires, e.g. by way of raising taxes. Our conceptual system thus provides criteria for deciding 'what' should be steered by the state. (For 'how' the state can intervene see Kaufmann-Hayoz/Brohmann et al. in this volume.)

The example of living space shall illustrate how this could be specified: The living space (in m^2) per person has, for instance in Germany, roughly doubled since the Second World War. In order to investigate whether the state is allowed to intervene with regard to living space, it has to be first established how far living space is an external condition of a good life and at what point the actual surface area of living space represents subjective desires or individual ideas about the extent of satisfaction of objective needs. It thus becomes a question of establishing both a minimum level (in the example: of living space) as a condition of a good life (as defined in a specific historical and cultural context), and a maximum level, above which the indefensibly subjective desires and individual ideas are situated. (Cruz 2011, O'Neill 2011 and Rauschmayer et al. 2011, for instance, propose that there is something like a 'minimum level', an 'optimal level' and an 'upper boundary' of external conditions with regard to objective needs.) The exact determination of lower and upper boundaries is dependent on how a good life is specified in an actual context, meaning that the answer is inevitably dependent on the historical and socio-cultural context. To define such boundaries is thus a matter of societal negotiation. (Rauschmayer et al. 2011, for instance, also proceed from a necessity to negotiate boundaries.) It should be noted that the determination of boundaries will in practical terms always be characterised by the fact that knowledge is uncertain and incomplete. Once the boundaries are defined, everything above the lower boundary can be assigned to subjective desires; there is no obligation to fulfil these, but equally, they need not be curtailed. Everything above the maximum level can be regarded as 'too much', and as something that should be prevented (current discussions e.g. about the poverty line and bankers' bonuses show that such questions are not beyond academic and public debate). Such a conception of lower and upper boundaries might be usefully employed by the state in steering consumer behaviour. (The legitimacy of strong state intervention would increase from the lower boundary towards the upper boundary.)

Such a specification of the normative content of the conceptual system touches on another basic issue within sustainability in consumption – a social justice issue that is often conveniently ignored by 'affluent countries in the North': debates often centre on what is 'too much' and pay no attention to what is 'not enough' in their own national context. However, what ultimately should be achieved is that the satisfaction of subjective desires by the wealthy and powerful does not hinder, even less preclude, the less privileged from fulfilling their objective needs. If the principle was strictly implemented to fulfil subjective desires only if this did not impinge on the satisfaction of objective needs elsewhere, and if boundaries such as those described above were used as criteria for state intervention, the state would not only be legitimised, but it would be under an obligation to intervene whenever this happens. In this way, the decisive criteria for sustainability in consumption would be applied to the current generation.

References

Bartiaux F. (2003): A socio-anthropological approach to energy related behaviours and innovations at the household level. St. Raphael: ECEEE 2003 Summer Study Proceedings.

Baumeister R. F., Leary M. R. (1995): The need to belong: Desire for interpersonal attachments as a fundamental human motivation. In: Psychological Bulletin 117 (3): 497–529.

Behrens T., Hoffmann E., Lindenthal A., Hage M., Thierfelder B., Siebenhüner B., Beschorner Th. (2005): Institutionalisierung von Nachhaltigkeit. Eine vergleichende Untersuchung der organisationalen Bedürfnisfelder Bauen & Wohnen, Mobilität und Information & Kommunikation. Marburg: Metropolis.

Brohmann B., Fritsche U., Hartard S., Schmied M., Schönfelder C., Schütt N., Roos W., Stahl H., Timpe C. (2002): Nachhaltige Stadtteile auf innerstädtischen Konversionsflächen: Stoffstromanalyse als Bewertungsinstrument. Darmstadt, Freiburg, Berlin: Forschungsbericht für das BMBF.

Buchert M., Fritsche U., Jenseit W., Rausch L., Deilmann C., Schiller G., Siedentop S., Lipkow A. (2004): Nachhaltiges Bauen und Wohnen in Deutschland. Berlin: Umweltbundesamt. [http://www.umweltdaten.de/publikationen/fpdf-l/2600.pdf, http://www.oeko.de/service/cities; 16.07.2012]

Chakrabarty D. (2000): Provincializing Europe: Postcolonial Thought and Historical Difference. Princeton: Princeton University Press.

Costanza R., Fisher B., Ali S., Beer C., Bond L., Boumans R., Danigelis N. L., Dickinson J., Elliott C., Farley J., Elliott Gayer D., MacDonald Glenn L., Hudspeth T., Mahoney D., McCahill L., McIntosh B., Reed B., Turab Rizvi S. A., Rizzo D. M., Simpatico T., Snapp R. (2007): Quality of life: An approach integrating opportunities, human needs, and subjective well-being. In: Ecological Economics 61: 267–276.

Cruz I. (2011): Human needs frameworks and their contribution as analytical instruments in sustainable development policy-making. In: Rauschmayer F., Omann I., Frühmann J. (eds.): Sustainable development. Capabilities, needs, and well-being. London, New York: Routledge. 104–120.

Di Giulio A. (2008): Ressourcenverbrauch als Bedürfnis? Annäherung an die Bestimmung von Lebensqualität im Kontext einer nachhaltigen Entwicklung. In: Wissenschaft & Umwelt INTERDISZIPLINÄR 11, Energiezukunft: 228–237.

Di Giulio A. (2004): Die Idee der Nachhaltigkeit im Verständnis der Vereinten Nationen – Anspruch, Bedeutung und Schwierigkeiten. Münster, Hamburg, Berlin, London: LIT.

Di Giulio A., Defila R., Kaufmann-Hayoz R. (2010): Gutes Leben, Bedürfnisse und nachhaltiger Konsum. In: Nachhaltiger Konsum, Teil 1. Umweltpsychologie 27 (14): 10–29.

FEST (2000): Forschungsstelle der evangelischen Studiengemeinschaft. Indikatoren im Rahmen einer Lokalen Agenda 21. In: ZIT/TU Darmstadt (eds.): Strategien Nachhaltiger Entwicklung – Dokumentation zur Fachtagung am 14.11.2000 in Darmstadt.

Fichter K. (2005): Interpreneurship. Nachhaltigkeitsinnovationen in interaktiven Perspektiven eines vernetzenden Unternehmertums. Marburg: Metropolis.

Fischer D., Michelsen G., Blättel-Mink B., Di Giulio A. (in this volume): Sustainable consumption: how to evaluate sustainability in consumption acts.

Giddens A. (1984): The Constitution of Society: Outline of the Theory of Structuration. Cambridge: Polity Press.

Ismaier F. (2000): Konzeptionelle und methodische Anforderungen an Nachhaltigkeitsindikatoren. In: TU München Nachhaltigkeitsindikatoren. Studien zur Raumplanung Nr. 3. München.

ISO 7730 (2006-05): Ergonomie des Umgebungsklimas – Analytische Bestimmung und Interpretation der thermischen Behaglichkeit durch Berechnung des PMV- und des PPD-Indexes und der lokalen thermischen Behaglichkeit.

Jackson T., Jager W., Stagl S. (2004): Beyond insatiability – needs theory, consumption and sustainabilty. In: Reisch L., Røpke I. (eds.): The ecological economics of consumption. Cheltenham, Northampton: Edward Elgar. 79–110.

Kaufmann-Hayoz R., Bamberg S., Defila R., Dehmel C., Di Giulio A., Jaeger-Erben M., Matthies E., Sunderer G., Zundel S. (in this volume): Theoretical perspectives on consumer behaviour – attempt at establishing an order to the theories.

Kaufmann-Hayoz R., Brohmann B., Defila R., Di Giulio A., Dunkelberg E., Erdmann L., Fuchs D., Gölz S., Homburg A., Matthies E., Nachreiner M., Tews K., Weiß J. (in this volume): Societal steering of consumption towards sustainability.

Klann U., Nitsch J. (1999): Der Aktivitätsfelderansatz – ein Ansatz für die Untersuchung eines integrativen Konzepts nachhaltiger Entwicklung. STB-Bericht Nr. 23. Köln.

Koch A., Huber A., Avci N. (2008): Behaviour Oriented Optimisation Strategies for Energy Efficiency in the Residential Sector. ICEBO'08 – 8th International Conference for Enhanced Building Operations. Berlin.

Lavers T. (2008): Reconciling the needs and wants of respondents in two rural Ethiopian communities. In: Social Indicators Research 86 (1): 129–147.

Lee E. (1995): Human Rights and Non-Western Values. In: Davis M. (ed.): Human Rights and Chinese Values, Oxford: Oxford University Press. 72–90.

Leßmann O. (2011): Sustainability as a challenge to the capability approach. In: Rauschmayer F., Omann I., Frühmann J. (eds.): Sustainable development. Capabilities, needs, and well-being. London, New York: Routledge. 43–61.

Manstetten R. (1996): Zukunftsfähigkeit und Zukunftswürdigkeit – Philosophische Bemerkungen zum Konzept der nachhaltigen Entwicklung. In: GAIA 5 (6): 291–298.

Max-Neef M. (1991): Human scale development: conception, application and further reflections. London, New York: The Apex Press.

McAllister F. (2005): Wellbeing concepts and challenges. Discussion paper for the Sustainable Development Research Network.

Michaelis L. (2000): Ethics of consumption. Oxford: Oxford Centre for the Environment, Ethics & Society.

Mogalle M. (2001): Management transdisziplinärer Forschungsprozesse. Basel: Birkhäuser Verlag.

Nussbaum M. C. (1992): Human Functioning and Social Justice: In Defense of Aristotelian Essentialism. In: Political Theory 20 (2): 202–246.

Nussbaum M. C., Sen A. (eds.) (2002): The Quality of Life. 6th Edition. Oxford: Clarendon Press.

O'Neill J. (2011): The overshadowing of needs. In: Rauschmayer F., Omann I., Frühmann J. (eds.): Sustainable development. Capabilities, needs, and well-being. London, New York: Routledge. 25–42.

Papastefanou G. (2007): Variatio delectat? – Verbreitung und sozialstrukturelle Differenzierung der Konsumvariabilität. In: Jäckel M. (ed.): Ambivalenzen des Konsums und der werblichen Kommunikation. Wiesbaden: VS Verlag für Sozialwissenschaften. 235–260.

Rauschmayer F., Omann I., Frühmann J. (2011): Needs, capabilities and quality of life. Refocusing sustainable development. In: Rauschmayer F., Omann I., Frühmann J. (eds.): Sustainable development. Capabilities, needs, and well-being. London, New York: Routledge. 1–24.

Røpke I. (1999): Some themes in the discussion of the quality of life. In: Köhn J., Gowdy J., Hinterberger F., van der Straaten J. (eds): Sustainability in Question. The Search for a Conceptual Framework. Cheltenham, Northampton: Edward Elgar. 247–266.

Schneidewind U. (1997): Wandel und Dynamik in Bedürfnisfeldern – Wesen und Gestaltungsperspektiven. Eine strukturationstheoretische Rekonstruktion am Beispiel der Ökologisierung des Bedürfnisfeldes Ernährung. Diskussionsbeitrag Nr. 2 des IP Gesellschaft I. St. Gallen: Institut für Wirtschaft und Ökologie.

Schramek E.-R., Recknagel H., Sprenger E. (2008): Taschenbuch für Heizung und Klimatechnik 09/10. Oldenburger Industrieverlag.

Soper K. (2006): Conceptualizing needs in the context of consumer politics. In: Journal of Consumer Policy 29: 355–372.

Staub C., Ott W., Heusi F., Klingler G., Jenny A., Häcki M. (2011): Indikatoren für Ökosystemleistungen: Systematik, Methodik und Umsetzungsempfehlungen für eine wohlfahrtsbezogene Umweltberichterstattung. Umwelt-Wissen Nr. 1102. Bern: Bundesamt für Umwelt.

TU München (2000): Nachhaltigkeitsindikatoren. Studien zur Raumplanung Nr. 3. München.

UBA (2006): Wie private Haushalte die Umwelt nutzen – höherer Energieverbrauch trotz Effizienzsteigerungen, Umweltbundesamt.

Van Vliet B. (2002): Greening the Grid. The Ecological Modernisation of Network-bound Systems. Wageningen University: PhD-Thesis.

Van Vliet B., Chappels H., Shove E. (2005): Infrastructures of Consumption. Environmental Innovation in the Utility Industries. Earthscan: London.

Veenhoven R. (2000): The Four Qualities of Life. Journal of Happiness Studies 1 (1): 1–39.

Daniel Fischer, Gerd Michelsen, Birgit Blättel-Mink, Antonietta Di Giulio[1]

3 Sustainable consumption: how to evaluate sustainability in consumption acts

There is no question that the term 'sustainable consumption' has enjoyed a remarkable career in recent history. Numerous privately initiated, industry-led and governmental conferences, strategies, projects and publications have been devoted to the promotion of sustainable consumption. They have all come up against the initial challenge of drawing a distinction between 'sustainable' and 'non-sustainable' acts of consumption. The objective of this article is to shed some light on this distinction by finding ways of assessing sustainable and non-sustainable acts of consumption.

The main focus of this article is on individual consumers and individual acts of consumption. However, this focus should not be taken to imply any particular position with regard to the "privatisation of sustainability" (cf. Grunwald 2010 and 2011), in the sense that individual consumers would be held responsible for achieving the overall goal of sustainability. Furthermore, it is not the intention of this article to provide a list of indicators that enable individual acts of consumption to be ticked off as either 'sustainable' or 'non-sustainable'. Instead, the aim is to identify *possible approaches* for assessing the sustainability of *acts* of consumption.[2]

This article is divided into three parts. The first part consists of a brief introduction to the discourse of sustainable consumption. The second part is initially concerned with the questions of which acts of consumption are to be evaluated, and then goes on to derive, from the key elements of the idea of sustainability, criteria for carrying out

---

1 Discussion participants were Marlen Arnold (User Integration), Dirk Dalichau (Consumer/Prosumer), Rico Defila (SÖF-Konsum-BF), Benjamin Diehl (User Integration), Elias Dunkelberg (ENEF-Haus), Jürgen Gabriel (Heat Energy), Felix Groba (Seco@home), Ulrich Hamenstädt (Transpose), Ingo Kastner (Change), Pia Laborgne (Heat Energy), Harald A. Mieg (BINK), Malte Nachreiner (BINK), Claudia Nemnich (BINK), Ralf-Dieter Person (Change), Marc Requardt (User Integration), Martina Schäfer (LifeEvents), Dirk Thomas (Change), Claus J. Tully (BINK).

2 Our focus on exploring possible approaches for assessing the sustainability of individual acts of consumption differs from the aim of Di Giulio et al. in this volume, who elaborate on a conceptual system that allows to investigate the relation between sustainability and consumption on the basis of the concept of needs.

such an evaluation. On that basis, two possible ways for evaluating sustainability in consumption will be considered. In the third and final part, these two possibilities will be illustrated and discussed using two actual projects within the focal topic.

3.1 The discourse of 'sustainable consumption'

Although environmentally and socially oriented considerations of consumption patterns, mainly Western consumption patterns, date back much further, the summit of Rio de Janeiro is generally regarded as the birthplace of 'sustainable consumption'. In one of the summit's outcome documents, Agenda 21, an entire chapter (Chapter 4) was devoted to "changing consumption patterns" – and yet, Agenda 21 failed to describe in detail what was meant by sustainable consumption. A concrete programme of action was only initiated ten years later at the Sustainable Development Summit in Johannesburg (Marrakech Process). A fundamental difficulty in defining sustainable consumption in concrete terms is that most definitions (including the one in Agenda 21) tend to derive it *ex negativo*, i.e. relationally, by contrast to non-sustainable consumption.

Over the years, various attempts have been made to better understand and conceptualise sustainable consumption – each attempt having its own different priorities (for German-speaking regions see Schrader/Hansen 2001; Scherhorn/Weber 2003, and Belz/Bilharz 2005; for an international overview see Jackson 2006). A widely recognised definition came out of a round table organised by the Norwegian government in 1994:

> *"Sustainable Consumption and Production can be defined as [...] the production and use of goods and services that respond to basic needs and bring a better quality of life, while minimising the use of natural resources, toxic materials and emissions of waste and pollutants over the life cycle, so as not to jeopardise the needs of future generations"* (Norwegian Ministry of the Environment 1994).

A year after its promulgation, the 'Oslo definition' of sustainable consumption was adopted by the Commission on Sustainable Development of the United Nations (UNCSD). Despite being one of the most frequently cited definitions, it is the object of some controversy in academic circles (see section 3.2.3).

In Germany, the debate around sustainable consumption was advanced by studies such as "Zukunftsfähiges Deutschland"[3] (BUND/Misereor 1996) and "Nachhaltiges

---

3 "Germany fit for the future" (translated by C. Holzherr).

Deutschland"[4] (UBA 1997). They were instrumental in classifying the current state of sustainable consumption. The study by the Federal Environment Agency (UBA) concludes that "at least 30 to 40 percent of all environmental problems [...] can directly or indirectly be attributed to prevailing patterns of consumer behaviour" (UBA 1997, p. 221; translated by C. Holzherr). New models of prosperity ('having a good life as opposed to having everything') identified in a study by the Wuppertal Institute gave new impetus to the debate around sustainable consumption.

An initial focus on case studies from a few specific sectors such as nutrition or housing, and an initial attempt to clarify different analytical levels (e.g. individual acts of consumption or broader patterns of consumption) (cf. Verbraucherzentrale Nordrhein-Westfalen 1997; Scherhorn/Reisch 1997; Günther et al. 2000; UBA 2001), have gradually given way to the development of indicators and the empirical assessment of sustainable consumption (cf. Lorek et al. 1999; Gebhardt et al. 2003; Baedeker et al. 2005). Two decades after Agenda 21, and despite a number of academic and non-academic initiatives, no binding definition of individual sustainable behaviour in consumption has been proposed. The considerations below are intended to fill this gap, by exploring ways of assessing the sustainability of individual consumer behaviours.

3.2 What is 'sustainable consumption'? Terminology and key concepts

Sustainability is a normative guiding principle. This section is concerned with clarifying the object of evaluation (i.e. which individual consumption acts should be assessed), and with finding ways of assessing these consumption acts on the basis of the normative idea of sustainability. In order to clarify what exactly should be assessed, we will, as a first step, take a closer look at the phenomenon of consumption. Next, with the aim of deriving assessment criteria, we will identify the values, norms and criteria that are related to the idea of sustainability. Finally, we will consider the relations between the assessment criteria and the objects of assessment (i.e. acts of consumption), and the impact of these relations on the type and manner of assessment.

3.2.1 Understanding 'consumption'

The key question to ask here is: what constitutes an act of consumption (the 'what' question). The determinants of acts of consumption, the 'why (do people consume)' question, fall under action theory (cf. Kaufmann-Hayoz/Bamberg et al. in this volume), and questions relating to steering consumer behaviour, the 'how to change' questions,

---

[4] "Sustainable Germany" (translated by C. Holzherr).

belong to the area of policy instruments and intervention techniques (cf. Kaufmann-Hayoz/Brohmann et al. in this volume).

In the literature the term consumption is used in a number of ways. Narrowly defined, it signifies the use of goods or services; in a broader sense, it refers additionally to the selection, acquisition and disposal or passing on of goods (cf. Campbell 1998).

It now becomes necessary to clarify the terms 'goods' and 'services'. Goods can be classified in terms of their materiality (material vs. immaterial), in terms of their life span (quickly used up goods vs. long-lasting goods), intended use (consumer goods vs. producer goods) or ownership (private vs. public). (For an informative outline from an economic perspective see Bea et al. 1994.) Consumer goods are tangible products, traded on the open market, exchanged privately and used by individuals. The apparently clear-cut distinction between production and consumption in the above definition does, however, need to be complemented with the notions of 'co-production' and 'prosumption'. The latter term was coined by Alvin Toffler (1983; cf. for example Blättel-Mink/Hellmann 2010), who – at the time when traditional industrial societies were changing into more service-oriented economies – recognised the role of private agents as contributors to the creation of economic value. He described an increasing hybridisation of *production* and *consumption*, citing examples of self service shops or home improvement stores, where many products are not conceived as end products to be consumed, but as components of a further production cycle. Such a further production cycle can consist of re-sale, personal use or co-production in the sense of adding value to a product. Consumption is thus to be understood as a "complex, multi-level process of acquisition, use and disposal of co-production and self-production" (Reisch 2011, p. 142). The customer becomes a "working customer" (Voß/Rieder 2005; translated by C. Holzherr), ceasing to consume goods as end products, and – as a co-producer – contributing to their further processing in the final link of the production chain (cf. Reichwald/Piller 2009). In contrast to goods, services (despite on-going difficulties in defining them) are to be understood as either "non-material or non-storable products" (Jacobsen 2010, p. 206; translated by C. Holzherr), although – with technological input – they can acquire material properties.

In the following paragraphs, goods and services are regarded as objects of individual acts of consumption and will be subsumed under the term 'consumer goods'. This term is to be understood not in a material, but in a referential sense. In terms of individual acts of acquisition, use, disposal, production and co-production, consumer goods are all functional, i.e. they are capable of satisfying needs. According to Di Giulio et al. (in this volume), the needs people seek to satisfy by means of consumer goods ('satisfiers') can be divided into objective needs and subjective desires: objective needs refer to those individual constructs of wanting that, if satisfied, allow individuals to develop those (non-negotiable) elements of a good life that are considered universally valid. By contrast, subjective desires are individual constructs of wanting that cannot claim to relate to universally acknowledged elements of a good life.

Against the above background, acts of consumption can thus be seen to consist in selecting, acquiring, using, as well as disposing of, recycling and co-producing consumer goods for the purpose of satisfying objective needs and subjective desires.

3.2.2 The 'sustainability' frame of reference and its normative implications

In this section, the question to be considered is what happens if consumption is included in the normative discourse of sustainable development, i.e. if one attempts to relate consumption phenomena to another social construction, namely the idea of sustainability. The concept of sustainability and its inherent ethical norms will briefly be explored before being applied to acts of consumption.

A frequent starting point for discussing sustainable development is the so-called Brundtland definition promulgated by the United Nations (WCED 1987, p. 54):

"Sustainable development is development that meets the needs of the present without compromising the ability of future generations to meet their own needs."

Sustainability is framed above as a normative concept of a world as it ought to be. Central to such a conceptualisation is the so-called "equity principle" (Kopfmüller et al., 2001, p. 31; translated by C. Holzherr) – with its inter- and intra-generational dimensions (cf. Ott/Döring 2004). In the discourse of the United Nations, this definition thus has a specific sense that differs from how it is used in other contexts, e.g. in everyday language or in forestry (cf. Di Giulio 2004). The Brundtland definition – often cited in the above, abbreviated, version – goes on to invoke two key concepts (WCED 1987, p. 54):

- "the concept of 'needs', in particular the essential needs of the world's poor, to which overriding priority should be given; and
- the idea of limitations imposed by the state of current technology and social organization on the environment's ability to meet present and future needs."

As for the question of which needs should be satisfied, Di Giulio et al. (in this volume) argue that only in the case of objective needs is there a duty to provide external conditions that allow all people (also and in particular people from the South) to satisfy their objective needs in order to (individually) lead a good life. These external conditions may additionally be subject to material constraints. For instance, ecosystems have finite carrying capacities and a failure to protect them will result in irreversible damage (cf. Rockström et al. 2009).

A definition of sustainable acts of consumption can thus be formulated along the following lines: a primary goal of sustainable development is to ensure that both now and in the future, all people should be given the opportunity to satisfy their objective needs in order to be able to develop universally valid human characteristics and capabilities (e.g. to be socially interactive, to be healthy and to be able to move from one location to another; cf. Di Giulio et al. in this volume). A key criterion in judging whether an individual act of consumption is sustainable is therefore the extent to which individual consumption acts contribute to creating or maintaining the external conditions that allow humans to satisfy their objective needs.

Such a definition has a number of advantages over the Oslo definition of sustainable consumption:

- Being based on a solid and explicit conceptual system, it goes beyond an everyday definition of 'basic needs' and 'quality of life'.
- By referring to external conditions for the satisfaction of objective needs it provides a superordinate evaluative criterion that is compatible with the idea of a good life and the Brundtland definition. By contrast, the Oslo definition mixes different levels, linking very basic and abstract notions on the one hand with (rather arbitrary and unfounded) specific pro-environmental measures (such as a reduction of toxic substances) on the other, while, at the same time, omitting economic and socio-cultural dimensions.
- It focuses on individual consumption. By contrast, the Oslo definition indiscriminately covers both production and consumption, with the consequence that acts of consumption are not sufficiently delineated from more general actions or behaviour.

The expectation that sustainable acts of consumption should contribute to satisfying the objective needs (within the constraints of our natural world) of present and future generations naturally requires further qualification. Questions to ask in this context include: What alterations to the ecosystem are tolerable, and which ones are not tolerable? What conditions have to be in place for people (which people?) to satisfy their needs (which needs?) and develop their capabilities (which capabilities?)? A variety of disparate proposals have been put forward in reply. For instance, in the context of strong versus weak sustainability, opinions are divided over the legitimacy and extent to which natural capital can and may be replaced by man-made artificial capital (cf. Ott/Döring 2004). The question of which needs have to be satisfied for a good life is another such controversial issue (cf. Di Giulio et al. in this volume; Costanza et al. 2007). Answers to these questions are not universally valid, but can only be given in relation to a specific historical context. In order to gauge the sustainability of an act of consumption, it is thus imperative to develop procedures and criteria whereby the key criterion proposed above is put into historical context and applied to specific acts of consumption.

What can be discussed more broadly is the object of evaluation, i.e. what aspects of an individual act of consumption should actually be evaluated by criteria yet to be specified. This will be the subject of the next section.

3.2.3 Evaluating sustainability in consumption in terms of *impact* and *intention*

In the context of sustainability and acts of consumption, the object of evaluation is not the acts of consumption per se, but their impacts or the intentions behind them. Building on this twofold focus will suggest distinguishing between an *impact-oriented* and an *intention-oriented* evaluation.

The assumption behind an *impact-oriented evaluation* is that the degree of sustainability of an act of consumption is determined by its consequences, which in turn "can only be defined in relation to a benchmark" (Scherhorn/Reisch 1997, p. 13; translated by C. Holzherr). Benchmarks could be ecological, economic and/or socio-cultural conditions that allow objective needs to be satisfied. From this perspective, an act of consumption is sustainable if the target-performance gap is closed (or narrowed). Naturally, such an approach requires a frame of reference in which specific targets for sustainable development are defined. These should reflect the overarching objective of allowing people – now and in the future – to satisfy their objective needs within the confines of our natural world.

Impact-oriented evaluation procedures only focus on those acts of consumption (selecting, acquiring, using etc. – see above) that have an effect on the external conditions for satisfying the objective needs of third parties (now and in the future). From an ecological perspective, for instance, the sustainability of the phases selection and acquisition of a given product can be evaluated in terms of how resource-intensive the product will be in the course of its entire life cycle (cf. for example Tukker et al. 2006). Looking at the phase of usage, life cycle assessments show that often usage accounts for most of the carbon footprint of a functional product (Grießhammer et al. 2004). This example illustrates that indicators relating only to the consumer good itself are necessary, but not always sufficient evaluation criteria. For an overall evaluation, all aspects of the act of consumption have to be included. Environmental impacts are thus related not only to the production of a product, but to its use, disposal, recycling and co-production. To refer, for instance, to the ecological life cycle assessment is indicative of the fundamental problem that there are relatively few examples of sophisticated proposals on how to operationalise and evaluate cultural and social aspects. Integrated evaluation instruments relating to the external conditions for the satisfaction of objective needs are still in their early stages of development (e.g. the Human Development Index – HDI).

An *intention-oriented evaluation* of acts of consumption addresses the objectives that individuals pursue through their actions. The normative notion of intra- and intergenerational equity means that not only one's own, but other people's and future generations' objectives come into the equation (cf. de Haan et al. 2008). In other words, both our generation and future generations must be entitled to satisfy their objective needs within the constraints of our natural world. The decisive criterion for an intention-oriented evaluation of the sustainability of an act of consumption is thus not its impact (or absence of impact), but solely the intention to act (or not to act) sustainably. Acts of consumption that are not intended to contribute to sustainable development should therefore be regarded as *neutral*, i.e. as neither sustainable nor unsustainable, but as 'not unsustainable'.

The above distinction is compatible with ethical approaches and distinctions, such as that drawn by Max Weber (2010) between *ethic of conviction* ("Gesinnungsethik") and *ethic of responsibility* ("Verantwortungsethik"), or more recent distinctions such as that between *intent*-oriented and *impact*-oriented research into individual consumer behaviour (Stern 2000). In its own right, our distinction contributes to a definition of sustainable consumption insofar as it links impact- and intention-oriented evaluation to individual acts of consumption, thereby overcoming Stern's purely ecological perspective by invoking the above-mentioned overarching external conditions for satisfying objective needs. Our proposal for evaluating the sustainability of acts of consumption thus not only allows for 'sustainable' and 'unsustainable', but additionally for *neutral* consumption. It also allows for a distinction between individual evaluation of individual consumer behaviour as well as societal evaluation of individual consumer behaviour, whereby the individual is explicitly not responsible for achieving sustainable development for the whole of society.

Ideally, individual acts of consumption are sustainable if they are oriented towards the objectives of sustainable development (intention-oriented), and actually contribute to their accomplishment (impact-oriented). In other words, they should be *both* intention-oriented *and* impact-oriented. This will be the case if an act of consumption is not only intended to narrow the target-performance gap in relation to the (specific) external conditions necessary for satisfying objective needs of current and future generations, but also contributes to actually achieving such a result. The idea behind sustainability is to strive for circumstances where all human beings are able to satisfy their objective needs and lead a good life. From a societal perspective, an impact-oriented evaluation is of more interest than an intention-oriented evaluation. The next section will – after illustrating both evaluation approaches – take a closer look at the relative importance of an intention-oriented evaluation.

3.3 Impact-oriented and intention-oriented evaluation in practice

Two projects from within the focal topic will now serve to illustrate the two evaluation approaches outlined above: the "Consumer/Prosumer" project, whose object it was to investigate the impacts of online second-hand consumer trading, and the "BINK" project group, who looked into what competencies people need in order to be able to consume more sustainably.

3.3.1 Online second-hand trading ("Consumer/Prosumer")

Online second-hand trading can, under certain circumstances, present opportunities for sustainable consumption. One of the objectives of the "Consumer/Prosumer" project was to gauge the environmental impacts of second-hand eBay trading (cf. Erdmann 2011). The investigation focused on impacts – although exclusively on ecological impacts. The social-ecological systems approach that was chosen for this investigation allowed for an analysis of primary, secondary and tertiary environmental impacts – at both product and society level. Private eBay users were questioned about their online trading, and a quantification of environmental impacts was undertaken with the help of a material-flow model. The advantage of such agent-centred analysis is that it tends to offset the weaknesses of a life cycle assessment. Both virtual (computer-based) and physical (packaging, shipping) activities were considered. It was found that online second-hand trading is environmentally beneficial only if it does not cause an acceleration of consumption (rebound effects) and if transport involved in trading is kept to a minimum (e.g. by trading regionally). Generally speaking, durable, high-quality products whose usage is extended by resale have positive environmental impacts. Second-hand trading can have a positive environmental impact if the second-hand products are used by the new owner for an extended period and if their usage does not require energy or water. 80 percent of eBay second-hand products fall into this category. With products requiring electricity and/or water for their usage, age and efficiency appear to dictate ecological impact. The improved efficiency of new products can, in fact, outweigh the longer usage of older products. Overall, it was found that, from an ecological perspective, second-hand online trading had, under certain circumstances, a beneficial impact on the environment.

In order to find out about people's motivations for eBay trading, an online survey was conducted (N=2,511) (Clausen et al. 2010). It emerged that motives such as saving money, not having to rely on opening hours, convenience and the thrill of online auctioning ranked higher than the desire to protect the environment. By means of a cluster analysis, five consumer types were identified. They included *environment-ori-*

ented second-hand purchasers (22 percent), *prosumers* (23 percent) and *price-conscious second-hand purchasers* (20 percent). Whilst the prosumers have a strong interest in reselling (this shows in their handling of products as well as their high motivation to resell), the price-conscious purchasers use eBay to buy goods that they could not afford to buy otherwise. Although both groups show relatively low environmental awareness, their online trading does contribute to the protection of the environment. Buying second-hand goods is thus an example of how acts of consumption can have sustainable *impacts* without people having sustainable *intentions*.

3.3.2 Education and sustainable consumption ("BINK")
In an intention-oriented evaluation of consumer behaviour the aim is to find out which competencies individuals must possess in order to be able to consume more sustainably and how they can be supported in their intentions to do so. The objective of the "BINK" project was to investigate what contribution educational institutions could make in this area.

The "BINK" project developed a framework capable of capturing concrete prerequisites for intentional sustainable consumption behaviour. The aim was to develop those competencies in individuals that would enable them to a) incorporate sustainability as a criterion for their consumer behaviour, b) plan their consumer behaviour accordingly, c) be able to critically evaluate the impacts of their behaviours, and d) reflect on how the culture they live in influences their own consumption practices. The ability to evaluate one's own consumption practices is crucial in this respect: if the focus of ethical scrutiny shifts from impacts to intentions, then self-evaluation comes to the fore. Faced with an abundance of different kinds of information about products (e.g. origin and quality), people have to be able to critically evaluate information and carefully choose between different sources of information. People are thus required to make a real effort to go out and gather information, as well as be ready to deal with the tensions and insecurities aroused by contradictory information.

In order to support people's good intentions with regard to sustainability, the "BINK" project put in place certain changes within the institutions (e.g. in catering). The objective was to observe how individuals acquired and used new competencies. The data from the empirical investigation showed, amongst other things, that – in comparison to an external control group – young people who were involved in the project were significantly more likely to incorporate sustainability criteria in their consumer practices. Moreover, they were significantly more likely to believe that their behaviour had a positive influence on the production process of a product. The findings indicate that involving students in an institution-wide sustainability campaign is a suitable strategy to support young people in consuming more sustainably (Barth et al. 2012).

Enabling people to become more conscious of their consumption patterns and consume more sustainably thus means equipping them with a wide range of competencies. Given the considerable cognitive and emotional demands made on people, and given the differences between individuals, it cannot be assumed that all people will acquire the necessary competencies to the same extent. It would therefore be unrealistic to expect educational institutions to be fully informed about their students' competencies in respect of sustainable choices, and to make good any deficiencies in this regard. Another challenge in this respect for future projects might be to investigate if and to what extent the right intentions lead to actual sustainable behaviours.

3.4 Conclusion and future perspectives

The idea that individuals should ideally evaluate their consumer behaviour from both an impact-oriented as well as an intention-oriented perspective is, in several respects, problematic. First, it is unreasonable to expect an individual to be fully aware of the consequences of his or her actions; consumer behaviour is not always reflected behaviour (see Kaufmann-Hayoz/Bamberg et al. in this volume). Second, lifestyle research has shown that consumption patterns which are sustainable in terms of their impacts need not be motivated by intentions to be sustainable. An action can thus result in a desired outcome without being sustainable from an intention-oriented perspective. Third, intentions to act sustainably do not necessarily lead to actions, and even if they do, these actions do not always have the desired effects. An act of consumption can thus be sustainable from an intention-oriented perspective and unsustainable from an impact-oriented perspective.

The examples above have demonstrated that the two approaches to evaluation are complementary rather than in competition to each other. An impact-oriented evaluation of individual consumer behaviour is highly complex and inevitably comes up against some gaps in information as well as unknown causal relations between means and ends. People cannot be expected to identify and weigh up all potential impacts before each act of consumption (e.g. far-away, future or unintended impacts). An exclusively impact-oriented evaluation would put undue stress on people – to the point where some might well refuse to take on (or be excused from) the responsibility for contributing to sustainable development. From an overall societal perspective, however, an impact-oriented analysis is indispensable. It is a prerequisite for judging which actions are conducive to sustainable development and which instruments might therefore be best suited to steer consumer behaviour in the desired direction (cf. Kaufmann-Hayoz/Brohmann et al. in this volume).

In summary, it can be said that it takes both evaluation approaches in order to achieve sustainable consumption on a long term basis. Intention-oriented approaches, which aim to reinforce the intentions and competencies necessary for sustainable consumer behaviour, are invaluable for creating the necessary capacities and the acceptance of a need for change in society towards more sustainable consumption. Impact-oriented approaches help to shed light on the impacts and interrelations between acts of consumption, which can then be steered in such a way that sustainable consumption becomes a feasible goal for society at large. The two approaches are thus complementary rather than in competition to each other. With regard to steering consumer behaviour, it is thus a question of weighing up both approaches: deciding when it is necessary to strengthen, or to awaken, intentions in people, and when it is sufficient to aim for, enact or prevent certain impacts.

References

Baedeker C., Liedtke C., Welfens J. M., Busch T., Kristof K., Kuhndt M., Schmidt M., Türk V. (2005): Analyse vorhandener Konzepte zur Messung des nachhaltigen Konsums in Deutschland einschließlich der Grundzüge eines Entwicklungskonzeptes. Endbericht. Wuppertal: Wuppertal-Institut.

Barth M., Fischer D., Michelsen G., Nemnich C., Rode H. (2012, accepted). Tackling the knowledge-action gap in sustainable consumption: insights from a participatory school programme. In: Journal of Education for Sustainable Development.

Bea F. X., Friedl B., Schweitzer M. (1994): Allgemeine Betriebswirtschaftslehre. Bd. 3: Leistungsprozess. Stuttgart: UTB.

Belz F.-M., Bilharz M. (2005): Nachhaltiger Konsum. Zentrale Herausforderung für moderne Verbraucherpolitik. München: Technische Universität München.

Blättel-Mink B., Hellmann K.-U. (eds.) (2010): Prosumer Revisited. Zur Aktualität einer Debatte. Wiesbaden: VS-Verlag für Sozialwissenschaften.

BUND & Misereor (1996): Zukunftsfähiges Deutschland. Ein Beitrag zu einer global nachhaltigen Entwicklung. Basel: Birkhäuser Verlag.

Campell C. (1998): Consuming Goods and the Good of Consuming. In: Crocker D. A. (ed.): Ethics of consumption. The good life, justice, and global stewardship. Lanham, Md: Rowman & Littlefield. 139–154.

Clausen J., Blättel-Mink B., Erdmann L., Henseling C. (2010): Contribution of Online Trading of Used Goods to Resource Efficiency: An Empirical Study of eBay Users. In: Sustainability 2 (6): 1810–1830.

Costanza R., Fisher B., Ali S., Beer C., Bond L., Boumans R., Danigelis N. L., Dickinson J., Elliott C., Farley J., Gayer D. E., Glenn L. M., Hudspeth T., Mahoney D., McCahill L., McIntosh B., Reed B., Rizvi S. A. T., Rizzo D. M., Simpatico T., Snapp R. (2007): Quality of life: An approach integrating opportunities, human needs, and subjective well-being. In: Ecological Economics 61 (2–3): 267–276.

Di Giulio A. (2004): Die Idee der Nachhaltigkeit im Verständnis der Vereinten Nationen. Anspruch, Bedeutung und Schwierigkeiten. Münster: LIT Verlag.

Di Giulio A., Brohmann B., Clausen J., Defila R., Fuchs D., Kaufmann-Hayoz R., Koch A. (in this volume): Needs and consumption – a conceptual system and its meaning in the context of sustainability.

Erdmann L. (2011): Quantifizierung der Umwelteffekte des privaten Gebrauchtwarenhandels am Beispiel von eBay. In: Behrendt S., Blättel-Mink B., Clausen J. (eds.): Wiederverkaufskultur im Internet. Chancen für nachhaltigen Konsum am Beispiel von eBay. Berlin: Springer. 127–158.

Gebhardt B., Dölle R., Weber C., Hugger C., Farsang A., Scherhorn G. (2003): Nachhaltiger Konsum im Spannungsfeld zwischen Modellprojekt und Verallgemeinerbarkeit. Abschlussbericht. Förderkennzeichen 07KON03/6. Stuttgart: Universität Stuttgart.

Grießhammer R., Bunke D., Eberle U., Gensch C.-O., Graulich K., Quack D., Rüdenauer I., Götz K., Birzle-Harder B. (2004): EcoTopTen – Innovationen für einen nachhaltigen Konsum. Pilot-Phase (final report). Freiburg im Breisgau: Öko-Institut.

Grunwald A. (2010): Wider die Privatisierung der Nachhaltigkeit. In: GAiA 19 (3): 178–182.

Grunwald A. (2011): On the Roles of Individuals as Social Drivers for Eco-innovation. In: Journal of Industrial Ecology 15 (5): 675–677.

Günther C., Fischer C., Lerm S. (eds.) (2000): Neue Wege zu nachhaltigem Konsumverhalten. Eine Veranstaltung der Deutschen Bundesstiftung Umwelt zur EXPO 2000. Berlin: Verlag Erich Schmidt.

Haan G. de, Kamp G., Lerch A., Martignon L., Müller-Christ G., Nutzinger H. G., Wütscher F. (2008): Nachhaltigkeit und Gerechtigkeit. Grundlagen und schulpraktische Konsequenzen. Berlin: Springer.

Jackson T. (ed.) (2006): The Earthscan reader in sustainable consumption. London: Earthscan.

Jacobsen H. (2010): Strukturwandel der Arbeit im Tertiarisierungsprozess. In: Böhle F., Voß G. G., Wachtler G. (eds.): Handbuch Arbeitssoziologie. Wiesbaden: VS Verlag für Sozialwissenschaften. 203–228.

Kaufmann-Hayoz R., Bamberg S., Defila R., Dehmel C., Di Giulio A., Jaeger-Erben M., Matthies E., Sunderer G., Zundel S. (in this volume): Theoretical perspectives on consumer behaviour – attempt at establishing an order to the theories.

Kaufmann-Hayoz R., Brohmann B., Defila R., Di Giulio A., Dunkelberg E., Erdmann L., Fuchs D., Gölz S., Homburg A., Matthies E., Nachreiner M., Tews K., Weiß J. (in this volume): Societal steering of consumption towards sustainability.

Kopfmüller J., Brandl V., Jörissen J., Paetau M., Banse G., Ceoenen R., Grunwald A. (2001): Nachhaltige Entwicklung integrativ betrachtet. Konstitutive Elemente, Regeln, Indikatoren. Berlin: Ed. Sigma.

Lorek S., Spangenberg J. H., Felten C. (1999): Prioritäten, Tendenzen und Indikatoren umweltrelevanten Konsumverhaltens. Endbericht des Teilprojekts 3 des Demonstrationsvorhabens zur Fundierung und Evaluierung nachhaltiger Konsummuster und Verhaltensstile. Im Auftrag des Umweltbundesamtes, FE Vorhaben: 209 01 216/03. Wuppertal: Wuppertal-Institut.

Norwegian Ministry of the Environment (1994): Symposium Sustainable Consumption. Oslo.

Ott K., Döring R. (2004): Theorie und Praxis starker Nachhaltigkeit. Marburg: Metropolis Verlag.

Reichwald R., Piller F. (2009): Interaktive Wertschöpfung. Wiesbaden: Gabler.

Reisch L. A., Bietz S. (2011): Communicating Sustainable Consumption. In: Godemann J., Michelsen G. (eds.), Sustainability Communication: Interdisciplinary Perspectives and Theoretical Foundation. Berlin: Springer. 141–150.

Rockström J., Steffen W., Noone K., Persson Å., Chapin F. S., Lambin E. F., Lenton T. M., Scheffer M., Folke C., Schellnhuber H. J., Nykvist B., Wit C. A. de, Hughes T., van der Leeuw S., Rodhe H., Sörlin S., Snyder P. K., Costanza R., Svedin U., Falkenmark M., Karlberg L., Corell R. W., Fabry V. J., Hansen J., Walker B., Liverman D., Richardson K., Crutzen P., Foley J. A. (2009): A safe operating space for humanity. In: Nature 461 (7263): 472–475.

Scherhorn G., Reisch L. A. (1997): Wege zu nachhaltigen Konsummustern. Überblick über den Stand der Forschung und vorrangige Forschungsthemen. Marburg: Metropolis Verlag.

Scherhorn G., Weber C. (eds.) (2003): Nachhaltiger Konsum. Auf dem Weg zur gesellschaftlichen Verankerung. München: oekom.

Schrader U., Hansen U. (eds.) (2001): Nachhaltiger Konsum. Forschung und Praxis im Dialog. Frankfurt am Main: Campus Verlag.

Spangenberg J. H., Lorek S. (2002): Environmentally sustainable household consumption: from aggregate environmental pressures to priority fields of action. In: Ecological Economics 43 (2–3): 127–140.

Stern P. (2000): Toward a Coherent Theory of Environmentally Significant Behavior. In: Journal of Social Issues 56 (3): 407–424.

Toffler A. (1983): Die dritte Welle. Zukunftschance. Perspektiven für die Gesellschaft des 21. Jahrhunderts. München: Goldmann Sachbuch.

Tukker A., Huppes G., Guinée J., Heijungs R., de Koning A., van Oers L., Suh S., Geerken T., Van Holderbeke M., Jansen B., Nielsen P. (2006): Environmental Impact of Products (EIPRO). Analysis of the life cycle environmental impacts related to the final consumption of the EU-25. Main report. Seville.

UBA – Umweltbundesamt [Federal Environment Agency] (1997): Nachhaltiges Deutschland. Wege zu einer dauerhaft umweltgerechten Entwicklung. Berlin: Verlag Erich Schmidt.

UBA – Umweltbundesamt [Federal Environment Agency] (ed.) (2001): Aktiv für die Zukunft – Wege zum nachhaltigen Konsum. Dokumentation der Tagung der Evangelischen Akademie Tutzing in Kooperation mit dem Umweltbundesamt vom 3.–5. April 2000. Berlin.

Verbraucherzentrale Nordrhein-Westfalen [Consumer advice centre for North Rhine-Westphalia] (ed.) (1997): Herausforderung Sustainability. Konzepte für einen zukunftsfähigen Konsum; Dokumentation einer Fachtagung am 11. + 12. November '96 in Aachen. Düsseldorf.

Voß G., Rieder K. (2005): Der arbeitende Kunde. Wenn Konsumenten zu unbezahlten Kunden werden. Frankfurt a. M.: Campus Verlag.

WCED – World Commission on Environment and Development (1987). Our common future. Oxford: Univ. Press.

Weber M. (2010): Politik als Beruf. Berlin: Duncker & Humblot.

Ruth Kaufmann-Hayoz, Sebastian Bamberg, Rico Defila, Christian Dehmel,
Antonietta Di Giulio, Melanie Jaeger-Erben, Ellen Matthies, Georg Sunderer, Stefan Zundel[1]

4 Theoretical perspectives on consumer behaviour – attempt at establishing an order to the theories

4.1 Introduction

4.1.1 Issue and objective of this article

The articles in this volume examine individual consumption[2], which is understood to be the utilisation of goods (products, services, infrastructure) by individuals. Such 'utilisation' is expressed in the form of consumer *behaviour*, i.e. in the acts of selecting, acquiring, using/consuming and disposing of or forwarding on consumer goods. It is therefore indicated that theories of action are key to the scientific analysis of questions of sustainable consumption from a theoretical point of view. Firstly, action theories allow us to describe, understand and explain phenomena of consumer behaviour. Secondly, they help determine and interpret the prerequisites which must be met and the difficulties which may arise in changing such behaviour towards greater sustainability. Thirdly, they allow for targeted and promising interventions to change consumer behaviour based on relevant theories.

---

[1] Discussion participants were Tanja Albrecht (ENEF-Haus), Marlen Arnold (Nutzerintegration), Matthias Barth (BINK), Frank-Martin Belz (User Integration), Birgit Blättel-Mink (Consumer/Prosumer), Bettina Brohmann (Seco@home), Henriette Cornet (User Integration), Dirk Dalichau (Consumer/Prosumer), Benjamin Diehl (User Integration), Elisa Dunkelberg (ENEF-Haus), Stefan Engeser (User Integration), Daniel Fischer (BINK), Dorika Fleissner (Intelliekon), Doris Fuchs (Transpose), Jürgen Gabriel (Heat Energy), Sebastian Gölz (Intelliekon), Konrad Götz (Intelliekon), Felix Groba (Seco@home), Ulrich Hamenstädt (Transpose), Christine Henseling (Consumer/Prosumer), Andreas Homburg (BINK), Katy Jahnke (Heat Energy), Angelika Just (LifeEvents), Martin Kesternich (Seco@home), Marian Klobasa (Intelliekon), Andreas Klesse (Change), Andreas Koch (Heat Energy), Michal Kohlhaas (Seco@home), Pia Laborgne (Heat Energy), Gerd Michelsen (BINK), Harald Mieg (BINK), Joachim Müller (Change), Ralf-Dieter Person (Change), Klaus Rennings (Seco@home), Marc Requardt (User Integration), Martina Schäfer (LifeEvents), Joachim Schleich (Seco@home), Volker Schneider (Transpose), Immanuel Stieß (ENEF-Haus), Susanne Steiner (User Integration), Kerstin Tews (Transpose), Thure Traber (Seco@home), Claus Tully (BINK), Victoria van der Land (ENEF-Haus), Sandra Wassermann (Heat Energy), Julika Weiß (ENEF-Haus), Wiebke Winter (Consumer/Prosumer), Daniel Zech (Heat Energy).

[2] Different to collective consumption of companies and public procurement, cf. Baedeker et al. 2005.

Technical-functional and symbolic-communicative functions of goods and their use are closely intertwined in individual consumer behaviour, with individuals' actions always integrated in socio-cultural processes.[3] This means that individual consumption is a complex and multi-faceted phenomenon, which can be examined from various perspectives in terms of (action) theories. To mention just two fundamentally different theoretical approaches: the object of study can be an individual's act of consumption as such, abstracted from its context for analytical purposes, or the main interest can rather be on the incorporation of actions in social, cultural and material contexts. The decisive question is not which action theory is better, but which action theory is more suitable in which case. Different perspectives and therefore different action theories are meaningful depending on the context and the question which is to be answered and depending on the problem which is examined. The question posed in this article is therefore which action theories are particularly suitable in examining which phenomena of consumer behaviour.

Answering this question requires a structure of the phenomena of consumer behaviour from the perspective of action theory, i.e. establishing an order for these phenomena, to which all of the relevant action theories from various disciplines can be related. Therefore, the first objective of this article is creating a structure of the multi-faceted phenomena of consumer behaviour which offers connection points to various disciplines, with the second objective to correlate established action theories to show which consumer behaviour phenomena these theories examine and which they are more likely to ignore. The point in this is not to conclusively clarify any potential discrepancies between various theoretical approaches. Rather, by assigning theories to phenomena, we would like to make visible where various approaches compete to explain the same phenomena and where any deficits in the theories may be located. Such an allocation is intended to enable experts – both in science and practice – who are well versed in partial areas of (sustainable) consumer behaviour:

- to find suitable action theories for the consumer behaviour phenomena they deem interesting and to identify groups of theories which examine the same areas of phenomena;
- starting with the difficulties of achieving sustainable consumption, to find relevant action theories, which can be used to justify and design targeted interventions (naturally, this requires the normative clarification of the objective of 'sustainability in consumer behaviour'; cf. Fischer et al. in this volume);
- to identify compatibilities and divergences between various theories.

---

3 The four-volume collection edited by Alan Warde (2010) gives a good insight into the social scientific research into consumption in the past few decades.

The fact that it is not possible to examine all of the existing action theories within the framework of a single article is self-evident. Limiting the scope of this article to the action theories presented in more detail was the result of pragmatic considerations: theories were selected that were used in the focal topic, as well as theories that are particularly popular in the context of sustainability according to the experiences of the researchers involved.

Following a few preliminary remarks regarding terminology, section 4.2 will present a structure of the phenomena of individual consumer behaviour. Having shown the structure, we will then present important 'families' of action theories that have been used in the focal topic (section 4.3), before section 4.4 points out the areas of consumer behaviour phenomena for which the various theories provide explanations.

4.1.2 Preliminary remarks regarding terminology

'Acting', 'action', 'behaviour'
'Acting' is generally understood as intentional human activity with a specific attributed meaning. An 'action' or 'act' is a specific unit of activity which has been singled out for analysis and description and which is organized in terms of content and time and focussed on a given objective. This may also include omitted actions. The exact way in which the analytical units are determined varies according to the various theories, with some theories using labels other than 'action' or 'act' (e.g. 'practice' as an analytical unit for bundles of activities structured in time and space; cf. section 4.3.6).

The term 'behaviour' is used differently according to different theoretical and disciplinary contexts. It is either used – as in many psychological theories – as a superordinated concept, which encompasses the entirety of human activities and thus includes 'acting' in the sense described above. Or it is distinguished from 'acting' to designate any human activities which are not intentional and are therefore performed without subjective meaning, e.g. reflex-like movements or errors caused by inattentiveness (such a usage is more common in sociology, cf. the definition according to Max Weber in Kaesler 2002).

Human activities in connection with individual consumption are generally performed within contexts, i.e. they have individual and social significance. Even habitual purposes or routine uses can be understood within such contexts: they are not reflex-like or coincidental, even if such routines or parts of the routines are largely performed automatically. Consumer activities are therefore to be understood as 'acting' as outlined above. Here, we generally use the term consumer behaviour in its superordinated meaning, and consumer actions or acts of consumption to designate specific parts or analytical units of consumer behaviour. However, we hereby do not tie ourselves down to a certain specific theory.

'Action theory'

By 'action theory' we mean all theories which can deal with phenomena of individual consumer behaviour as outlined above. This does not just include the theories which label themselves as action theories, but also those which are labelled as, e.g., behavioural, decision or practice theories. The decisive factor is whether a theory relates to individual behaviour (i.e. on micro level). The social and cultural integration may be examined or the individual behaviour may be viewed in aggregate form (e.g. as the behaviour of households or lifestyle groups). In contrast, theories which refer to the meso or macro level and any changes occurring there – e.g. theories on how institutional frameworks are created or how organisations 'act' – are not examined under the scope of this article, even if they label themselves as 'action theories' (such as system-functional action theories in the tradition of Talcott Parsons or other more modern theories of action inspired by evolutionary biology). Naturally, this does not mean that such theories may not also be of importance for the topic of sustainable consumption.

'Consumption', 'consumer behaviour'

Following Di Giulio et al. and Fischer et al. (both in this volume), consumption is understood as the utilisation of goods (products, services, infrastructures) to satisfy objective needs and subjective desires (cf. Di Giulio et al. in this volume for more on this distinction). Correspondingly, we refer to consumer actions as the actions such as selection, acquiring, use or consumption, as well as disposal or recycling and co-producing of goods.

The fact that consumer actions affect third parties and that consumer actions also indirectly utilise or 'consume' components of nature (resources, states, processes) is – in terms of action theories – an (often unintended) side-effect of consumer actions. It is less important to the explanation or understanding of consumer behaviour than for its assessment as more or less sustainable (cf. Fischer et al. in this volume).

4.2 Establishing a structure to the phenomena of consumer behaviour

An order or structure of the phenomena of consumer behaviour to which essentially all of the relevant action theories should be able to relate must itself be as 'free from theory' as possible. Admittedly, this is not something which can be completely achieved, as it is impossible to refrain from making distinctions and using concepts which are not largely predicated on theory. The structure for the phenomena of consumer behaviour which we propose below merely claims to be free from theory in the sense that it is not rooted in a single theory of action.

The order suggested here for the phenomena was developed following several rounds of discussion among the researchers of the SÖF focal topic of sustainable consumption. It represents a synthesis which incorporated the various disciplinary backgrounds and theoretical approaches shaping the perspectives to the phenomena of consumer behaviour taken by those involved. The suggested structure is not the only possible structure, but allows connections to be made to a relatively broad spectrum of scientific approaches.

The structure of the phenomena of consumer behaviour relevant to consumer theory was established from three perspectives: Firstly from the perspective of differentiating between various types of consumer actions from an individual point of view (4.2.1), secondly from the perspective of the social and cultural incorporation of the consumer behaviour (4.2.2) and thirdly from the perspective of the change of individual consumer behaviour over time (4.2.3).

4.2.1 Characteristics of consumer actions

From an individual perspective, individuals' consumer acts can be characterised according to three general dimensions – namely (1) how *consciously* a consumer action is carried out, i.e. the degree of attention it is accompanied by; (2) how *significant* the consumer good or consumer act is for the agent; and (3) the *degree of freedom* with regard to the consumer act, i.e. the degree to which the consumer action is pre-structured. These three characteristic dimensions are described below, with typical examples used to illustrate them in practice.

Degree of consciousness of the consumer act: reflected vs. non-reflected

An individual may be more or less aware or conscious when performing actions. With regard to consumer actions, it seems sensible to label the poles at the two ends of this dimension as 'reflected' and 'non-reflected'. Consumer acts that are near to the 'reflected' pole are those which a person consciously elects to take after considering all of the options available to them. They require a great deal of attention, and there can be several reasons for this (e.g. novelty of the action or situation, rarity, significance due to connection with high costs). In contrast, consumer acts that are near to the 'non-reflected' pole are performed with little attention paid to them, and there is no (renewed) conscious decision and check. Often such actions have the character of routines, i.e. they are performed out of habit due to frequent repetition, meaning that they require little conscious control.

Table 1: Typical examples of consumer acts from an individual perspective, situated in different areas of the three dimensions: degree of consciousness, significance and pre-structuring. The typical examples were selected regardless of whether the corresponding consumer act would be judged sustainable or non-sustainable.

Degree of consciousness	Signifi-cance	Pre-structuring	Typical examples
Reflected	Non-essential	High degree of freedom	**Procuring new windows for an 'ordinary' house** ◆ As the windows of a house only need to be replaced every few decades and the hypothetical person only owns the one house, this is a *reflected* action. ◆ Replacing the windows or specific product features does not have a great significance beyond the functional use, meaning that the consumer action is *non-essential*. ◆ As the windows offered in the price range that the hypothetical person can afford are of various makes and different frame colours and materials, different hinges, heat and noise insulation qualities etc., which all correspond to the structural and legal circumstances, there is a relatively *high degree of freedom* regarding the type of the consumer action and the type of product.
		Low degree of freedom	**Installation of digital TV/internet** ◆ The installation is one-off – at least over a time horizon of a few years – meaning that it is a *reflected* action. ◆ The installation is the means to an end for the hypothetical person and does not have any significance beyond the functional use, making it *non-essential*. ◆ The hypothetical person sometimes works from home, but they can only do their work with a suitable internet connection. They enjoy watching TV and their favourite TV programmes can now only be received with a digital reception. There is only one telecommunications provider. All of this means that the consumer action has a *low degree of freedom*.
	Essential	High degree of freedom	**Participating in a 3-month trekking expedition in the Himalayas** ◆ The trip is a unique experience in the life of the hypothetical person, making it a *reflected* action. ◆ In making the trip, the hypothetical person is fulfilling a lifetime dream, which is therefore an *essential* element in their idea of a full life. ◆ There is a *high degree of freedom* for the hypothetical person for this consumer act. The person's social environment neither forces nor prevents them from performing the action, and there are various providers, who offer such trips at comparable quality at an attainable price.
		Low degree of freedom	**Procuring new windows for a 'listed' building** ◆ As the windows for such a building are only very rarely replaced, and the hypothetical person only owns this property, it is a *reflected* action. ◆ Maintaining the house is of great significance to the hypothetical person, as they are convinced of its value in terms of cultural history. Replacing the windows is an *essential* action for them. ◆ The provisions for the protection of listed buildings only permit a highly specific design of the windows, and as some of the windows are leaky, they need to be replaced urgently. Therefore, there is a *low degree of freedom* for the action.

Theoretical perspectives on consumer behaviour – attempt at establishing an order to the theories

Degree of consciousness	Significance	Pre-structuring	Typical examples
Non-reflected	Non-essential	High degree of freedom	**Buying milk** ◆ The hypothetical person buys milk at least once a week at the supermarket, without thinking too much about it, and normally grabbing the same product. It is a habitual purchase, meaning that the action is *non-reflected*. ◆ The action does not have any particular significance for the hypothetical person beyond its functional use – covering the household's milk requirements. The person does not look into how the milk has been produced and processed. The consumer action is *non-essential* as long as the product fulfils its functional use. ◆ The supermarket offers around a dozen different types of milk with different fat contents, best before dates and various packaging, as well as a few special milk types, and there is the option in the village to purchase milk directly from the farmer, so there is a relatively *high degree of freedom*.
Non-reflected	Non-essential	Low degree of freedom	**Flushing the toilet** ◆ Flushing the toilet is an automatic action, which is performed several times throughout the day. It is a *non-reflected* action. ◆ The action does not have any specific significance for the hypothetical person. It is *non-essential*. ◆ It would not really be socially acceptable not to flush the toilet. The technical sanitary installation prescribes the type of use, and the hypothetical person has little or no influence on what type of and how much water is used with each flush. There is a *low degree of freedom*.
Non-reflected	Essential	High degree of freedom	**Buying clothes** ◆ The hypothetical person is constantly adding to and renewing their wardrobe. Finding and trying on clothes is an almost weekly activity for them. Buying clothes is therefore a *little reflected* action. ◆ The hypothetical person places a high value on the effects of their clothing, meaning that clothing is a way for them to display their personality. Choosing and buying clothing is therefore *essential* to them for their quality of life. ◆ The hypothetical person has a good income, lives in an area with a wide array of clothes on offer, without restrictive guidelines on clothing. There is therefore a *high degree of freedom* in buying clothes.
Non-reflected	Essential	Low degree of freedom	**Preparing thanksgiving dinner for the family** ◆ The hypothetical family celebrates thanksgiving dinner every year, so there is little thought as to whether and how it is to be organised. It is therefore a *little reflected* consumer action. ◆ Observing the thanksgiving tradition has great significance for the family. It is part of their national identity and helps strengthen the family bond. Preparing the traditional meal is therefore an *essential* consumer action. ◆ The guests expect that the traditional dishes are served, particularly the stuffed turkey. Therefore, there is a *low degree of freedom* for the action.

Significance of the consumer good or act: essential vs. non-essential
As social beings, humans are always part of a social world or a social order, in which consumption is laden with meaning and significance, which also enables social interaction and affiliation. Owning and using/consuming consumer goods therefore often have a more or less strong symbolic social or emotional significance in addition to its functional benefit. In each individual's search for what they deem a fulfilled life, with their idea of quality of life realised, consumption does not just play a role due to its functional use, but also due to such additional significance.

In this dimension, the poles result from the degree of significance an individual attaches to a specific activity, a specific personal objective or ownership. We will label the poles of this dimension 'essential' and 'non-essential'. Consumer acts which have strong significance and are indispensible to an individual's idea of a good life (e.g. wearing certain brands of clothing, playing certain sports, enjoying certain drinks) are found near the 'essential' pole. Consumer acts which are substitutable in terms of their significance in this regard are found near the 'non-essential' pole. The functional benefit of such consumer acts can naturally be important to an individual, but they can or could be replaced by another consumer act, for example another product (e.g. concluding health insurance, purchasing washing powder).

Degree of pre-structuring of the consumer act: high vs. low degree of freedom
Human behaviour is pre-structured to varying extents by economic, political, sociocultural and physical-material framework conditions, situational contexts (e.g. the requirements of organisation of everyday life) and personal traits and abilities (personal resources). We have labelled the poles in this dimension as 'high degree of freedom' and 'low degree of freedom' to express that the individual freedom in deciding for or against a consumer act, in choosing a consumer good or how to use it can be more or less extensive. Consumer acts located near the 'high degree of freedom' pole are such where the individual is free to perform the actions or not, based on the various context levels and their personal traits and abilities, or where they can choose from a rich pallet of consumer goods, or where the forms of use of a consumer good can be freely designed. Consumer acts located near the 'low degree of freedom' pole are such where a person is practically forced to take them or where they cannot or may not perform such actions, where only one single consumer good is available or where the use of the consumer good is strictly prescribed (e.g. if there are binding clothing rules, nutritional guidelines or cleanliness standards or where a consumer action is a mandatory component part of a higher-level complex action).

Typical examples for the description of consumer acts

Table 1 shows typical examples of consumer actions from an individual's perspective, which are situated differently with respect to the three dimensions. The examples highlight the fact that specific consumer acts can be placed in an order or structure in terms of whether they are nearer to one pole or another in the three dimensions. However, the typical examples also make it clear that the characterisation of a specific consumer action is not possible without knowledge of the social, cultural and material context in which it is performed. The description of the typical examples is therefore 'contextualised', i.e. it shows under which assumptions regarding the person, situation, framework conditions etc. the respective example can be related to the three dimensions in the manner shown.

4.2.2 The incorporation of individual consumer acts in social, cultural and material contexts

The fact that an outwardly identical consumer action can be represented differently with regard to the three aforementioned dimensions, depending on a person, situation and context, indicates the need to incorporate social, cultural and material contexts of individual consumer behaviour into the order for the phenomena. A distinction can be made between various levels, which cannot be separated in this form in general life, but which can be individually focussed on for the purposes of analysis.

The level of (everyday) social interaction

Consumer actions are mostly performed through interaction and cooperation with others. For consumer actions within a household this means that material or spatial changes, the shared use of equipment or the form of daily nutrition become the subject of negotiations. Outside the household, situations arise such as interaction or cooperation, for example with sales staff or colleagues during lunch breaks. The routine nature of actions for several everyday activities is based on the cooperation of the social environment. If the environment remains stable, it contributes to establishing habits; if it changes or cooperation is refused (e.g. during a train strike), routines are interrupted and short-term or long-term alternative solutions are needed. Even on this level, therefore, consumer actions also have social significance, for example as a symbol of togetherness or as the subject of negotiations. Interaction contexts and relationship structures also directly contribute to the pre-structured nature of behaviour.

The level of affiliation to social groups and milieux

Consumer actions also help people define themselves on a more abstract level. To a certain extent, specific consumer actions are communicated and reproduced as a label within lifestyle groups or social milieux, allowing people to express their affiliation and to identify this in others (see e.g. Reusswig et al. 2005). In addition to this demonstrative reproduction of the symbolic significance of consumer actions (see e.g. Schulze 1992), the term 'habitus' rather designates the unconscious reproduction of social affiliation (primarily to social classes), for example, through a certain taste in food, the choice of clothing or the purchase of certain objects (Bourdieu 1982). On this level, the production and reproduction of socio-cultural significances through consumer behaviour are particularly clear.

The level of institutional embeddedness
Consumer actions are institutionally embedded in the various areas of daily life, but also with regard to various stages of life in different ways. This includes, for example, clothing guidelines and regulated break times at work or guidelines for use of leisure facilities. However, institutional influences can also be found in what is supposedly the private life. For example, upon the birth of their first child, parents become part of a system of family policy. They become the target market for institutional offers and have to follow certain rules (both official guidelines such as medical examinations, vaccinations, schooling and unofficial guidelines on diet or upbringing), which are commonly accompanied by highly specific consumer actions. Consumer actions are pre-structured by way of rules and guidelines, but also have a specific social significance due to their sometimes implicitly conveyed social expectations.

The level of cultural embeddedness
On this level, consumer actions are connected with culturally anchored values, in turn representing and reproducing these values. For example, the definitions of hygiene, comfort (this also includes temperature and lighting conditions expected or regarded as comfortable), luxury, healthy nutrition and flexibility are culturally specific (see e.g. Shove 2005). These definitions are communicated and acquired through processes of socialisation and (re)produced through dialogue, but also have an effect on which standards, for example in terms of cleanliness and climatisation in buildings, are implemented and how these are then experienced. Cultural definitions and allocations of meaning also shape how actions are organised in terms of space and time, contributing to the pre-structured nature of actions.

The level of socio-technical and socio-spatial embeddedness
Consumer actions can particularly be observed with regard to their incorporation in socio-technical and socio-spatial contexts and therefore in terms of their material pre-structure. Even actions which are performed very individually, for example using one's

own bath, are socially embedded inasmuch as 'the social' is also 'imprinted' in technology and space. Both the design and method of use of baths are culturally influenced, and the existence and use of a bathtub is also connected to the social definition of cleanliness and/or comfort (see e.g. Thiemann 2006). Technical and spatial arrangements may favour the automation of actions. The more technology determines the process of the action or the more the room design prescribes specific ways or methods of use, the less the action needs to be consciously controlled and managed. At the same time, routines can also be interrupted through changes or disruptions of such arrangements, allowing for some reflection.

4.2.3 Phenomena of changing consumer behaviour

Looking at whether and how consumer behaviour changes over time and which processes are involved in such changes, a distinction must be made as to whether the changes as such are to be examined (what is changing?) or whether their facilitating

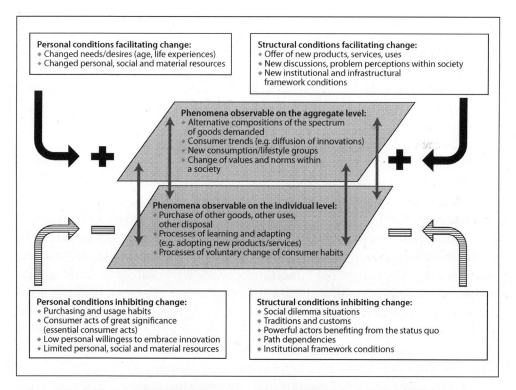

Figure 1: Structure of the phenomena in connection with changes in consumer behaviour: phenomena identifiable on individual and aggregate level, as well as personal and structural conditions which can facilitate or inhibit changes

and inhibiting conditions are to be observed (what promotes changes and what prevents them?). The distinctions shown in Figure 1 are discussed below.

What is changing?

On *individual* level, consumer actions can change with regard to all of the characteristics outlined above. Reflected actions may become non-reflected routines (and vice-versa), the significance of consumer acts may change (from essential to non-essential and vice-versa) and their degree of freedom may become greater or lesser. For example, specific changes can be detected in the fact that less, more or different consumer goods are being purchased, that they are being sourced differently, that the consumer goods are more or less frequently/intensively used or used in another way, that goods are used for longer or shorter periods and that the way in which a no longer used good is disposed of. Furthermore, such changes cannot only be viewed as differences over time, but also as dynamic processes (e.g. learning or adaptation processes for different life circumstances or as processes of voluntary behavioural changes).

On a *collective* (aggregate) level, changes of individual consumer behaviour can be detected if they occur in the same way in a large proportion of the population. In this case, they appear in the changed composition of the spectrum of goods in demand, social consumer trends (cf. for example Bosshart et al. 2010), in the formation of new lifestyle groups (cf. for example Rauch/Kirig 2007) or in changes of values and norms (cf. for example Inglehart 1997). If the individual change is the adoption of innovations (e.g. new ideas, products or methods of use), the corresponding social phenomenon can be described as a diffusion of innovations (cf. Rogers 2003).

What facilitates and what inhibits changes?

Looking at people as essentially autonomous subjects, who possess creativity in terms of ideas, actions and solutions to problems and who are able to set and implement their own change objectives – also in their role as consumers – it is clear that the non-determined nature of human behaviour also represents the fundamental *personal condition facilitating change*. For consumer behaviour it is possible to be more specific in that the course of an individual's life in itself promotes change, as objective needs and subjective desires may change as the person gets older or in connection with life events (e.g. the birth of a child may result in a desire for a bigger apartment). Equally, changes to personal, social and material resources may facilitate changes in consumer behaviour. For example, newly acquired knowledge of the meaning of a product label may lead to a decision to pay attention to such labels when making a purchase; a new partnership with a vegetarian may cause an individual to question and subsequently reduce his or her own meat consumption; a higher income facilitates the purchase of a new car. Such changes linked to the individual course of life become relevant for the phenomena observable on a collective level in the case of changes in the age composition or a general increase in the level of education or prosperity across a population.

On the other hand, some *inhibiting conditions primarily established in individuals* can stabilise the status quo of consumer behaviour or slow down or prevent change processes. Less reflected habits for purchases or uses represent such a condition, with the advantage for the individual of achieving an easier daily life; it is possible to break down these habits, but this is associated with uncertainty and effort. Consumer actions which carry great significance for an individual's idea of quality of life (essential consumer actions) would be rather more resistant to changes as they are, by definition, indispensible. A low personal willingness to embrace innovation, i.e. a person who is generally not prepared to accept new products and services ("innovativeness", cf. Rogers 2003) and limited personal, social or material resources (e.g. lack of time, financial resources) are considerable conditions which inhibit change.

A clear *structural condition promoting change* to individual consumer behaviour is the offer of new consumer goods (or the removal of such goods). These are more or less willingly accepted by individuals, which in turn expresses itself on a collective level in the speed of diffusion of the innovation. Newly arising social debates and problem perceptions can also promote change by prompting people to question their consumer habits or change their attitudes towards products and services, which in turn will be expressed as a new consumer trend on aggregate level – if this happens among a large proportion of the society. Finally, new institutional and infrastructural framework conditions may also prompt a change of individual consumer behaviour, e.g. the introduction (or repeal) of bans of certain types of use, the expansion (or reduction) of disposal infrastructures or changed default regimes[4]. Such changes may influence the degree of pre-structure of an individual's behaviour in that they may, for example, have more or less control on how an action is performed, or in that the changes represent opportunity structures for a desired or enforced change of consumer behaviour.

In general, prolonged institutional and socio-technical conditions and structures set boundaries in terms of willingness and possibilities to change and therefore represent *structural conditions which inhibit change*. Namely, this includes firmly established traditions and customs (such as the consumer frenzy at Christmas), but also social dilemmas, powerful actors as beneficiaries of the status quo (e.g., manufacturers have a vested interest in their products being replaced in short intervals and design their

---

4 'Default' is used to label a (technical) presetting (e.g. "normal programme" on a washing machine) or in general a rule which applies if nothing else is specified (e.g. general rights of way for pedestrians in a specific traffic zone).

offers accordingly, cf. the phenomenon of planned obsolescence, Slade 2006), path dependencies (e.g. in traffic and settlement policy, cf. Haefeli 2008) and institutional frameworks.

4.3 Action theories for the research into sustainable consumption

In this section, we will present short profiles of various action theories or families of theories that have a certain significance for the research into sustainable consumption and have been used in the focal topic. The basic assumptions, scope and conceptual structure are briefly outlined, with key literature stated. The objective of the outline is to show the various approaches of the theories to describing and explaining human behaviour to a certain extent 'at a glance' and thus to create the starting situation to discuss their suitability for the study of sustainable consumption below.

The following six theories or families of theories were selected:
- The theory of consumer preference (originally developed in economics),
- Rational choice theories (originally developed in economics and sociology),
- The theory of planned behaviour (originally developed in psychology),
- The norm activation model (originally developed in psychology),
- Stage models of voluntary behavioural changes (originally developed in psychology),
- Practice theories/theories of social practices (originally developed in sociology).

This selection in no way claims to be complete. It is primarily justified by the fact that all of the theories shown have been used in one or more projects within the focal topic and that they have a certain level of prominence and circulation. The selection also indicates the range of disciplines in which action theories have been developed that today appear relevant to the research into sustainable consumption. It is not surprising that such theories have their origins predominantly in disciplines such as economics, sociology and psychology. In turn, this does not mean that there are no productive theoretical approaches in other areas or that such approaches could not be developed through inter-disciplinary research.

In the course of the representations, we will start with the theory of consumer preference not just because this approach has a long tradition and is regarded as a highly successful programme for explaining consumer actions, but also as several more modern action theories used to explain (sustainable) consumption were developed, stressing their critical differences to this approach. One first step towards emancipation could be the liberalisation of a few basic assumptions of the theory of consumer preference in the family of rational choice theories. For this reason, from a systematic perspective, the theory of consumer preference could also be regarded as the narrowest form of the rational choice theories (see section 4.3.2). In this way, the theory families, which follow on from this in the following portrayal, are characterised among other things by an ever increasing distance from the basic assumptions in the theory of consumer preference.

4.3.1 The economic theory of consumer preference
(written by Stefan Zundel)

Basic assumptions
The economic theory of consumer preference acts on the basic assumption that the actors can assign prices to their behavioural options. Given the supplementary condition of a given income and given prices, the option which provides the greatest possible benefit is selected from a given set of actions. It is further assumed that the actors assess the alternative actions based on their consequences (consequentialism), they are aware of the properties of the optional actions (bundles of goods and their qualities), they can compare and assess all of the options in pairs on the basis of their consequences (forming a complete order of preferences) and that this order of preferences fulfills some axioms (assumption of rationality), primarily transitivity (consistency).

Scope of application
The original scope of application of the economic theory of consumer preference is variations of the manifested affordable consumer demand for goods by large groups. Actions (explanandum) are therefore essentially understood as purchase actions. The scope of application of the theory was expanded to actions and their consequences, where the costs are not identifiable in the form of a market price and the supplementary condition under which the benefit is maximised need not necessarily be a restriction on income (Becker 1982).

Conceptual structure
The concept of revealed preference, which signals an evaluative relationship between bundles of goods, is central to the conceptual structure of the economic theory of consumer preference. If the bundle of goods A is favoured over the bundle of goods B, the language of the theory is: A is preferred to B. If the bundle of goods A is preferred to B, it is assigned a higher value index. This standard for representation results in a utility function which represents this value index. Correspondingly, the assumption of benefit maximisation means that the actor always chooses the option which is preferred to all others.

Key literature
- Developing the axiomatic principles: Arrow/Debreu 1954.
- Describing the principles of the theory in a textbook: Varian 2007.
- Expanding the theory to issues beyond economic contexts: Becker 1982; Lancaster 1966.
- Critical arguments and further developments: Friedman 1953; Kahnemann 2002; McFadden 1999; Simon 1957.

4.3.2 Rational choice theories (RC theories)
(written by Georg Sunderer)

Basic assumptions
The classic RC theory is oriented around three basic assumptions:
1. Preference assumption: Human behaviour is target-oriented and is therefore influenced by preferences, objectives or motives.
2. Restriction assumption: A party's potential actions are limited by behavioural restrictions.
3. Utility maximisation assumption: Actors attempt to realise their preferences, objectives or motives to the greatest possible extent, considering the restrictions on actions.

Some RC theorists replace the assumption of utility maximisation with a more general assumption of a decision-making rule. It simply requires that the theory contains a decision-making rule which indicates the action an actor will take. This allows RC models to be created, which assume limited rationality (e.g. "satisficing behaviour").

Scope of application
In principle, the RC theory claims to be a general action theory, but primarily relates to actions which are connected with a (conscious) decision. If it is to be used for non-reflected actions, additional assumptions are required (for example: the non-reflected actions are results of earlier conscious decisions or of unconscious inner processes, which are performed according to the decision-making logic assumed in the RC model). Narrow RC approaches only analyse material preferences and objective restrictions, while broad approaches may also include non-material preferences and subjectively perceived restrictions.

Conceptual structure
The formalisation of the decision-making rule is key to all RC approaches. This cannot simply be taken from the basic assumptions – it requires additional assumptions. In narrow variations of the RC theory, it is assumed that actors have exclusively selfish,

material preferences and these preferences are almost identical for all actors. Furthermore, the assumption is made that the actors have all of the relevant information. As a result, it is exclusively objective (hence actually existing) material restrictions which are relevant for explaining behaviour (e.g. income, time or opportunity structures). In broad variations, all types of preferences are regarded as relevant, i.e. this includes moral motives, with the assumption of complete information forgone, and it is assumed that different actors may have different preferences.

Key literature
- Introduction and overview: Diekmann/Voss 2004; Kunz 2004; Opp 1999.
- Sample applications in the environmental field: Bamberg et al. 2008; Best 2006; Diekmann/Preisendörfer 1998.

4.3.3 The theory of planned behaviour (TPB)
(written by Christian Dehmel and Stefan Zundel)

Basic assumptions
TPB assumes that human actions are always based on intentions. These are influenced by various internal and external factors, but are essentially based on rational decisions.

Scope of application
Like the RC theories, TPB primarily examines reflected, planned activities which are connected with (conscious) decisions to take actions.

Conceptual structure
The concept of intention is key. Beliefs regarding negative or positive consequences and the effectiveness of an action, a resulting attitude to the action and subjective norms held by a person all influence intention. Together with the perceived control over one's own behaviour and its consequences, the specific action is explained. The perceived control is based on the assessment of internal (e.g. capabilities, knowledge) and external resources (e.g. time, money) and influences both the intention to perform the action and, more directly, the implementation of the action.

Key literature
- Original theory (Theory of Reasoned Action): Fishbein/Ajzen 1975.
- Further development to TPB: Ajzen 1988, 1991.
- Important additions to the model: Armitage 1999; Charng et al. 1988; Krömker/Werner 2009; Krömker/Dehmel 2010; Sparks/Shepherd 1992, 2002.

- Integration of TPB with the norm activation model (NAM) by Schwartz and Howard (1981) and Rogers' protection motivation theory (Rogers/Prentice-Dunn 1997) to form a new overall model: Krömker 2004, 2008.

4.3.4 The norm activation model (NAM)
(written by Ellen Matthies)

Basic assumptions
The theories of norm activation act on the assumption that humans can act on moral, altruistic motives to help others or contribute to the common good. They therefore imply an unselfish view of human nature.

Scope of application
The focus in NAM is on altruistic actions (e.g. helping, purchasing fair-trade products), but also the support of political steering measures for the common good.

Conceptual structure
The original form of NAM postulated that altruistic actions are based on the activation of personal norms which are experienced as feelings of moral obligation. The activation of personal norms is conditional on four key variables: Firstly, the awareness of the need or problem, i.e. the perception of a problematic situation, which requires corrective action. Secondly, the ascription of responsibility for negative consequences, which can result from non-moral actions. Thirdly, the perception of optional actions to remedy the situation (outcome efficacy). And fourthly, the perception of one's own abilities.

The further development of the model to the theory of moral decisions assumes that moral motivation in the decision-making process is weighed against other, potentially conflicting motivations. Another assumption is that, where a personal norm is activated that is not followed by a corresponding action a new cognitive assessment is performed, which may lead to a denial of the perceived personal responsibility or the problem as a whole. In further developments of the model, the aspect of habitualisation was also taken into consideration. Where there is a strong habitualisation, personal norms remain without consequence, as actions are performed automatically, i.e. without perceiving the consequences or responsibilities.

Key literature
- Original theory (norm activation theory, NAM): Schwartz 1977.
- Further development into the Theory of Moral Decision Making: Schwartz/Howard 1981.

- Further developments in the environmental field: Value-Belief-Norm theory of environmentalism (VBN): Stern 2000.
- Overview of applications in the environmental field: Steg/Nordlund 2012.
- Expansions of the model, considering social/subjective norms and habits: Bamberg et al. 2007; Hunecke et al. 2001; Klöckner/Matthies 2004.

4.3.5 Stage models of voluntary behavioural changes
(written by Sebastian Bamberg)

Basic assumptions

The models of voluntary behavioural changes assume that human behaviour is completely determined by neither social nor material conditions. Humans have the capacity of motivational self-direction, i.e. they can determine their own personally desired changes and invest their mental and physical energies into achieving these changes. A central assumption is that voluntary behavioural changes are not a linear process, but that different stages with different qualities can be identified within the process, which represent typical motivational and volitional difficulties people encounter in their attempt to change their previous behaviour.

Scope of application

The theory family focuses on voluntary behavioural changes on an individual level. At the heart of the theory lie the motives for change and the time-dependent process of changing what are often habitually performed patterns of actions.

Conceptual structure

The process of behavioural change is characterised as passing through several stages. In each stage, the person is confronted with a specific task. During the pre-contemplative stage, the person assesses the consequences of their behaviour to this point in the light of the standards or norms they deem critical. At the contemplative stage, they select and assess potential alternative strategies for action. During the preparation/testing stage, the task is preparing and implementing the new alternative actions selected in the previous stages, with the experiences of the new action evaluated during the maintenance stage, ultimately forming a new habitual action. Three specific types of intention mark the successful solution of these stage-specific tasks, with the formation of the intention influenced by stage-specific socio-cognitive constructs.

Key literature
- Original models: Achtziger/Gollwitzer 2008; Prochaska/Velicer 1997; Weinstein/Sutton 1998.
- Further development, primarily relating to the research of sustainable consumption: Bamberg 2007, 2012; Bamberg et al. 2011.

4.3.6 Practice theories / theories of social practices (TSP)
(written by Melanie Jaeger-Erben)

Basic assumptions
Theories of social practices are not a defined theory programme, but a number of various approaches which contain similar basic assumptions. They assume that most human activities are a component of social practices and these are ontologically more fundamental than individual motives or intentions. A 'social practice' is the smallest analytical unit, regarded as a combination of spatially and temporally contextualised verbal and physical action patterns and motivational, normative and affective elements. Social practices emerge from an interplay between structure and agency, creating order and thus comprehensibility and normality. Routines are carriers of social order and social identity, constantly reproducing both elements. Practice theories particularly consider the role of materials and socio-technical contexts. Following this, objects or artefacts do not only carry symbolic meaning, but are integral and self-constituting components of practice. Social structures are not just reproduced by agents in the form of practices, but are also adapted, modified and transformed.

Scope of application
Theories of social practices were established as "social theories", which essentially attempt to explain how social behaviour and social order arise and what is the nature of social change. In principle, therefore, all actions, in particular everyday routines, can be regarded as part of social practice. TSP focuses on the actions or practices themselves and their structure, not necessarily on the acting individual.

Conceptual structure
Social practices are the central analytical unit. Social practices are 'logically' organised bundles of activities (doings) and supporting 'materials' (artefacts, technical or other media), which have a spatial and temporal structure. Parts of these bundles are normative structures, which organise the activities among one another and further combine certain norms, social expectations, roles etc. The normative structures enable practices to be embedded in social contexts and only in this way can they become significant.

Key literature
- Theories of social practices in general: Reckwitz 2002, 2003; Schatzki 1996, 2002.
- Application in the field of consumption: Brand 2010; Brunner 2007; Gram-Hanssen 2010; Jaeger-Erben 2010; Røpke 2009; Shove 2003; Spaargaren/Oosterveer 2010; Warde 2005.

4.4 Combination of the structure of phenomena and the action theories

In section 4.2, we provided a structure of the consumer behaviour phenomena relevant for action theory, considering three perspectives: Firstly, the perspective of differentiating between various types of consumer actions from an individual point of view (section 4.2.1), secondly, the perspective of the social and cultural incorporation of the consumer behaviour (section 4.2.2) and thirdly, the perspective of the change of individual consumer behaviour over time (section 4.2.3). The subsequently demonstrated theories or theory families from section 4.3 show that no theory encompasses all of the phenomena of consumer behaviour. Which phenomena of consumer behaviour a theory fundamentally attempts to understand or explain, i.e. the phenomena for which it is particularly suitable, follows from their stated subject area and from the perspective (i.e. the assumptions and conceptual strucuture) from which they view that subject area. Below, we will review the six theories once again, stressing the relationship between the respective theory and the phenomena. Table 2 shows an overview of this.
- The *economic theory of consumer preference* does not make a distinction between reflected and non-reflected actions as it is generally assumed that actors have full information and act on this basis or at least act as such. At most, the distinction between essential and non-essential is reflected in the assumed order of preferences in which essential consumer acts are naturally located right at the top. In the theory of consumer preference, the degrees of freedom of the consumer act are simply shown in the number and quality of the optional actions. However, the textbook version does not provide subjective restrictions of choice, and institutional restrictions, which could be responsible for a limitation of optional actions, are at best framework conditions, but not the object of the formation of hypotheses. In general, the incorporation of consumer behaviour in social and cultural contexts is not the genuine object of the theory. These may, however, be integrated into the theory if they are assigned shadow prices or if non-monetary restrictions are constructed, for example time restrictions. In contrast, material contexts are directly addressed in the form of price, income, assets etc. Personal conditions which promote change are not addressed in the theory – the focus is placed exclusively on structural condi-

tions that facilitate change. In other words: From the point of view of the theory, it is always the situation, in which the actor finds himself, that changes, but never the actor himself or his preferences.

- *Rational choice theories* are particularly suitable for reflected consumer actions. Given the basic assumptions, it is questionable whether they can also be used for non-reflected consumer actions. Broad RC models can be used regardless of the significance or degree of freedom of the consumer act, while narrow RC approaches do not allow for this. The latter are rather unsuited to consumer acts which are essential to a person due to non-material preferences. The same applies for consumer actions with a great degree of freedom with regard to material aspects, as explanations regarding material restrictions would then be less promising.

  Effects of change processes can only be shown for the change of objective material restrictions (e.g. prices) in narrow variations. In contrast, broader variations also allow for an explanation of changes in consumption, which are the result of changed non-material conditions (e.g. change of environmental awareness). The same is true for aspects of social embeddedness: narrow RC variations only consider material aspects, while broad variations also consider all types of non-material aspects.

  RC approaches are tried-and-tested, having been used in numerous studies of sustainable consumption. The role of material factors and attitudes towards the environment was particularly intensively examined. However, the fact that there are often no positive material incentives for sustainable consumption is problematic for the use of narrow RC variations. In such cases, narrow models are unable to explain sustainable consumption. Despite this, narrow models can still be useful for such consumer acts, for example when investigating which behavioural changes can be achieved as a result of changes to objective material conditions (Diekmann/Voss 2004, p. 19). In contrast to the more restrictive, narrow variation, the broad version has a larger spectrum of application in the area of sustainable consumption. In any case, broad models are to be preferred to narrow models if it is to be assumed that non-material aspects have a relevant influence on consumer behaviour. The low-cost hypothesis (Diekmann/Preisendörfer 1998) indicates that a narrow RC model could better explain behaviour given rising material costs (Diekmann/Voss 2004, p. 20).

- The scope of application of the *theory of planned behaviour* reaches as far as reflected decisions. The research into routines is viewed rather critically by one of the two creators of the theory (Ajzen 2002). The theory does not distinguish between essential and non-essential consumer acts. The degrees of freedom of the consumer actions are shown in the theoretical construction of perceived control, which shows the extent to which actors believe that they can perform certain actions and influence the results of such actions. As an individualistic action theory, the primary focus of TPB is not on interactions and the incorporation of consumer acts in social, cultural and material contexts. However, the phenomenon is indirectly considered, with the theoretical construction of the "attitude" influencing the intention to perform the action. The subjective attitude is understood as an overall balance of valuations, which is formed on the basis of the envisaged beliefs about the effectiveness and the advantages and disadvantages of an action. The more positive the attitude towards a specific action, the stronger the intention and the more likely the action is to be performed. Equally, the theoretical construction of "social norms" can be understood as a type of internal reflection of group affiliation. The theory of planned behaviour is not a theory that explicitly describes the changes in theoretical terms, nor is it a dynamic conception. However, drivers or obstacles to sustainable consumption can be identified by way of statistical comparisons with the help of the theory.

- The *norm activation theories* can be applied to reflected actions associated with a certain degree of freedom, which may potentially be perceived as morally relevant (the degree of freedom of the consumer action can be partially shown using the "abilities" construct). The model therefore acts as a point of connection to normative aspects of sustainable consumption or sustainable development (e.g. intergenerational equity). The explanatory power of the norm activation model was impressively displayed for the area of environmentally friendly consumption in a number of empirical studies (cf. Steg/Nordlund 2012). The empirical findings of the expanded model indicate that, in addition to the altruistic motivation, other motives can also play an important part in consumer behaviour.

  The social, cultural and material embeddedness of actions is not the primary focus of the norm activation theory as an individualistic action model. However, there are model variations which more closely consider these aspects. For example, in applications of the value-belief-norm theory (Stern 2000), it is assumed that personal norms have their roots in culturally imparted, enduring values. The theory of moral decision making (Schwartz/Howard 1981) assumes that, in addition to the moral motivation, a social motivation (in the sense of concrete expectations of the social environment) also influences the moral decision. Further developments also consider restrictions on behaviour (Hunecke et al. 2001). Thus, the norm activation model or its further developments within the model family can also be used to develop interventions.

- The *stage models of voluntary behavioural changes* focus on the self-initiated change of consciously reflected decisions. The model starts with the pre-contemplative stage, i.e. the non-scrutinised performance of routine actions (e.g. the everyday car journey to work). According to the model, the motivation to make changes arises in the pre-contemplative phase, if a person becomes aware of a discrepancy between the negative individual and/or collective consequences of their current action and any standards or values which are central to the person's self-identity (or their ideal self). Furthermore, the voluntary behavioural change is predicated on the fact that the person is convinced that there are, in principle, alternatives to the current behaviour (degree of freedom). The model stresses the influence of social norms and cultural values on creating the motivation to change. Furthermore, stage models focus on the temporal dimension of change processes. Behavioural changes are designed as a non-linear process of passing through four different stages with different qualities. People can move through these stages quickly, but they may also get 'caught up' in a given stage or return to earlier stages.

- Following the classification of the *practice theories or theories of social practices* (TSP) proposed by Merton (1962), they are to be labelled as universal theories rather than 'middle range theories'. They attempt to provide a fundamental explanation of social behaviour. In principle, therefore, all of the phenomena described in the order can be incorporated into theoretical considerations relevant for practice, as consumer actions are understood as part of social practice or social practices (cf. Røpke 2009; Warde 2005). The reconstruction of social practices can be regarded both as an 'entanglement' of a consumer act within a spatial and temporal context and as a point of connection with 'social fields', which makes the action part of the social order and therefore 'socially charged' or 'socialised'. The room for manoeuvre involved in the actors' performance can be seen through the reconstruction of the structure of social practices. TSP allow the integration of the macro and micro (or individual and structural) level of consumption by looking at the interrelationship between consumers and the available systems of provision.

  TSP are therefore not suitable for approaches based on deductive logic. To this point there has been no systematic application in quantitative studies on sustainable consumption. Instead, they offer heuristics, which can be used to reconstruct and contextualise the structure of social phenomena, which in turn allows us to determine the most suitable places for implementing interventions. Are these found on the level of resources and competences or do certain structures or socio-cultural significances promote a non-sustainable practice? As the approach to consumer actions is embedded in ever-changing structures and social reforms, predictions can be made on the development of consumption, for example in the event of continuing technological advances. However, TSP is a relatively young concept in the area of sustainable consumption, meaning that there is still potential for development in this area (cf. Spaargaren et al. 2010).

4.5 Conclusion and outlook

This overview has highlighted several points. Firstly, it is clear that, in light of the multi-faceted and complex phenomenon of individual consumer behaviour, a variety of different theories is not just desired, but is rather unavoidable. There is no 'best' theory for examining consumption and sustainable consumption – instead, the phenomena can and must be observed from a variety of perspectives. Depending on the issue at hand, the objectives and questions of a planned research project or practice task, different theories are suitable. Those theoretical approaches have to be chosen which are most suitable for the phenomena that are in the focus of a specific project.

Furthermore, it should be stated that there are not only complementary elements within the theories portrayed, but also divergences. These are evident in the distinctions and differences they make or do not make, and they can also be seen in the basic assumptions upon which the individual theories are based – in other words their view of human nature. Whether consumers are viewed as benefit-maximisers, determined by psychological factors and processes or controlled by their incomes and prices, as autonomous agents or as embodiments of social practices, influences the way in which a theory describes and explains, or reconstructs and understands, the phenomena of consumer behaviour. No scientific determination can be made on which view of human nature is 'correct'. Instead, it is a question of the patterns of meaning and values prevalent in a culture and is therefore subject to changes.

Finally, there is a specific advantage in the structure of the phenomena we have presented here. Namely, it shows where there appears to be a relative dearth of theories with regard to the phenomena of consumer behaviour. Going through the entries in the middle column of Table 2, it becomes apparent that the majority of the action theories portrayed and used within the focal topic are suitable for describing and explaining *reflected* consumer acts and the accompanying (conscious) decisions, but are not so suitable for less reflected consumer actions, such as everyday routines. The phenomena of the (symbolic) significance of consumer actions and the dynamic nature of change processes remain largely overlooked or are only implicitly addressed. Ultimately, it is striking that all of the theories focus either on personal or structural conditions of behaviour and on how these conditions change. Admittedly, a tendency

Table 2: Overview of the portrayed action theories: references to the phenomena of consumer behaviour and suitability for research into sustainable consumption

Theory	Phenomena upon which the theory focuses	Suitability of the theory in researching sustainable consumption
Economic theory of consumer preference	Reflected consumer acts (generally acts of purchasing) Material contexts (institutional and socio-technical level): Price, income, wealth etc. Structural conditions facilitating and inhibiting change	Explanation of the variation of demand (on aggregate level) for a given bundle of goods based on price and income variations
Rational choice theories	Reflected consumer acts (primarily decisions) Narrow variations: Material aspects of personal resources and contexts (institutional and socio-technical level) and their role in facilitating or inhibiting change Broad variations: Additional non-material aspects of personal resources and contexts (on all levels) and their role in promoting or preventing change	Heuristics for explaining situation-specific actions Estimation of the effects of measures Narrow variations can only be used in the case of material aspects Broad variations have a variety of uses and are primarily preferred where a relevant influence of non-material aspects on consumer behaviour must be assumed
Theory of planned behaviour (TPB)	Reflected consumer acts (primarily decisions) Subjective representation (perception) of pre-structuring, as well as significance and phenomena of embedding in social, cultural and material contexts Personal conditions facilitating and inhibiting change	Explanatory power for largely different fields of action Heuristics for identifying starting points for promising interventions
Norm activation model (NAM)	Reflected consumer acts with a certain degree of freedom, which are perceived as morally relevant Psychological process for triggering the feeling of an obligation	Starting point for normative aspects of sustainable consumption as the model refers to the significance of moral motives Can be used to analyse individual consumer actions or as a heuristic for developing interventions
Stage models of voluntary behavioural changes	Process of voluntary changes of less reflected and repeated consumer actions Personal reasons for the success or failure of the desired changes	These models are particularly suitable for theory-based development of person-focused intervention concepts
Practice theories / theories of social practices	Social, cultural and socio-technical contexts of consumer behaviour, which form the pre-structure for social behaviour and daily routines relevant to consumption Structural conditions facilitating and inhibiting change	Heuristics for the contextualisation of consumer acts as part of a dynamic relationship between actors and structures Derivation of methods for interventions, considering that the action is entangled in social and structural contexts

to overcome this dichotomy of perspectives has been observed in more recent times. Practice theories claim to combine both perspectives. More recent developments of the theory of planned behaviour and the norm activation model, as well as the broad RC approaches, have also shown attempts at theoretically reflecting the processes acting between individuals and structures (e.g. the relationship between attitudes, personal norms or perceived control and socio-cultural and structural conditions).

Thus, certain deficits in terms of theory appear with regard to non-reflected consumer actions, the significance of consumer goods and acts, the dynamic nature of change processes and the mediation between individual and structural factors and processes. Of course, this statement must be qualified in various regards: For one thing, it is a consequence of the selection of theories and disciplines made here.[5] Furthermore, a theoretical discussion to tackle the latter deficit is currently being established around practice theories. Finally, this does not mean that there were no valuable theoretical approaches, which could be used to develop a better understanding of the phenomena that have tended to be overlooked in the social-scientific research into sustainable consumption. To suggest some possibilities:

1. Classic consumption research – research into both marketing and sociology of consumption – might provide concepts and theories which are suitable for describing and explaining non-reflected consumer actions and the aspect of the significance of such actions. The themes of moral and political consumption could be particularly relevant for sustainable consumption (cf. for example Warde 2010; Jäckel 2007).
2. The approach of system dynamics could yield significant potential for describing and modelling dynamic change processes in complex systems (cf. for example Ulli-Beer et al. 2010; Müller/Ulli-Beer 2010).
3. The approaches used in ecological psychology and cultural psychology, which have been a small niche of psychology for decades, could help bring about a helpful theoretical framework to more precisely define the area between personal (individual) and structural (aggregate) factors and processes (cf. for example Boesch 1980; Lang 1992; Kaminski 2008; Kaufmann-Hayoz 2006; Kaufmann-Hayoz et al. 2010).

Looking beyond the limits of classical social scientific research into sustainable consumption raises one final point to be discussed here: The focus placed on action theories of the social sciences and the extent to which they are empirically verifiable in this contribution could lead to a one-sided socio-technocratic view of the big project 'sustainable consumption' in society – put simply, the idea or hope that 'people could be made' to consume sustainably if it were only possible to flick the correct switch in the

---

5 Anyhow, other authors have found a similar need for action, e.g. Hunecke 2008.

gears of society (although the models of voluntary behavioural changes and the essentially sceptical practice theories explicitly distance themselves from such an idea that people can be steered). The disciplines and research fields which are more affiliated to cultural sciences or humanities (e.g. philosophy, research on the history of ideas, legal studies or educational sciences) might provide interesting supplementary approaches to the perspectives taken here regarding phenomena of individual (sustainable) consumer behaviour, which would be indispensible to a more extensive assessment. The recurring debate on the good life (cf. Di Giulio et al. in this volume) or the closely connected revitalisation of epoch theories are probably not a coincidence. Precisely, the inter-disciplinary context of research into sustainable consumption would provide the ideal field for illuminating the theoretical deficits as part of inter-disciplinary cooperation and to establish a more comprehensive theoretical understanding of individual consumer behaviour and the conditions of changes in such behaviour. In this way, the theoretical debate would be a departure from the often partially concealed assumption that the sole purpose of theories in the field of sustainable consumption was to create the cleverest socio-technocratic intervention on the basis of action theories.

References

Achtziger A., Gollwitzer P. M. (2008): Rubikonmodell der Handlungsphasen. In: Brandstätter V., Otto J.H. (eds.): Handbuch der Allgemeinen Psychologie: Motivation und Emotion. Göttingen: Hogrefe. 150–156.
Ajzen I. (1988): Attitudes, personality and behaviour. Milton Keynes: Open University Press.
Ajzen I. (1991): The theory of planned behavior. In: Organizational Behavior and Human Decision Processes 50: 179–211.
Ajzen I. (2002): Residual Effects of Past on Later Behavior: Habituation and Reasoned Action Perspectives. In: Personality and Social Psychology Review 6 (2): 107–122.
Armitage C. J., Conner M. (2001): Efficacy of the theory of planned behaviour: a meta-analytic review. In: British Journal of Social Psychology 40: 471–499.
Armitage, C. J., Conner M. (1999): Efficacy of the theory of planned behavior: assessment of predictive validity and perceived control. In: British Journal Social Psychology 38: 35–54.
Arrow K. J., Debreu G. (1954): Existence of an Equilibrium for a Competitive Economy. In: Econometrica 22: 265–290.
Baedeker C., Liedtke C., Welfens J. M. (2005): Analyse vorhandener Konzepte zur Messung des nachhaltigen Konsums in Deutschland einschließlich der Grundzüge eines Entwicklungskonzeptes: Endbericht. Wuppertal: Wuppertal Institut für Klima, Umwelt, Energie.
Bamberg S. (2007): Is a stage model a useful approach to explain car drivers' willingness to use public transportation? In: Journal of Applied Social Psychology 37: 1757–1783.
Bamberg S. (2012): Processes of change. In: Steg L., van den Berg A. E., de Groot J. I. M. (eds.): Environmental Psychology: An Introduction. Wiley-Blackwell. 267–280.

Bamberg S., Davidov E., Schmidt P. (2008): Wie gut erklären "enge" oder "weite" Rational-Choice-Versionen Verhaltensveränderungen? Ergebnisse einer experimentellen Interventionsstudie. In: Diekmann A., Eichner K., Schmidt P., Voss T. (eds.): Rational Choice: Theoretische Analysen und Empirische Resultate. Wiesbaden: VS Verlag. 143–170.

Bamberg S., Fujii S., Friman M., Gärling T. (2011): Behaviour theory and soft transport policy measures. In: Transport Policy 18: 228–235.

Bamberg S., Hunecke M., Blöbaum A. (2007): Social context, personal norms and the use of public transportation: Two field studies. In: Journal of Environmental Psychology 27: 190–203.

Becker G.S. (1982): Der ökonomische Ansatz zur Erklärung menschlichen Verhaltens. Tübingen: Mohr Siebeck.

Best H. (2006): Die Umstellung auf ökologische Landwirtschaft als Entscheidungsprozess: Eine Anwendung der Theorie rationalen Handelns. Wiesbaden: VS Verlag.

Boesch E. E. (1980): Kultur und Handlung. Einführung in die Kulturpsychologie. Bern, Stuttgart, Wien: Hans Huber.

Bosshart D., Muller C., Hauser M. (2010): European Food Trends Report. Rüschlikon: Gottlieb Duttweiler Institut, Studie Nr. 32.

Bourdieu P. (1982): Die feinen Unterschiede – Kritik der gesellschaftlichen Urteilskraft. Frankfurt a. M.: Suhrkamp.

Brand K.-W. (2010): Social Practices and Sustainable Consumption: Benefits and Limitations of a New Theoretical Approach. In: Gross M., Heinrich H. (ed.): Environmental Sociology: European Perspectives and Interdisciplinary Challenges. In: Springer Business and Media: 217–235.

Brunner K.-M. (2007): Ernährungspraktiken und nachhaltige Entwicklung – eine Einführung. In: Brunner K.M., Geyer S., Jelenko M., Weiss W., Astleithner F. (eds): Ernährungsalltag im Wandel – Chancen für Nachhaltigkeit. Wien: Springer. 1–38.

Charng H. W., Piliavin J. A., Callero P. L. (1988): Role identity and reasoned action in the prediction of repeated behavior. In: Social Psychology Quarterly 51: 303–317.

Diekmann A., Preisendörfer P. (1998): Umweltbewusstsein und Umweltverhalten in Low- und High-Cost-Situationen. In: Zeitschrift für Soziologie 27: 438–453.

Diekmann A., Voss T. (2004): Die Theorie rationalen Handelns. Stand und Perspektiven. In: Diekmann A., Voss T. (eds.): Rational-Choice-Theorien in den Sozialwissenschaften. Anwendungen und Probleme. München: Oldenbourg. 13–29.

Di Giulio A., Brohmann B., Clausen J., Defila R., Fuchs D., Kaufmann-Hayoz R., Koch A. (in this volume): Needs and consumption – a conceptual system and its meaning in the context of sustainability.

Fischer D., Michelsen G., Blättel-Mink B., Di Giulio A. (in this volume): Sustainable consumption: how to evaluate sustainability in consumption acts.

Fishbein M., Ajzen I. (1975): Belief, attitude, intention and behavior: An introduction to theory and research. Reading, MA: Addison-Wesley.

Friedman M. (1953): Essays in Positive Economics. Chicago: Chicago Press.

Gram-Hanssen K. (2010): Standby consumption in households analysed with a practice theory approach. In: Journal of Industrial Ecology 14 (1): 150–165.

Haefeli U. (2008): Verkehrspolitik und urbane Mobilität. Deutsche und Schweizer Städte im Vergleich 1950–1990. Stuttgart: Franz Steiner Verlag.

Hunecke M. (2008): Möglichkeiten und Chancen der Veränderung von Einstellungen und Verhaltensmustern in Richtung einer nachhaltigen Enwicklung. In: Lange H. (ed.): Nachhaltigkeit als radikaler Wandel: die Quadratur des Kreises? Wiesbaden: VS Verlag für Sozialwissenschaften. 95–122.

Hunecke M., Blöbaum A., Matthies E., Höger R. (2001): Responsibility and environment: Ecological norm orientation and external factors in the domain of travel mode choice behavior. In: Environment and Behavior 33: 830–852.

Inglehart R. (1997): Modernization and Postmodernization: Cultural, Economic and Political Change in 43 Societies. Princeton, NJ: Princeton University Press.

Jäckel M. (ed.) (2007): Ambivalenzen des Konsums und der werblichen Kommunikation. Wiesbaden: VS Verlag für Sozialwissenschaften.

Jaeger-Erben M. (2010): Zwischen Routine, Reflektion und Transformation. Die Veränderung von alltäglichem Konsum durch Lebensereignisse und die Rolle von Nachhaltigkeit – eine empirische Untersuchung unter Berücksichtigung praxistheoretischer Konzepte. Dissertationsschrift. Berlin: Technische Universität Berlin.

Kaesler D. (ed.) (2002): Max Weber: Schriften 1894–1922. Stuttgart: Alfred Kröner Verlag.

Kahnemann D. (2002): Maps of Bounded Rationality. Nobel Prize Lecture. [http://nobelprize.org/nobel_prizes/economics/laureates/2002/kahnemann-lecture.pdf; 24.07.2012]

Kaminski G. (2008): Das Behavior Setting-Konzept – Entstehungsgeschichte und Weiterentwicklungen. In: Lantermann E. D., Linneweber V.H. (eds.): Grundlagen, Paradigmen und Methoden der Umweltpsychologie. Göttingen: Hogrefe. 333–376.

Kaufmann-Hayoz R. (2006): Human action in context: A model framework for interdisciplinary studies in view of sustainable development. In: Umweltpsychologie 10 (1): 154–177.

Kaufmann-Hayoz R., Bruppacher S., Harms S., Thiemann K. (2010): Einfluss und Beeinflussung externer Bedingungen umweltschützenden Handelns. In: Lantermann E.D., Linneweber V.H., Kals E. (eds.): Spezifische Umwelten und umweltbezogenes Handeln. Göttingen: Hogrefe. 697–734.

Klöckner C.A., Matthies E. (2004): How habits interfere with norm directed behavior – A normative decision-making model for travel mode choice. In: Journal of Environmental Psychology 24: 319–327.

Krömker D. (2004): Naturbilder, Klimaschutz und Kultur. Weinheim: Beltz.

Krömker D. (2008): Globaler Wandel, Nachhaltigkeit und Umweltpsychologie. In: Lantermann E. D., Linneweber V.H. (eds.): Grundlagen, Paradigmen und Methoden der Umweltpsychologie. Göttingen: Hogrefe. 715–747.

Krömker D., Dehmel C. (2010): Einflussgrößen auf das Stromsparen im Haushalt aus psychologischer Perspektive. Kassel: Transpose Working Paper No. 6. [http://www.uni-muenster.de/Transpose/en/publikationen/index.html; 24.07.2012]

Krömker D., Werner J. (2009): Interventionen für den Klimaschutz im Bau- und Sanierungsbereich: eine Bewertung aus handlungstheoretischer Sicht. In: Umweltpsychologie 13 (1): 10–34.

Kunz V. (2004): Rational Choice. Frankfurt/Main, New York: Campus Verlag.

Lancaster K. (1966): A new approach to consumer theory. In: Journal of Political Economy 174: 15.

Lang A. (1992): Kultur als "externe Seele". Eine semiotisch-ökologische Perspektive. In: Allesch Ch., Billmann-Mahecha E., Lang A. (eds.): Psychologische Aspekte des kulturellen Wandels. Wien: Verlag des Verbandes der wissenschaftlichen Gesellschaften Österreichs. 9–30.

McFadden D. (1999): Rationality for Economists. In: Journal of Risk and Uncertainty 19 (1–3): 33.

Merton R. K. (1962): Social Theory and Social Structure. Glencoe: The Free Press.

Müller M., Ulli-Beer S. (2010): Policy Analysis for the transformation of Switzerland's stock of buildings. A small model approach. Proceedings of the 28th International Conference of the System Dynamics Society. Seoul, Korea. [http://www.systemdynamics.org/conferences/2010/proceed/papers/P1172.pdf; 24.07.2012]

Opp K. D. (1999): Contending Conceptions of the Theory of Rational Action. In: Journal of Theoretical Politics 11: 171–202.

Prochaska J. O., Velicer W. F. (1997): The Transtheoretical Model of health behavior change. In: American Journal of Health Promotion 12: 38–48.

Rauch C., Kirig A. (2007): Zielgruppe LOHAS. Kelkheim: Zukunftsinstitut GmbH.

Reckwitz A. (2002): Towards a Theory of Social Practice: A development in culturalist theorizing. In: European Journal of Social Theory 5 (2): 245–265.

Reckwitz A. (2003): Grundelemente einer Theorie sozialer Praktiken. Eine sozialtheoretische Perspektive. In: Zeitschrift für Soziologie 32 (4): 282–301.

Reusswig F., Lotze-Campen H., Gerlinger K. (2005): Changing global lifestyle and consumption patterns: The case of energy and food. In: Radhakrishna G. (ed.): Consumer Behaviour: Effective measurement tools. Hyderabad, India: The ICFAI University Press. 197–210.

Rogers E. M. (2003): Diffusion of innovations. New York u.a.: Free Press. (5th edition).

Rogers R. W., Prentice-Dunn S. (1997): Protection Motivation Theory. In: Gochman D. S. (ed.): Handbook of Health Behavior Research, Vol. I: Personal and Social Determinants. New York, USA: Springer. 113–132.

Røpke I. (2009): Theories of practice – New inspirations for ecological economic studies. In: Ecological Economics 68: 490–497.

Schatzki T. (1996): Social Practices: a Wittgensteinian Approach to Human Activity and the Social. Cambridge: Cambridge University Press.

Schatzki T. (2002): The Site of the Social: A Philosophical Account of the Constitution of Social Life and Change. Penn State University Press.

Schulze G. (1992): Die Erlebnisgesellschaft – Eine Kultursoziologie der Gegenwart. Frankfurt a. M.: Campus Verlag.

Schwartz S. H. (1977): Normative influences on altruism. In: Advances in Experimental Social Psychology 10: 221–279.

Schwartz S. H., Howard J.A. (1981): A Normative Decision-Making Model of Altruism. In: Rushton J.P., Sorrentino R.M. (eds.): Altruism and Helping Behavior. Hillsdale: Erlbaum. 189–211.

Shove E. (2003): Comfort, cleanliness and convenience: the social organization of normality. Oxford, New York: Berg Publishers.

Shove E. (2005): Changing human behaviour and lifestyle: a challenge for sustainable consumption? In: Røpke I., Reisch L. (eds.): The Ecological Economics of Consumption. Cheltenham: Elgar. 111–132.

Shove E., Watson M., Hand M., Ingram J. (2007): The design of everyday life. Oxford and New York: Berg Publishers.

Simon H. (1957): Models of Man. New York: John Wiley.

Slade G. (2006): Made to Break: Technology and Obsolescence in America. Cambridge, Mass.: Harvard University Press.

Spaargaren G., Oosterveer P. (2010): Citizen-Consumers as Agents of Change in Globalizing Modernity: The Case of Sustainable Consumption. In: Sustainability 2 (7): 1887–1908.

Sparks P., Shepherd R. (2002): The Role of Moral Judgements within Expectancy-Value-Based Attitude-Behavior Models. In: Ethics and Behavior 12 (4): 299–321.

Sparks P., Shepherd R. (1992): Self-identity and the theory of planned behavior – assessing the role of identification with green consumerism. In: Social Psychology Quarterly 55: 388–399.

Steg L., Nordlund A. (2012): Models to explain environmental behaviour. In: Steg L., van den Berg A.E., de Groot J.I.M. (eds.): Environmental Psychology: An Introduction. Wiley-Blackwell. 185–196.

Stern P. C. (2000): Toward a coherent theory of environmentally significant behavior. In: Journal of Social Issues 56: 407–424.

Thiemann K. (2006): Neue Dinge für eine nachhaltige Entwicklung. Ansätze zu einer Kulturpsychologie nachhaltigen Produktdesigns. Dissertation an der Philosophisch-humanwissenschaftlichen Fakultät der Universität Bern, Schweiz. [http://www.ikaoe.unibe.ch/forschung/nomixwc; 24.07.2012]

Ulli-Beer S., Gassmann F., Bosshardt M., Wokaun A. (2010): Generic structure to simulate acceptance dynamics. In: System Dynamics Review 26 (2): 89–116.

Varian H. (2007): Grundzüge der Mikroökonomik. München: Oldenbourg. (7th edition).

Warde A. (2005): Consumption and Theories of Practice. In: Journal of Consumer Culture 5 (2): 131–153.

Warde A. (ed.) (2010): Consumption. Los Angeles: Sage. (4 volumes).

Weinstein N. D., Rothman A. J., Sutton S. R. (1998): Stages theories of health behavior. In: Health Psychology 17: 290–299.

Ruth Kaufmann-Hayoz, Bettina Brohmann, Rico Defila, Antonietta Di Giulio,
Elisa Dunkelberg, Lorenz Erdmann, Doris Fuchs, Sebastian Gölz, Andreas Homburg,
Ellen Matthies, Malte Nachreiner, Kerstin Tews, Julika Weiß[1]

5 Societal steering of consumption towards sustainability

5.1 Introduction

It is not just often stated in science and politics that the prevalent individual patterns of consumption were not sustainable, it is also a general assumption that sustainability in consumption will not appear 'of its own accord', but rather it requires societal or political steering. At the same time, the question of such steering is a tricky one as it not only touches on the aspect of the effectiveness of the steering, but also contains the question of the legitimacy of such steering measures – that is not to forget the question of how and by whom the objective of such steering measures will be determined. Based on the experiences and findings from the project groups of the focal topic and the discussion across the project groups, we aim to provide a structure to map the debates regarding the topic of societal steering of consumption and in particular of individual consumer behaviour across various scientific disciplines, as well as in the political arena. In doing so, we consider one aspect of the steering in particular – namely the possibilities and effects of exerting an influence on individual consumer behaviour. It goes without saying that there are several important questions which can only be touched on briefly and not tackled in detail.

In the introduction, we will discuss a few fundamental questions on the legitimacy of steering and on what we understand steering to mean, followed by an overview of the main types of instruments to which measures used to steer individual consumer behaviour can be assigned. Finally, we will discuss various questions and problems regarding the assessment of the effectiveness of steering measures.

---

1 Discussion participants were Martin Achtnicht (Seco@home), Tanja Albrecht (ENEF-Haus), Sebastian Bamberg (LifeEvents), Jens Clausen (Consumer/Prosumer), Dorika Fleissner (Intelliekon), Katy Jahnke (Heat Energy), Angelika Just (LifeEvents), Martin Kesternich (Seco@home), Marian Klobasa (Intelliekon), Joachim Müller (Change), Klaus Rennings (Seco@home), Joachim Schleich (Seco@home), Sandra Wassermann (Heat Energy), Daniel Zech (Heat Energy).

5.1.1 The question of legitimacy

The question of whether the societal steering of individual consumption and in particular state interventions are legitimate at all is itself not without controversy. One popular belief is that individual consumption completely belongs in the private sphere and the freedom of consumers may not be limited in a liberal state. For example, Grunwald (2010) speaks against morally charging private behaviour with the normative political idea of sustainability. However, a more exact observation of the phenomena of individual consumer behaviour shows that consumer acts always have a symbolic-communicative function in addition to their technical-functional role. This includes, among other things, allocations of meaning and significance which, in turn, are always associated with cultural values and societal norms (cf. Kaufmann-Hayoz/Bamberg et al. in this volume; Warde 2010). Furthermore, consumption is regarded as the central driver of economic growth (see e.g. Røpke 2010) and also has socio-cultural and ecological effects both in the present and looking towards the future. Individual consumer behaviour is therefore not a purely private matter and cannot be morally neutral (cf. Fischer et al. in this volume).

The conceptual analysis performed by Di Giulio et al. (in this volume) can help to provide further clarification on the legitimacy of societal or political steering. The authors have developed a concept of needs, which is compatible with the notion of consumption and the idea of sustainability, as well as the notion of a good life. They further suggest a conceptual system which relates individual constructs of wanting, consumer goods and components of nature to one another. They distinguish between four different constructs of wanting: objective needs, subjective desires, demands made of consumer goods and ideas regarding the degree and breadth of satisfaction of needs and desires. Furthermore, they show that objective needs cannot be contested on ethical grounds, but the other constructs of wanting certainly can. Following this line of reasoning, the idea of sustainability leads to an obligation on the part of the community to provide the individuals with any external conditions which allow them to satisfy their objective needs and lead what they deem to be a good life. This includes offering indispensible consumer goods in the sense of a basic provision and the required quantity and quality of components of nature (natural resources, processes and states). This community obligation provides legitimacy for the steering of individual consumer behaviour in two respects. For one thing, societal steering is legitimate when and to the extent that it is proven to be necessary to create the external conditions to satisfy objective needs for present or future generations. For another thing, the steering is legitimate to prevent undesired effects of individual consumer behaviour; beyond that which is necessary to satisfy objective needs (therefore in the area of subjective desires, the demands made of consumer goods and ideas regarding the degree and breadth of satisfaction of requirements and desires), a steering yet limiting intervention into individual consumer behaviour would be legitimate if, without such steering, the guarantee of the external conditions for satisfying objective needs for present or future generations would be compromised (e.g. if one portion of people satisfy their subjective desires to an extent and in a way which prevents others from satisfying their objective needs).[2]

5.1.2 The question of the objective of the steering

Based on these considerations, we act on the assumption that the normative political objective of sustainability fundamentally legitimises a controlling influence on individual consumer behaviour. Individual consumer actions in their entirety should be such that the external conditions for satisfying at least the objective needs of all people should be guaranteed today and in future. Admittedly, this general objective must be substantiated when setting steering objectives. In doing so, the idea of a liberal society, which includes the autonomy of individual ways of life and the accompanying consumer actions, must also be considered. If steering consumption towards sustainability were understood as enforcing a certain lifestyle with specified concrete patterns of consumer behaviour, this would hardly be compatible with such an idea.

Assuming the legitimacy of the steering in the sense outlined above, the question arises as to which objective the steering is to be oriented around. Answering this question requires the clarification of how individual consumer actions are to be evaluated with a view to sustainability. Fischer et al. (in this volume) make a distinction between an *impact-oriented* and an *intention-oriented* evaluation. In an intention-oriented perspective a consumer action is said to be sustainable if the underlying intention is (at least in part) to reduce the difference between the actual state and the desired state in establishing the external conditions for satisfying the objective needs of present and future individuals. In an impact-oriented perspective, a consumer action is said to be sustainable if the consequences of the action are proven to reduce the difference between the actual and desired state (consumer actions which maintain the difference between the actual and desired state at a constant are therefore equally unsustainable as those which increase the difference). Ideally, a consumer action should meet both criteria.

---

2 As stated by Di Giulio et al. (in this volume), the society is to negotiate and determine an upper and lower boundary as to what needs to be ensured to satisfy the objective needs. Furthermore, the distinction between objective needs and subjective desires requires a further specification and operationalisation in a given historic and societal context in order to be useful for (empirical) research into sustainable consumption and for the steering.

In terms of the impact-oriented evaluation, two points are of particular importance with regard to the societal steering of consumer behaviour. Firstly, the impact-oriented evaluation requires prior qualitative or quantitative statements regarding the external conditions, which are to be ensured to satisfy the objective needs of all people. This implies value judgements, which need to be made in the course of the negotiations process within a society. These statements would need to be substantiated in the form of target values to determine specific steering objectives (desired states), to identify the need for steering (calculating differences between actual and desired states) and to evaluate the success of such steering. Secondly, in performing an impact-oriented evaluation, the aggregated level, considering the entire society, is of greater importance than the individual level. This is because in the societal steering of consumer behaviour – in contrast to the evaluation of consumer actions in terms of individual ethics – the focus is on achieving states of sustainability within societies rather than ensuring that each single consumer action performed by an individual meets the stated criteria of sustainability. Therefore it makes sense to define the objectives of the steering and to construct the corresponding indicators on the aggregate level of a society or in parts of the society (e.g. in certain fields of consumption).

With regard to the intention-oriented evaluation of consumer behaviour in view of the societal steering, it is of particular significance that all consumer actions are regarded as neutral in terms of sustainability, i.e. they are neither non-sustainable nor sustainable where the individual is not aware of their relevance in terms of sustainability (e.g. as the individual cannot have any knowledge of the sustainability discourse). Therefore, the key question for societal steering of consumer behaviour in this case is what enables people to form corresponding intentions (cf. Barth et al. in this volume) and how the development of these abilities can be promoted. The imparting of knowledge regarding the consequences of consumer actions in terms of the impact-oriented evaluation should naturally be part of such an empowerment. The objective of the steering is therefore primarily oriented on the competences to be promoted, with the aim of equipping the people within a society with such competencies (e.g. by anchoring the topic of sustainability in education).

5.1.3 The understanding of 'steering'

Societal steering can generally be paraphrased as the goal-oriented and intentional structuring of (framework) conditions within societies for the purpose of achieving societal objectives. To the extent that social objectives are politically determined, the term political steering can be used synonymously. Admittedly, societies are not as easily steered as a car. Individuals and societies can bring forth new ideas at any time, and people can react to new situations in unanticipated ways. In this context, a mechanistic model would also be unsuitable due to the multi-faceted and complex nature of consumer behaviour and its changes. We are acting on the assumption that consumer behaviour cannot in fact be steered in a narrow, technical sense, but can be influenced so that desired behaviour becomes more likely – and that scientifically sound knowledge of their effectiveness is necessary for such steering interventions (cf. Kaufmann-Hayoz/Gutscher 2001).

Particularly in connection with complex steering tasks, such as steering with regard to sustainable development, the concept of political steering is mostly understood in a broad sense – namely as the interplay of controlling influences by state as well as non-state actors on the development of the common good. Such an understanding of societal or political (self-)steering and coordination has long since been connected with the concept of *governance*.

The term governance primarily emerged from research within political science and adjacent disciplines into questions of the political steering of a society in the widest sense. The key idea of the governance approach is the distinction between *governance* in the aforementioned sense and *government* as classic state steering in policy cycles. Its precursors in the discourse include the steering debate, the policy networks approaches, the debate on the "joint decision trap" and the literature on regimes. These debates examined questions of the steering capacity of increasingly complex societies and problems faced across societies. At the same time, the state's ability to steer was questioned, with the presence and actions of an entire range of non-state actors in political processes and structures particularly stressed (Mayntz/Scharpf 1995).

The governance approach places the primary focus on the role of non-state actors (e.g. the private sector, associations, non-government organisations) and political networks, along with the interaction between state and non-state actors for coordinating solutions to problems within a society and problems associated with the distribution of resources (Benz 2004; Kohler-Koch/Eising 1999). On an international level, global governance was correspondingly defined as "multi-actor, multi-level decision-making", stressing the development of new structures and processes for solving problems in global politics (Rosenau/Czempiel 1992). Again, the focus is mainly placed on increasing regulatory initiatives and controls of non-state, supra-state and sub-state actors (Cutler et al. 1999; Fuchs 2007).

When we describe various (steering) instruments in the rest of this paper, we assume a broad understanding of political steering in the sense of the governance approach, i.e. we assume that certain instruments can be used by various actors – not just state actors – to influence individual consumer behaviour. In doing so, we in no way view governance as a 'non-political' problem-solving process, not influenced by existing interests (cf. Hewson/Sinclair 1999; Fuchs 2002). This means that governance structures and processes do not put an end to the existence of conflicting interests, the continued competition for the distribution of resources and the (often highly asymmetrical) distribution of power in this competition. Governance must therefore be understood as steering by determining, applying and observing rules and as creating and changing practices which may be directed by interests, but are not necessarily performed in the interests of the society as a whole. Governance structures and processes must therefore always be observed not only with regard to their effectiveness, but also in terms of their effects on democratic legitimacy and social justice.

5.2 Instruments of societal steering

Essentially, behind the question of how consumer behaviour can be steered lies the following question: Which actors have which (legitimate) opportunities to influence the behaviour of other actors so as to make individual consumer behaviour (more) sustainable, and what prompts the actors to use such opportunities? The concept of (steering) instruments has become established for labelling such possible actions in the sense of fundamental procedures, through which the actors in society attempt to steer the behaviour of others explicitly towards objectives within a society.

The use of the term 'instruments' is inconsistent and is determined, among other things, by disciplinary traditions. The term 'political instruments' is widely used in political science research (even if an instrument is not used by state actors, cf. the remarks on the governance approach above), while 'techniques' (for purposes of behaviour change) are more commonly mentioned in environmental psychology research. In turn, research based largely on economics also talks of 'enterprise instruments' as a term for the steering possibilities available to the providers of consumer goods. These 'enterprise instruments' are differentiated from 'political instruments' (often narrowly understood in discussions within these fields as instruments used by state actors).

The term 'intervention' is also used in the context of steering behaviour, but this is also inconsistently understood. On one hand, it stands for any type of regulatory interference, in particular by the state into the market. On the other hand, 'interventions' in environmental psychology research designate programmes or 'field tests' limited in terms of time and area of application (e.g. a community or organisation), in which different instruments or techniques are generally used to change behaviour and are evaluated in terms of their effects. In this article, we will use the term 'instruments' to designate basic ways of steering the behaviour of addressees in society in the desired direction, and 'interventions' in the described environmental psychological sense.

Changes to consumer behaviour unfold in a dynamic framework of both structural and personal conditions which facilitate and inhibit change (cf. Kaufmann-Hayoz/Bamberg et al. in this volume). Therefore, a distinction is made between *structure-focused* and *person-focused* instruments (cf. Steg/Vlek 2009; Mosler/Tobias 2007; Kaufmann-Hayoz et al. 2001). The former start from the structural conditions for behaviour, i.e. they primarily create changes in the consumer environment (e.g. in the offer of consumer goods or in institutional and infrastructural framework conditions) and thereby influence the behaviour of consumers. The latter start from the personal behavioural conditions or specific behavioural situations, i.e. they directly produce or support changes in consumer behaviour while the structural conditions remain unchanged. Essentially, the steering instruments always attempt to create or strengthen conditions which promote change and to eliminate or weaken conditions which inhibit change, with a possible distinction made between different instrument types depending on their 'mechanism of action'.

The literature on (steering) instruments contains several suggestions as to their typecasting, and various disciplines provide different typologies (cf. Tews 2009; Rubik et al. 2009; Kaufmann-Hayoz et al. 2001; Bemelmans-Videc et al. 1998). The distinction between regulatory (command and control) instruments, market-based instruments and instruments based on the provision of information (soft/informational instruments) is common mainly in the political science literature. Wolff und Schönherr (2011) have adopted these categories (which they label as 'regulatory', 'economic' and 'communicative' instruments) specifically for the field of sustainable consumption, adding a vaguely outlined remaining category of 'procedural instruments and instruments of societal self-regulation', in which they include instruments such as the provision of infrastructure (for example for the separate collection of waste), voluntary agreements between producers, activities performed by companies in the course of their CSR strategies or networks (e.g. roundtables).

Below we typecast the (steering) instruments without defining new instrument types, but combining approaches from various disciplinary contexts. We distinguish between four types:
- Regulatory instruments,
- Cooperative instruments,
- Economic instruments,
- Communicative instruments.

We will provide a brief outline of these types, referring to previous applications and experiences and illustrating each type using one or two examples of applications in the field of sustainable consumption. In doing so, we will only provide a short description of the regulatory, cooperative and economic instruments, as the distinctions between these types is commonly known and widely used. In contrast, we will discuss communicative instruments in more detail as this allows for a less known perspective to be integrated into the approach which explicitly focuses on behavioural changes.

As with any typological distinction, this primarily represents a way of facilitating analysis. In practical applications, the boundaries are not static, and individual measures can be allocated to various instrument types, depending on the perspective. For example, the mandatory requirement on manufacturers and retailers to place energy labels on devices is a regulatory instrument from the perspective of the political decision-makers, which takes effect in the consumer environment. However, from the perspective of the end consumer, it is a communicative instrument, which helps support the action.

5.2.1 Regulatory instruments

Regulatory instruments place requirements or bans on the actions of certain target groups. This includes mandatory guidelines established by the state, which are binding on all people and, if necessary, can also be enforced by way of state sanctions. The decisive element of requirements and bans or prohibitions is their binding nature. They have a very direct influence on the scope for action of the parties they address, by directly creating obligations and imposing sanctions in the event that such obligations are violated. Provided that they can be enforced, their effectiveness can be relatively accurately predicted in advance. However, due to their intrusive nature, they are often associated with great controversy and potentially high costs for the decision-makers in a political sense. In the context of sustainable consumption, regulatory instruments often target actors in the consumer environment (namely producers or providers of goods), but they are also used to directly regulate individual consumer behaviour.

> *For example: Regulations for builders in the area of energy-efficient refurbishment*
>
> The key regulatory instruments for controlling the room heating demand of buildings in Germany are the Energy Saving Ordinance (EnEV) and the Renewable Energy Heat Act (EEWärmeG). In this case, the regulatory legislation is aimed directly at consumers in their role as homeowners or building owners/developers.

- The Energy Saving Ordinance obliges building owners to observe legal limits in terms of energy consumption. New builds must therefore not exceed a specific primary energy demand. In contrast, there are few requirements placed on existing buildings, which are not tied to the case of refurbishment. However, the owners of buildings subject to more extensive renovation performed on the building envelope must implement energetically effective measures (e.g. facade/roof insulation).
- The Renewable Energy Heat Act places further requirements on the manner in which rooms are heated and obliges the use of renewable energies if compensatory measures are not taken (e.g. improving the building's energy standard). These guidelines apply exclusively to new builds throughout the country. Only in Baden-Württemberg is there a state law which prescribes the use of renewable energies also for existing buildings.

An analysis of these two regulatory instruments, in particular with regard to the energy-efficient refurbishment of single family and two-family houses, showed that, in addition to the limited scope due to the specific requirements, the insufficient enforcement represents the main problem of these regulatory instruments. One important reason for this is the lack of information available to the authorities relating to the existing buildings, as the authorities generally need not be informed of renovations, although the majority of requirements apply to this case. Furthermore, the example also shows that effectively checking the enforcement of the instruments is associated with great effort due to the number of different actors involved. However, as the authorities are likely to have little interest in such an effective control of the enforcement – even to avoid conflicts with the local residents – there is little checking with regard to the actual implementation of the requirements of renovations on existing buildings (Weiß/Vogelpohl 2010).

However, if there is to be a nationwide energy-efficient refurbishment of existing buildings in Germany, regulatory law is still to be regarded as an important part of the mix of instruments, as a significant group of homeowners have shown only little interest in the idea of energy-efficient refurbishment, meaning that it will be hard to reach them through instruments such as providing advice and incentives. It may be possible to partially remedy the insufficient enforcement through obliging sampling regimes or by choosing requirements that can be easily checked (cf. Weiß/Vogelpohl 2010).

5.2.2 Cooperative instruments

Cooperative instruments act as alternatives to the hierarchical forms of regulatory intervention, instead placing the emphasis on the self-regulation of actors within society. Common forms include (voluntary) agreements between target groups and (not always) the state, as well as commitments by industries to achieve specific objectives or take measures in areas or sectors which are not subject to regulatory approaches, or to achieve objectives that go beyond the existing regulation. The key themes discussed in the literature include the intention of filling gaps in state regulation to help solve problems within a society, as well as pre-emptive strategies by actors in the private sector, i.e. entering into voluntary agreements to avoid more stringent state regulation (Gibson 1999; Vogel 2005). This type of instrument focuses on changing the consumer environment and thereby indirectly influences consumer behaviour.

Empirical studies have shown that cooperative instruments are mainly effective when they are legally binding (i.e. where they act as quasi regulatory instruments) and sanctions are enforced for failure to observe the cooperative instruments. If the objectives of the companies involved are in line with those of the policies, they can even be more effective than regulatory instruments. Even non-binding guidelines may be effective if they are used effectively by pressure groups in the public. However, if these conditions are not met, it can be assumed that these instruments will have little effect, despite being highly popular in politics due to their low costs. For example, there is often a significant lack of correspondence between the objectives of the private sector or companies and societal objectives (Fuchs 2006). Also, limited financial resources and the limited potential to attract a society's attention hinder the effective use of non-binding guidelines by pressure groups in the public.

Regulatory instruments are increasingly talked of as a necessary legal framework of a so-called "controlled self-regulation" (Hey et al. 2005). The talk is therefore of "hybrid instruments", which combine the advantages of regulatory legislation and self-regulation or, in other words, of regulatory and cooperative instruments (cf. Tews 2009).

For example: Smart metering as a hybrid instrument

The legislation within Europe and Germany has incorporated smart metering as an instrument for increasing energy efficiency in climate protection policies. This innovation allows households to receive information on electricity consumption within far shorter intervals than when using the mechanical counters previously used.

> The legally-binding introduction of intelligent measurement systems represents one example of a hybrid instrument. In their role as measurement service providers, the network operators or the new market actor "metering point operator" (*Messstellenbetreiber*) will in future have to implement the measurement systems all over the country. However, it is left to them to decide how they will actually implement this requirement in the liberalised measurement and metering technology market, which was only established in 2008. The concept of the pertinent Ordinance on Access to Meters (*Messzugangsverordnung*) focuses on establishing the greatest possible competition, without limiting freedom of consumers and companies as far as possible. The fact that the connection users are able to choose the metering point operator or measurement service provider is key to opening up the market. To this point, the ordinance has not consciously prescribed technical standards (including protocol formats), minimum technical requirements or equipment details of smart counters or a schedule for an extensive rollout of any kind. Instead, the end user is granted the right to demand a semi-annual, quarterly or monthly bill from their energy supplier in place of the annual bill, which was the previous standard. It can be assumed that a more economically efficient implementation of these requirements requires the use of digital meters, which can be remotely monitored. As there has been no noteworthy market activity to this point, the amendment of this competitive, liberalised legislation, which is currently being passed, is largely displaced by classical regulatory measures (obligation to incorporate smart meters and exchange all of the old counters at the end of the calibration period, no possibility for the connection user to contradict the reading due to the legally governed data protection profile).[3]

5.2.3 Economic instruments

Using pricing signals, economic instruments attempt to change the individual assessment of various options for the pursuit of a given objective. They can be used both to directly influence individual consumer behaviour and also to influence actors within the consumer environment. One group of economic instruments attempts to prevent undesired actions by increasing the price, for example of resource-intensive goods (such as fuel). This includes taxes, (incentive) duties and fees. Another group of instru-

---

[3] Feedback using smart metering systems is a communicative instrument from the perspective of households (cf. section 5.4.4 below and Sunderer et al. in this volume on the effectiveness of the instrument).

ments creates financial incentives – such as subsidies, reduced interest-rate loans or tax write-offs – for desired behaviour, often for decisions to invest. Examples of such instruments include bonus schemes for purchasing energy-efficient devices and subsidies or reduced interest-rate loans for the energy-efficient refurbishment of residential buildings. The mechanism of financial incentives for promoting desired behaviour is also used regardless of such state measures, for example, by companies in introducing new 'sustainable' products onto the market.

Economic instruments primarily became popular in political and scientific discussions in the 1980s and 1990s against the backdrop of strengthening (neo)liberal perspectives and the accompanying focus on market forces and discrediting direct state intervention. They were often praised for what was perceived to be a greater degree of effectiveness. However, in reality, the effectiveness of these instruments can only be predicted to a certain extent as it is not always known in advance how and when individual people and companies will react to pricing signals. In fact, trends may be declared with the help of economic simulation models, but reliable predictions are impossible to make. However, the determination of economic incentives (in particular for increases of pricing for undesired behaviour) at an amount which would make the instruments effective is in fact again connected with high political costs. This means no determination of the appropriate amount is realised or the presumed political advantage of the economic instruments fades away (Fuchs 1995).

For example: Progressive electricity tariffs as an incentive for more frugal electricity consumption

In the 1970s, progressive electricity tariffs were introduced in Italy, California and a few other countries to promote energy-saving consumer behaviour. Progressive electricity tariffs – a kilowatt hour price which progressively increases along with increasing electricity consumption – are an example of an economic instrument that attempts to produce both of the aforementioned effects. More thrifty electricity consumption is more significantly 'rewarded' than under the conventional electricity tariff consisting of the basic rate and the price per kilowatt hour. In contrast, higher consumption is more significantly 'punished' than under a unit price per kilowatt hour.

The progression may be found in the various components of the total electricity price or may affect the electricity provision itself. An example for the former would be a comparison between the current situation in Italy and California. In Italy, the progres-

> sion exists for all households in the network transit fees and taxes. In California, the progression is established in the proportion of the total electricity price made up by production costs. An example for the latter can be seen in Italy, where a distinction is made between contracts with a limitation on output to three kW (which means, for example, that two high-consumption household appliances cannot be used at the same time) and contracts with a higher capacity (cf. Brohmann/Dehmel et al. in this volume).

5.2.4 Communicative instruments

Communicative instruments attempt to change the psychological factors which influence the formation of intentions to perform actions and the performance of the actions itself. These include knowledge, values, attitudes, social or personal norms, the perception of physical and social reality and the perception of opportunities for actions as well as one's own possible actions. Communicative instruments often directly target individual consumer behaviour, but target audiences may also be actors within the consumer environment.

Product and consumer information, such as labels, stickers or various feedback, may be regarded as communicative instruments, as well as training and advisory services, information and awareness-raising campaigns and the establishment of networks and platforms.

Communicative instruments are highly popular in politics, particularly in the area of sustainable consumption, as they are only slightly intrusive and the provision of information, for example, has essentially positive connotations. This is primarily the case where the parties are not obliged to provide the information. However, the more potential the offering party in question deems the instrument to have in changing (consumer) actions contrary to their interests, the more political controversy and costs are associated with its introduction. For example, this can be seen in the EU's attempts at introducing a mandatory traffic light labelling system for foodstuffs[4] or making the energy efficiency label for household devices more expressive.

The effectiveness of communicative instruments is not always easy to assess and calculate, resulting in different estimates of their effectiveness. Communicative instruments, which are used as part of campaigns or interventions, can have a rapid effect,

---

4 A familiar red-yellow-green signal was to be used to show the fat, saturated fat, sugar and salt quantities of food at a glance. The information should always refer to 100 g or 100 ml and was to be highlighted on the packaging in the respective colour. Red stands for a high amount, yellow for a medium amount and green for a low amount.

but this is initially limited to the corresponding context in time and space. Long-term and wide-reaching effects would be expected if it were possible to use the communicative instruments to trigger or enhance desired trends within a society. As communicative instruments, or rather the messages communicated by such instruments, need to compete with a number of other messages (in consumption primarily advertising), they will only be heeded if they are easily accessible, understandable and resilient. Consumers rarely have the time and resources to obtain additional information which is not directly available when they are making their purchase decision, nor do they have the time to compare the quality of various labels. There are a number of studies which show that economic and regulatory instruments tend to be more effective than their communicative counterparts (ASCEE team 2008; Lorek et al. 2008; Rehfeld et al. 2007).

Research of the past two decades on the use of communicative instruments has produced a great number and sophisticated findings. It quickly became clear that communicative instruments must in no way be purely a matter of providing (product) information if they are to bring about behavioural changes. Argumentative and affective persuasion, creating states of tension and reflexion, memory aids, self-management, reporting the consequences of actions and stimulating social dissemination processes are equally important for an effective communicative instrument (Mosler/Tobias 2007; Steg/Vlek 2009; Vandenbergh et al. 2010; Flury-Kleubler/Gutscher 2001). In recent years, the view has been increasingly accepted that individual behavioural changes need to be understood as processes that consist of various stages, with different communicative instruments necessary depending on the stage at hand (cf. Bamberg 2007 and Kaufmann-Hayoz/Bamberg et al. in this volume). If there is no awareness of the problem, or if there are no known alternatives to an action and changing the behaviour is not considered at all, *motivational* instruments are required. If, on the other hand, a person intends to change their consumer behaviour, *supportive* instruments are often required in order that the intention is translated into actual behaviour. If the primary objective is breaking down dilemma situations within defined social systems and encouraging as many people as possible to adopt a new behaviour, additional *diffusion-focused* instruments are advisable. This differentiation between the various communicative instruments is briefly explained below.

Motivational instruments

Motivational communicative instruments attempt to encourage consumers to reflect on their behaviour up to that point and set a target of changing that behaviour. This type of instrument includes any attempts at stimulating an individual to reflect on their previous behaviour and create a changed awareness of the problem, as well as providing knowledge of alternative options for actions and their effectiveness with regard to sustainability objectives. Motivational instruments would also aim at a change of general norms and values, which may help establish subjective feelings of obligation, and changed perceptions of social norms (cf. Kaufmann-Hayoz/Bamberg et al. in this volume). Such instruments also have the potential to make symbolic meanings conscious and thus open a possibility for changing them.

The array of motivational instruments varies widely and comprises general education provided on the subject of sustainability and consumption, as well as targeted information on specific topics (e.g. brochures or internet sites with product information), but also various forms of persuasion within the scope of campaigns. Motivational instruments seem primarily suited to encourage individuals with previously non-reflected consumer behaviour (both in the sense of unconscious and socially preformed, cf. Kaufmann-Hayoz/Bamberg et al. in this volume) to enter the early stages of changing that behaviour – first setting an objective or target to change and then creating the specific intention towards a certain action. Empirical studies have shown that these instruments in fact produce such effects, but that they do not necessarily result in the behaviour change actually being achieved (cf. Abrahamse et al. 2005 relating to the field of energy use).

Supportive instruments
Supportive instruments attempt to support the (first) implementation of changed behaviour and help maintain the new form of action. This means that they presuppose that a specific intention already exists. Supportive instruments include techniques such as direct feedback (e.g. on current consumption of fuel or electricity), as well as prompts which (again) make the individual aware of the specific intention during the situation when the action is to be performed. Highly visible product labels which point out the opportunity during the purchase situation are also classified as supportive communicative instruments. Procedures involving private or public self-commitment and certain methods of self-management would also be regarded as supportive instruments.

For example: Feedback through smart metering as a supportive communicative instrument

The implementation of a smart metering system offers the opportunity to link the measurement system to feedback regarding energy consumption. Detailed information on consumption should help the consumers to identify potential for savings in their household, to set specific targets for change and to implement and maintain new forms of action. A broad field study of over 2000 households investigated two instruments of feedback on the basis of smart metering (web portal and monthly written information) (cf. Sunderer et al. in this volume). The results show that the information provided by the feedback was largely positively welcomed (informative, useful, helpful, user-friendly, fun and motivational to tackle the topic of energy savings). The majority of households were clearly oriented around saving energy, and the feedback helped support this attitude, meaning that the feedback system was useful as an instrument of supporting this energy saving in practice. However, pure information provided promptly and well represented does not automatically lead to a significant reduction in consumption for everyone. A total energy saving of 3.5 percent was achieved in German cities and Linz. It seems that displaying specific options is important as a quarter of the users of the web portal were very interested in the energy-saving tips provided on the portal. However, a third of users only used the portal once, and for a short time. The reasons for this have not yet been established.

Diffusion-focused instruments

Diffusion-focused instruments target the diffusion of social norms, motives, attitudes or new practices within social systems (population groups, organisations). Such instruments are partially used under the scope of environmental psychology interventions (see Gutscher et al. 2001 for a specific example of this in the area of mobility). In doing so, establishing networks and platforms and designing participative processes within communities and organisations have become a relatively self-contained area of change management and the diffusion of social innovations (cf. Schweizer-Ries 2010; Kristof 2010).

For example: Energy-saving advice as a bundle of communicative instruments

Energy-saving advice has been spread as an instrument to stimulate and support people in planning and implementing energy-efficient refurbishments. An analysis of the various offerings of such advice defined six types of advice: Energy check,

> energy advice at hardware stores, office-based initial advice, initial advice provided on-site, orientation advice provided on-site and concept-oriented advice. Progressing through the list, each type increases the amount of time taken by the parties providing the advice, the relevance of the building and the depth of content (cf. Dunkelberg/Stieß 2011). The way in which contact is established between the parties providing the energy-saving advice and the recipients/customers is the main determinant of which type of communicative instrument the respective energy-saving advice is allocated to. If the customers are actively addressed by the party providing the advice, such as during energy checks, the offer of the advice generally serves the purpose of raising awareness and is therefore primarily motivational. If the parties providing the advice offer a technically competent expert as a contact person but the contact is made by the house owner themselves, the energy-saving advice is rather more of a supportive, action-focused instrument.
>
> Accordingly, the effects of energy-saving advice vary: The advice may have a positive effect on the renovation decision by triggering energy-efficient refurbishments (e.g. IFEU 2005; IFEU 2008; Stieß/Birzle-Harder 2010). If homeowners are already in the course of planning a specific energy-efficient measure, energy-saving advice can have a positive influence on the refurbishment, for example a more demanding implementation in the form of greater insulation (e.g. Stieß/Birzle-Harder 2010). The example of energy-saving advice shows that communicative instruments often fulfil more than one function and a clear allocation to specific types is not always possible.

5.3 Questions and problems in assessing the effects of steering measures

A central point of interest is having knowledge of the effectiveness of social steering of consumer behaviour and particularly the discussed instruments, i.e. knowledge of whether using such instruments will bring about a change in consumer behaviour towards sustainability. However, there are numerous challenges facing research into the effectiveness of steering instruments. Hereinafter, we will briefly discuss the most important of these challenges.

5.3.1 Causality and empirical ascertainability

According to the impact-oriented evaluation of the sustainability of consumption (cf. Fischer et al. in this volume), the steering must ultimately cause a change in the actual state towards the desired state. Depending on the consumer field and specific sustainability objective, this may be, for example, lesser consumption of resources, fair working conditions or the sufficient provision of food for all. The ultimately intended effect is widely labelled the 'impact', with its scientific investigation labelled 'impact evaluation' (e.g. Wolff/Schönherr 2011), or as the 'distal impact' in the evaluation research into interventions (Rossi et al. 2004).

However, these impacts can only indirectly be achieved through instruments, which alter key personal and/or structural conditions (e.g. newly acquired knowledge of the effects of high meat consumption or the removal of energy-inefficient household appliances from the market), which may in turn lead to a change of consumer behaviour (e.g. to reduced meat consumption or increased purchases of energy-efficient appliances). The direct effect of the steering instrument on the addressees or target audience is known as the 'outcome' or the 'proximal effect', whereby – depending on the number of phases in the assumed causal chain of events – a distinction is made between primary, secondary outcomes and so on (Wolff/Schönherr 2011; United Way of America 1996).

Given this, there are two essential challenges facing the research:

- *The problem of causality:* Each investigation into effects examines hypotheses regarding the causal connection between a steering measure (i.e. the use of a steering instrument) and the phenomena which they are to effect. The hypotheses are explicitly or implicitly based on an "impact theory" (cf. Rossi et al. 2004), which makes statements on the short-term, medium-term and long-term, direct and indirect, intended and unintended causal consequences of the measures. An empirically based action theory is used to explain the effects on human behaviour, while corresponding theories from natural science, economics and social science are needed for the relationships between the behaviour and its ecological, economic and social consequences. There is no comprehensive, all-in-one impact theory, which covers all elements of societal steering. Rather, before each individual investigation, it must carefully be checked which theories can be used to create the most precise possible impact hypotheses for the instruments and consumer actions in question and their consequences.

  Strictly speaking, the causal effect of using steering instruments can only be concluded if actual experiments are performed, isolating and controlling the variation of the factors/variables whose effects are to be examined. However, this is only possible very selectively and approximately in so-called field experiments (see e.g. Leung et al. 2011). When performing impact studies in the real world, there are always greater or lesser uncertainties about the causal links, due to uncontrollable, context-dependent influencing factors, which may enhance or inhibit the effectiveness of a specific instrument. Often it is not a case of investigating the effect of just one instrument, but of entire bundles of instruments, making it even harder to make statements regarding the specific effectiveness of individual instruments.

As a whole, due to the complex cause-effect relationship, it is nigh on impossible to clearly attribute changes of the actual state to the use of specific steering instruments.

- *The problem of empirical ascertainability:* If effects are to be empirically demonstrated and causally attributed to individual consumer actions, the phenomena in question (i.e. both the behaviour of consumers and their ecological, economical and social effects) would need to be reliably determined and must be valid. However, this is often highly difficult, if not (currently) impossible. Determining behavioural changes on an individual level, which may affect electricity consumption, for example, would not only require the availability of suitable measuring devices (which have been correctly fitted and function properly) – it would also be necessary to exactly distinguish between the various uses (such as heating, showering, lighting, use of the appliances etc.) (cf. Klesse et al. in this volume). In the area of nutrition, it would need to be clarified which actions should be covered at all and which methods would be suitable for doing so (methods such as recording behaviour or analysing receipts often entail a great effort). The same question arises in recording meaningful and valid data on an aggregate level.

5.3.2 Influence of political trends in assessing the effectiveness of steering instruments

Not least due to the large gaps in the body of knowledge and the aforementioned challenges facing scientific research, generalised statements on the effectiveness of steering instruments are often based more on political and societal trends than on scientific analysis and therefore need to be critically assessed. In this way, the research into the perceived legitimacy of norms indicates that the legitimacy and acceptance increases along with the extent to which the norm fits into the general normative context within a society. To that extent, it is hardly surprising that, in the era of the neoliberal zeitgeist, economic instruments were often assigned particular effectiveness[5], even if this has not been proven (Fuchs 1995). Similarly, the preference received by communicative instruments in recent decades can probably be more assigned to their simpler political implementation rather than to the scientific proof of their greater effectiveness. On the contrary – comparative analyses into the effectiveness of instruments in the area of policies regarding sustainable consumption have reached the conclusion that regula-

---

5 A clear distinction is to be made between effectiveness and efficiency. Market-based instruments would in fact often be more efficient, as they allow the target audience to make a more flexible reaction, which is suitable to the respective cost-benefit conditions. However, efficiency alone may not be a criterion for the effectiveness of political instruments, as it provides no information as to the level to which the intended objective has been achieved.

tory and economic instruments tend to be more effective (ASCEE 2008; Lorek et al. 2008; Schönherr et al. 2010).

The first prerequisite for the effectiveness of instruments is naturally the introduction of the instrument. For this reason, the political feasibility of adopting and implementing a specific instrument is often considered in discussions as to the effectiveness of that instrument. In this respect too, there has been a tendency in previous decades to regard communicative and economic instruments as more easily achieved than regulatory instruments. However, such assessments can also be critically questioned as it has been shown that the introduction of communicative and economic instruments 'with bite' – i.e. those which run contrary to vested interests if they are successful – encounter similar political obstacles as the introduction of regulatory instruments. This is also one reason why, for example, as outlined above, the traffic light system for foods in the EU failed.

The acceptance, expectations and specific design of instruments and value systems within a society therefore interact and influence the 'theoretical' effectiveness of steering instruments.

5.3.3 Different traditions and perspectives of research into the effectiveness of steering instruments

The difficulty of making statements on the effectiveness of various types of instruments, which can be generalised, consequently results in the need to analyse the effect of individual instruments or bundles of instruments in specific political and societal contexts. Such analyses can be developed from different perspectives. For example, there is evaluation research in political science and sociology which investigates the effects and effectiveness of the political measures introduced (cf. Stockmann/Meyer 2009). On the other hand, scientifically evaluated interventions have produced findings on the effectiveness of various instruments on individuals and their behaviour. Such findings, for example from the field of environmental psychology, are not only useful in terms of analysing the effectiveness of communicative instruments, but also in terms of estimating the (probable) effects and optimising the design and adjustment of regulatory, economic or cooperative instruments. To that effect, Krömker and Dehmel (2010) have identified the relevance of various psychosocial factors for different target actions in the field of saving electricity in private households, which highlights the fact that steering instruments need to be tailored to the desired target and the desired context.[6]

---

6 The natural scientific approaches of analysing material flow (material flow analysis, MFA) in calculating sustainability effects represent a third research perspective. They have recently been further developed to a hybrid assessment tool by connecting them to approaches in political sciences.

In the focal topic, research was performed from both perspectives. Below, we will provide three examples, which help show what we can learn with regard to the effectiveness of instruments. The first two examples are political and sociological evaluation research and are based on qualitative comparative investigations into the effectiveness of similar instruments in different countries. The third example stems from research into scientifically evaluated interventions and is an example of an intervention in the field of environmental psychology.

Bonus schemes for energy-efficient household appliances

In the past, bonus schemes have been used in various countries around the world as an economic instrument for steering consumer purchasing decisions in favour of household appliances with high energy efficiency, particularly in order to promote the early replacement or better replacement of household appliances. The programmes examined in Austria, Denmark and the Netherlands were discussed regarding differences in terms of the financing of the programmes and their ecological effects in addition to the expected differences in terms of the amount of the bonus, the change of the promoted efficiency classes and types of appliances and the duration of the programmes (Dehmel 2010). Essentially, the assessments on the effectiveness in terms of promoting the market penetration of more efficient appliances and the accompanying electricity savings and reductions in emissions have been positive. Thereby, different design details imply interesting conditions for the success of their introduction and effectiveness. In terms of the introduction of these instruments, the key questions seem to relate to their financing and the constellations of interests among the relevant actors. For instance, the necessary funds were already available in Austria due to the existing consumer obligation to pay for the disposal of cooling appliances at the time they were purchased. Furthermore, the actors responsible for making the decision as to their use primarily represented the interests of trade and recycling. In Denmark and the Netherlands, the bonus schemes were financed by returning ecotaxes to consumers or reducing the tax liability for trade, meaning that more politically attractive arguments and financing structures were used than would have been the case in placing the burden directly on the budget.[7]

However, the most important differences would have to lie in the ecological effects of the programmes. For example, the old appliance only had to be returned in the

---

[7] Furthermore, it became evident that it was always small, relatively independent organisations who were implementing the programme that would lead to reduced transaction costs. At the same time, the previous assumption of the limited duration of a programme as a condition for its success was not confirmed. Ultimately, there were differences in how the programme was rounded off (i.e. how the economic instruments were supplemented by communicative instruments), whereby the provision of information on the pricing and availability of the best appliances on a website in Denmark was regarded as particularly effective.

Austrian bonus scheme. Although it is possible to criticise the lack of control or supervision of the return of the appliances, which only needed to be confirmed by way of a signature, the fact that this aspect was entirely overlooked in the other bonus schemes raises some large questions in terms of the calculated electricity savings and estimated reductions in emissions. In the worst case, these bonus schemes could simply subsidise putting a second appliance in households, thereby contributing to an increase in electricity consumption. Furthermore, all of the programmes examined neglected the necessity to consider trends towards larger appliances in connection with the exchange of household appliances. From the perspective of ecological effectiveness, one essential condition for the success of future bonus schemes would be to only pay the bonus up to a specific size of appliance or a specific maximum consumption (cf. Dehmel/Brohman et al. in this volume).

Progressive electricity tariffs

As shown in the section "Economic instruments", progressive electricity tariffs, i.e. tariffs where the price per kilowatt hour rises in steps along with consumption, are (have been) used in various countries, such as Italy, the USA on state level (California), Japan, South Korea and various developing countries. A closer look at the cases of Italy and California shows – despite immense differences in terms of the type of liberalisation and the structure of the electricity industry – interesting similarities and differences in terms of the conditions for introducing the tariff and how the tariffs work, as well as the design of the tariff itself (Brohmann/Dehmel et al. in this volume; Dehmel 2011; Tews 2011). The tariffs were each introduced or enhanced during energy crises both to ensure the supply of electricity through a reduction in demand induced in this way and to make electricity affordable for socially vulnerable households.

One fundamental prerequisite for realising potential electricity savings through tariff controls is the *binding nature* of the progressive tariff structure. This has been realised in different ways, depending on the electricity market. While today the electricity market in Italy has been liberalised, which enables clients a free choice of their electricity provider, there is no such liberalised electricity market in California (Tews 2011). In Italy, the existing progressive tariff structure has been transformed as a result of the liberalisation of the market, meaning that it is only binding on the parts of the electricity price which are not subject to competition – i.e. for network fees and taxes. Furthermore, a limit is placed on the output.[8] In contrast, the progressive tariff struc-

---

8 Since the 1970s, there have been contracts limited to 3 kW output in Italy, which are considerably cheaper than contracts with higher limits placed on output (generally 6 kW), and are still used by over 90 percent of Italy households.

ture in California is only binding for the clients of large electricity companies, who provide 77 percent of electricity for private households. Electricity clients can also not switch to another supplier as the market for end clients has not been liberalised. The progression is contained within the proportion of electricity costs attributable to production costs and is prescribed by the California Public Utilities Commission.

It is difficult to make an accurate estimate of the electricity savings achieved by progressive tariffs as there are a number of factors which influence the electricity consumption of private households. The effects of the tariffs also change depending on the specific design of the tariff. This makes it hard to make *generally valid* statements on the ecological effectiveness (i.e. the electricity savings), as it is the specific design of the tariff, which ultimately determines its effectiveness (Faruqui 2008, p. 24; cf. Tews 2011, pp. 12 ff.). American researchers have simulated the effects of introducing the last five-tier progressive tariff and developed broader model calculations on the electricity and costs saved by progressive tariffs. Depending on the assumptions in the model, a 6–10 percent reduction of the average annual electricity consumption for the households was calculated in these studies (Reiss and White 2004, p. 876; Faruqui 2008, p. 22). In the long-term, electricity consumption reductions of up to 20 percent and cost reductions of up to 25 percent could be achieved in the households (Faruqui 2008, p. 27).

The average consumption and trends in Italy and California both posted considerably lower values than in comparable countries or federal states. However, as electricity is a highly inelastic good, at least in the short-term, and therefore changes can only be achieved to a limited extent using price changes, supplementing this instrument with communicative instruments is of great importance (Hamenstädt 2009).[9]

Impact analysis of communicative instruments

In the focal topic, motivational and supportive intervention packages to promote energy-efficient user behaviour at the workplace were investigated (cf. Matthies et al. 2011 and Matthies/Thomas in this volume). On one hand, the idea was to investigate whether communication instruments alone could be at all effective in this context, i.e. whether they lead to energy saving (distal effect or secondary outcome). Beyond this, it was a question of whether attention-arousing instruments such as prompts (memory aids in the form of stickers on appliances) or commitments (feedback card with commitment) could more effectively bring about a change of user behaviour (proximal effect or primary outcome) than purely motivational instruments (placards and bro-

---

9 In this context, for example, the question of smart metering and other feedback strategies was constantly discussed.

chures with standard energy-saving tips). An impact analysis used three measurement values as indicators of the effects of the different types of intervention: energy consumption (only building-related), observed ventilation behaviour (also only building-related), and self-reported usage behaviour (on employee level). The indicators were calculated before and after the intervention and one year after the end of the intervention. The analysis included a control group, i.e. buildings where there was no intervention, with measurements taken and employees surveyed.

The distal effect of energy saving was shown by comparing buildings with and without the intervention in terms of their energy consumption before and after the intervention (average energy savings across the buildings of 7 percent in electricity and 1 percent in heating). The intervention group also displayed significant changes in the ventilation behaviour observed. The proximal effect of self-reported behavioural changes could be shown in a number of behaviours covered by the questionnaire. Significant changes were mainly found in the group with additional supportive communication instruments.

5.3.4 Conditions for acquiring effectiveness knowledge from interventions

In the focal topic, interventions to promote sustainable consumer behaviour were performed and analysed as part of several projects. In performing the evaluation research to assess and optimise the use of interventions (cf. Rossi et al. 2004), a number of particular challenges beyond the general challenges described above (in particular causality and empirical acertainability) came to light. These challenges and initial suggestions on dealing with the obstacles were explored as part of a survey of four project groups, which all used communicative instruments for the intervention.[10] The results, which partly confirm the experiences and findings gleaned from other projects and in the literature (cf. Balzer 2005), can be summarised in four requirements on trans-disciplinary evaluation research in the scope of interventions, formulated in the form of propositions:

1. *Develop fields of practice for sustainability innovations.* Evaluation research into interventions for sustainable consumption in trans-disciplinary research projects stringently requires fields of practice (schools, households, municipal utility companies etc.), where the actors are prepared to implement an innovation (such as a

---

10 Researchers from the project groups "LifeEvents", "Change", "BINK" and "Intelliekon" were surveyed. We would like to thank them for their support with the survey. Among other things, the following questions were posed: "What rather topic-specific challenges did you face in the course of the evaluation work within the SÖF focal topic?", "How did you deal with these challenges?", "What development tasks do you see for evaluations within the topic area of 'Transdisciplinarity/sustainable consumption'?"

new behaviour, a new technology). The partners must be highly committed and willing to take risks – this makes good arguments for winning over practice partners indispensible.
2. *Work out complex concepts.* Interventions which look to promote sustainable consumption rely on all of those involved having extensive background knowledge (such as "What is sustainable here?" or "Which measure is effective and how?"). A lack of knowledge can cost a lot of time during the goal-setting and planning phase of transdisciplinary work. A common understanding of concepts among all of the actors as a 'basis for the work' is not a given and must be established before the intervention. It is important to provide further education and tools with a practical relevance.
3. *Actively handle 'drop-out'.* Interventions rely on a long-term plan and the active cooperation of all of the partners. However, actors from fields of practice may only be active in the short-term or in phases. The motivation towards long-term and regular cooperation should therefore be actively promoted, for example by communicating the advantages of involvement (increased knowledge, new social relationships or expanding professional opportunities) (cf. Dovidio et al. 2006). At the same time, the possibility that interventions and the corresponding evaluation research may not be performed as planned should be considered as early as the planning phase, and possible ways of dealing with this must also be considered.
4. *The evaluation must be performed across various disciplines.* The importance of the evaluation is not always recognised by the practice actors and the evaluation is occasionally neglected due to limited time in favour of other activities (such as implementing the interventions). In case of extensive documentation and calculation requirements (see above: problem of empirical ascertainability), this problem situation may further deteriorate. Conveying the use of the evaluation and actively supporting the documentation and 'low-threshhold' survey options (such as telephone interviews) may help in this regard.

5.4 Conclusion

Scientifically sound knowledge (of their effectiveness) is necessary for the design and implementation of measures which attempt to change individual consumer behaviour. In the focal topic, the internal and external conditions of individual consumer behaviour were examined in various consumer fields, with the possibilities of highly different instruments and bundles of instruments explored from a number of different scientific perspectives (cf. the contributions in Part 2 of this volume). The questions relating to steering consumption towards sustainability probably represent the most important synthesis questions addressed to the interdisciplinary and transdisciplinary research community in sustainable consumption. The work of integrating existing findings on fundamental questions regarding the steering and the individual instruments, bundles of instruments and measures and their effectiveness and thereby generating useful knowledge for designing the socio-political steering of consumption towards sustainability is not yet finished and must be performed over and over again.

The instrument typology used in this article is not new, but combines – in particular by differentiating within the communication instruments – the approaches of different disciplines under a shared perspective. The same applies for the statements on assessing the effectiveness of the instruments, which are partially based on evaluation research in political science and sociology and partially on evaluation research in the context of scientifically accompanied interventions – two different research traditions, which are interrelated here. Outlining the different approaches as to the topic of the effects and effectiveness of (steering) instruments should provide a basis for classifying the different scientific and political debates relating to this question and should also highlight the key questions and limitations of research for the societal steering of consumption towards sustainability.

If it is possible to induce changes to individual consumer behaviour, it would also be possible – depending on the constellation and effective forces – that a development dynamic towards sustainability could arise in society as a whole due to the complex links between the effects and repercussions in the system. However, the desired reorganisation of the consumption system as a whole with regard to sustainable development is undoubtedly an enormous challenge in terms of the society's ability to transform and reform on a global, national and sub-national level. Whether and to what extent such a pervasive system transformation can be deliberately steered at all and the extent to which in particular steering interventions into individual consumer behaviour are legitimate are essential questions, which go beyond the specific design of instruments and must be critically discussed on a regular basis.

References

Abrahamse W., Steg L., Vlek C., Rothengatter T. (2005): A review of intervention studies aimed at household energy conservation. In: Journal of Environmental Psychology 25: 273–291.

ASCEE team (2008): Policy Instruments to Promote Sustainable Consumption. Brussels, Heidelberg, Oslo: IES, IÖW, SIFO.

Balzer L. (2005): Wie werden Evaluationsprojekte erfolgreich? Landau: Verlag Empirische Pädagogik.

Bamberg S. (2007): Is a Stage Model a Useful Approach to Explain Car Drivers' Willingness to Use Public Transportation? In: Journal of Applied Social Psychology 37: 1757–1783.

Barth M., Fischer D., Michelsen G., Rode H. (in this volume): Schools and their 'culture of consumption': a context for consumer learning.

Bemelmans-Videc M.-L., Rist R. C., Vedung E. (eds.) (1998): Carrots, Sticks and Sermons. Policy instruments and their evaluation. New Jersey: Transaction Publishers.

Benz A. (ed.) (2004): Governance – Regieren in komplexen Regelsystemen. Eine Einführung. Wiesbaden: VS Verlag für Sozialwissenschaften.

Brohmann B., Dehmel C., Fuchs D., Mert W., Schreuer A., Tews K. (in this volume): Bonus schemes and progressive electricity tariffs as instruments to promote sustainable electricity consumption in private households.

Brohmann B., Schönherr N., Wolff F., Fritsche U., Brunn C., Heiskanen E. (2011): Tracing the Pathways of Sustainable Consumption Policy Instruments and Measuring Sustainability: Results from an EU-wide Impact Evaluation Project. Paper at Konferenz Consumer 11 in Bonn, 18–20 July 2011.

Cutler C., Haufler V., Porter T. (ed.) (1999): Private Authority and International Affairs. Albany: State University of New York Press.

Dehmel C. (2010): Austausch von Kühlgeräten durch effiziente Neugeräte in privaten Haushalten – Die Trennungsprämie in Österreich im Vergleich zu ähnlichen Programmen in Dänemark und den Niederlanden. Münster: Transpose Working Paper No. 9. [http://www.uni-muenster.de/Transpose/en/publikationen/index.html; 24.07.2012]

Dehmel C. (2011): Der Einfluss von progressiven Tarifen auf den Stromkonsum in privaten Haushalten in Italien und Kalifornien. Münster: Transpose Working Paper No. 10. [http://www.uni-muenster.de/Transpose/ en/publikationen/index.html; 24.07.2012]

Di Giulio A., Brohmann B., Clausen J., Defila R., Fuchs D., Kaufmann-Hayoz R., Koch A. (in this volume): Needs and consumption – a conceptual system and its meaning in the context of sustainability in consumption acts.

Dovidio J. F., Piliavin J. A., Schroeder D. A., Penner L. A. (2006): The social psychology of prosocial behavior. Mahwah, NJ: Erlbaum.

Dunkelberg E., Stieß I. (2011): Energieberatung für Eigenheimbesitzer/innen. Wege zur Verbesserung von Bekanntheit und Transparenz durch Systematisierung, Qualitätssicherung und kommunale Vernetzung. Institut für ökologische Wirtschaftsforschung, Institut für sozialökologische Forschung. [http://www.enef-haus.de/index.php?id=6; 24.07.2012]

Faruqui A. (2008): Inclining Toward Efficiency. Is Electricity Price-elastic Enough for Rate Designs to Matter? Public Utilities Fortnightly, August: 22–27.

Fischer D., Michelsen G., Blättel-Mink B., Di Giulio A. (in this volume): Sustainable consumption: how to evaluate sustainability.

Flury-Kleubler P., Gutscher H. (2001): Psychological principles of inducing behaviour change. In: Kaufmann-Hayoz R., Gutscher H. (eds.): Changing things – moving people. Strategies for promoting sustainable development at the local level. Basel: Birkhäuser. 109–129.

Fuchs D. (1995): Incentive Based Approaches in Environmental Policy. Little We Know. Claremont, CA: The Claremont Graduate School. CPE Policy Paper.

Fuchs D. (2002): Globalization and Global Governance. Discourses at the Turn of the Century. In: Fuchs D., Kratochwil F. (eds): Transformative Change and Global Order. Reflections on Theory and Practice. Münster: LIT Verlag. 1–23.

Fuchs D. (2006): Business and Governance: Transnational Corporations and the Effectiveness of Private Governance. In: Schirm S. (ed.): Globalization. State of the Art and Perspectives. London: Routledge. 175–216.

Fuchs D. (2007): Business Power in Global Governance. Boulder: Lynne Rienner.

Gibson R. (ed.) (1999): Voluntary Initiatives. Peterborough: Broadview Press.

Grunwald A. (2010): Wider die Privatisierung der Nachhaltigkeit. In: Gaia 19 (3): 178–182.

Gutscher H., Mosler H.-J., Artho J. (2001): Voluntary collective action in neighborhood slow-down: Using communication and diffusion instruments. In: Kaufmann-Hayoz R., Gutscher H. (eds.): Changing things – moving people. Strategies for promoting sustainable development at the local level. Basel: Birkhäuser. 151–169.

Hamenstädt U. (2009): Stromsparen über den Preis? Ein Experiment. Münster: Transpose Working Paper No. 4. [http://www.uni-muenster.de/Transpose/en/publikationen/index.html; 24.07.2012]

Hewson M., Sinclair T. (ed.) (1999): Approaches To Global Governance Theory. Albany: SUNY Press.

Hey C., Volkery A., Zerle P. (2005): Neue umweltpolitische Steuerungskonzepte in der Europäischen Union. In: Zeitschrift für Umweltpolitik und Umweltrecht 1: 1–38.

IFEU [Institut für Energie- und Umweltforschung] (2005): Evaluation der stationären Energieberatung der Verbraucherzentralen, des Deutschen Hausfrauenbundes Niedersachsen und des Verbraucherservice Bayern. Heidelberg. [http://www.ifeu.de; 24.07.2012]

IFEU [Institut für Energie- und Umweltforschung] (2008): Evaluation des Förderprogramms "Energiesparberatung vor Ort". Heidelberg: Im Auftrag des Bundesministeriums für Wirtschaft und Technologie.

Kaufmann-Hayoz R., Bättig C., Bruppacher S., Defila R., Di Giulio A., Ulli-Beer S., Friederich U., Garbely M., Gutscher H., Jäggi C., Jegen M., Müller A., North N. (2001): A typology of tools for building sustainability strategies. In: Kaufmann-Hayoz R., Gutscher H. (eds.): Changing things – moving people. Strategies for promoting sustainable development at the local level. Basel: Birkhäuser. 33–107.

Kaufmann-Hayoz R., Bamberg S., Defila R., Dehmel C., Di Giulio A., Jaeger-Erben M., Matthies E., Sunderer G., Zundel S. (in this volume): Theoretical perspectives on consumer behaviour – attempt at establishing an order to the theories.

Kaufmann-Hayoz R., Gutscher H. (2001): Transformation toward sustainability: An interdisciplinary, actor-oriented perspective. In: Kaufmann-Hayoz R., Gutscher H. (eds.): Changing things – moving people. Strategies for promoting sustainable development at the local level. Basel: Birkhäuser. 19–25.

Klesse A., Müller J., Person R.-D. (in this volume): Achieving and measuring energy savings through behavioural changes: the challenge of measurability in the actual operation of university buildings.

Kohler-Koch B., Eising R. (eds.) (1999): The Transformation of Governance in the European Union. London: Routledge.

Kristof K. (2010): Models of change. Einführung und Verbreitung sozialer Innovationen und gesellschaftlicher Veränderungen in transdisziplinärer Perspektive. Zürich: vdf.

Krömker D., Dehmel D. (2010): Einflussgrößen auf das Stromsparen in Haushalten aus psychologischer Perspektive. Münster: Transpose Working Paper No 6. [http://www.uni-muenster.de/Transpose/en/publikationen/index.html; 24.07.2012]

Leung D., Panzone L., Swanson T. (2011): Improving the Design and Implementation of Sustainable Consumption Instruments and Strategies. EUPOPP Work Package 5, Deliverable 5.1.1. [http://www.eupopp.net/docs/d5.1improving_design_impl_sc_instr.pdf; 24.07.2012]

Lorek S., Giljum S., Bruckner M. (2008): Sustainable Consumption Policies Effectiveness Evaluation (SCOPE2) – Inventory and Assessment of Policy Instruments (final draft). Overath, Vienna: Sustainable Europe Research Institute.

Matthies E., Thomas D. (in this volume): Sustainability-related routines in the workplace – prerequisites for successful change.

Matthies E., Kastner I., Klesse A., Wagner H.-J. (2011): High reduction potentials for energy user behavior in public buildings: how much can psychology-based interventions achieve? In: Journal of Environmental Studies and Sciences 1 (3): 241–255.

Mayntz R., Scharpf F. (eds.) (1995): Gesellschaftliche Selbstregelung und politische Steuerung. Frankfurt a. M.: Campus.

Mosler H., Gutscher H. (1998): Umweltpsychologische Interventionsformen für die Praxis. In: Umweltpsychologie 2 (2): 64–79.

Mosler H., Tobias R. (2007): Umweltpsychologische Interventionsformen neu gedacht. In: Umweltpsychologie 11 (1): 35–54.

Rehfeld K.-M., Rennings K., Ziegler A. (2007): Integrated product policy and environmental product innovations: An empirical analysis. In: Ecological Economics, 61 (1): 91–100.

Røpke I. (2010): Konsum: Der Kern des Wachstumsmotors. In: Seidl I., Zahrnt A. (eds.): Postwachstumsgesellschaft. Konzepte für die Zukunft. Marburg: Metropolis-Verlag. 103–115.

Rosenau J., Czempiel E.-O. (eds.) (1992): Governance without Government: Order and Change in World Politics. Cambridge: Cambridge University Press.

Rossi P. H., Lipsey W. M., Freemann H. E. (2004): Evaluation. A Systematic approach (Seventh Edition). Thousand Oaks: Sage.

Rubik F., Scholl G., Biedenkopf K., Kalimo H., Mohaupt F., Söebech O., Strandbakken P., Turnheim B. (2009): Innovative approaches in European sustainable consumption policies. Assessing the potential of various instruments for sustainable consumption practices and greening of the market (ASCEE). Berlin: Schriftenreihe des IÖW 192/9.

Schönherr N., Brunn C., Wolff F. (2010): WP 3.2: In-depth analyses of SC instruments' impact on consumption patterns and of conditions of success or failure – Synthesis Report – EUPOPP Deliverable 3.2.1., October 2010.

Schweizer-Ries P. (2010): Umweltplanung, Gestaltung und Bewertung: Design, Umsetzung und Evaluation. In: Linneweber V., Lantermann E.-D., Kals E. (eds.): Spezifische Umwelten und umweltbezogenes Handeln. Enzyklopädie der Psychologie, Serie IX (Umweltpsychologie), Band 2. Göttingen u.a.: Hogrefe. 1031–1058.

Steg L., Vlek C. (2009): Encouraging pro-environmental behaviour: An integrative review and research agenda. In: Journal of Environmental Psychology 29: 309–317.

Stieß I., Birzle-Harder B. (2010): Evaluation der Kampagne "Gut beraten starten". [http://www.klimaschutz-hannover.de; 24.07.2012]

Stockmann R., Meyer W. (2009): Evaluation: Eine Einführung. Stuttgart: UTB.

Sunderer G., Götz K., Gölz S. (in this volume): The evaluation of feedback instruments in the context of electricity consumption.

Tews K. (2009): Politische Steuerung des Stromkonsums privater Haushalte. Portfolio eingesetzter Instrumente in OECD-Staaten. Münster/Berlin: Transpose Working Paper No. 2. [http://www.uni-muenster.de/Transpose/en/publikationen/index.html; 24.07.2012]

Tews K. (2011): Stromeffizienztarife für Verbraucher in Deutschland? Vom Sinn, der Machbarkeit und den Alternativen einer progressiven Tarifsteuerung. FFU-Report 05-2011, Forschungszentrum für Umweltpolitik, FU Berlin.

United Way of America (1996): Measuring program outcomes: A practical approach. Alexandria, VA.: United Way.

Vandenbergh M. P., Stern P. C., Gardner G. T., Dietz T., Gilligan J. M. (2010): Implementing the Behavioral Wedge: Designing and Adopting Effective Carbon Emissions Reduction Programs. In: Environmental Law Review 40: 10547–10554.

Vogel D. (2005): The Market for Virtue. The Potential and Limits of Corporate Social Responsibility. Washington: The Brookings Institution.

Warde A. (ed.) (2010): Consumption. Los Angeles: Sage. (4 volumes).

Weiß J., Vogelpohl T. (2010): Politische Instrumente zur Erhöhung der energetischen Sanierungsquote bei Eigenheimen – Eine Analyse des bestehenden Instrumentariums in Deutschland und Empfehlungen zu dessen Optimierung vor dem Hintergrund der zentralen Einsparpotenziale und der Entscheidungssituation der Hausbesitzer/innen. Berlin: Institut für ökologische Wirtschaftsforschung. [http://www.ioew.de/uploads/tx_ukioewdb/ENEF-Haus_2010_Instrumente.pdf; 24.07.2012]

Wolff F., Schönherr N. (2011): The impact evaluation of sustainable consumption policy instruments. In: Journal of Consumer Policy 34: 43–66.

Melanie Jaeger-Erben, Martina Schäfer, Dirk Dalichau, Christian Dehmel, Konrad Götz, Daniel Fischer, Andreas Homburg, Marlen Schulz, Stefan Zundel

6 Using 'mixed methods' in sustainable consumption research: approaches, challenges and added value

6.1 Starting point and objectives

The fact that nine out of ten projects in the SÖF focal topic "From Knowledge to Action – New Paths towards Sustainable Consumption" applied both qualitative and quantitative research methods[1] begs the question why a mixed method approach is so popular in this topic area.[2] Looking at the development of the two research traditions, mixing methodologies is by no means self-evident. While 'paradigm wars' of pitting one methodology against the other have largely been overcome (cf. Denzin 2010; Prein/Erzberger 2000), and a so-called "third methodological movement" (cf. Cresswell/Plano Clark 2007, p. 13) has even appeared on the research horizon, some authors remain sceptical (Symonds/Gorard 2010; Gorard 2007; Giddings 2006; Hammersley 2004). Over recent years, numerous projects and publications have looked into combining qualitative and quantitative research methods (some publications, such as the Journal of Mixed Methods Research, or Quality and Quantity, are specifically devoted to this field). While some areas of research have come to apply a combination of qualitative and quantitative methods (e.g. health and education research, cf. Cresswell 2009; O'Cathain 2009), a single method approach remains the general rule for most research projects (Bryman 2007).

---

[1] Given the variety in which methods were applied by the various projects, we have drawn a rather broad distinction between quantitative methodology referring to standardised research with numerically represented data and qualitative research referring to non-standardised methods where text or pictorially represented findings (mainly transcriptions from interviews and group discussions) are described and/or interpreted.

[2] Members from altogether seven project groups contributed to answering this question with both their own ideas and findings from their projects.

In the present article we will take a closer look at some of the projects of the focal topic, in terms of expectations of and experiences with specific combinations of methods. First, we will highlight some advantages described in the literature, some challenges and some different approaches to mixed methods research. Subsequently, we will look at how seven projects combined qualitative and quantitative methods. On that basis we will estimate the potential added value that mixed methods hold for research on sustainable consumption and sustainability.

6.2 Mixed methods: background, advantages and criticism

In the beginning, the main objective of combining qualitative and quantitative methods was to increase the validity of research findings. The original impetus came from the sociologist Norman K. Denzin (1970), who argued that a triangulation of methods (and of perspectives and theories), i.e. a multi-perspective approach to a research topic, was superior to applying one method only. Following in Denzin's footsteps, a number of researchers subsequently developed this approach. In the present article we will – not without giving due consideration to some more critical voices – focus on the advantages that a combination of methods can bring.

It is frequently argued that a mixed methods approach suits particular research issues, and it appears that situations which focus on interactions between 'structure' (social structures) and 'agency' (individual behaviour) lend themselves particularly well to mixed methods. There is, for instance, a series of studies that have applied quantitative and qualitative methods (e.g. surveys and interviews/focus group discussions) in their investigations of people's educational careers and life stories, investigating phenomena at both institutional and individual levels (cf. Wooley 2009; Prein et al. 1993). Health research is another case in point: as a field with a tradition for 'precise' scientific methods believed to be generated by standardised methods, an increasing number of health studies have started to combine methods and are increasingly incorporating qualitative methods (cf. O'Cathain et al. 2007). The inclusion of qualitative research can be particularly advantageous in projects investigating socially and politically complex issues, as well as circumstances and perspectives of specific target groups. In such contexts, a combination of methods can serve particular objectives or types of research, e.g. evaluation research (e.g. Greene 2007; Greene/Caracelli 1997) or intervention research (Nastasi et al. 2010). According to Nastasi et al. (2010), combining qualitative and quantitative methods allows for both participatory and representative research, capable of generating the development of culturally-sensitive and targeted group-oriented instruments.

In investigations that focus on research practice in the field of mixed methods, mention is often also made of the strategic advantages of combining methods. O'Cathain (2009) for example found that combined approaches had in some instances been applied because it was the required or at least the preferred approach by funding bodies. A further justification for their use appears to be the fact that they are currently rather 'fashionable' in the research community.

The possibility that mixed methods might be chosen, not as the best fit for a given research question, but rather for the sake of following a 'trend', has been one focus of recent criticism of this research strategy. Gorard (2007) and Giddings (2006), for example, call into question the assertion that combining methods is intrinsically superior to applying a single method. Investigations by Bryman (2007; 2006) indicate that project groups who used mixed methods possibly did so without giving adequate thought to a clear strategy. Having compared a wide range of research publications where mixed methods were applied, and drawing on interviews with social scientists, Bryman found that qualitative and quantitative procedures were often described separately in project reports, and researchers had failed to consider whether applying the two methods had amounted to more than simply conducting two concurrent investigations. Bryman (ibid.) was able to demonstrate that the difficulties of true method integration only partly lie in the methods themselves (e.g. incompatibility of different epistemological assumptions, different procedures). The reality that one method is often added as something of an 'afterthought' to make a mixed method project is a consequence of the fact that publications and publishers themselves (who are the target groups for research reports) often show a preference for certain methodological approaches, or researchers tending to be better acquainted with one approach.

So far, we have established that – given its fashionable appeal – 'mixed methods' may on occasion be used as a mere label. In such cases it is unlikely to be implemented to full effect during the project. In the light of such findings, Symonds and Gorard (2010) contest the usefulness of distinguishing between a qualitative procedure, a quantitative procedure and a supposedly superior 'third methodological movement'. Highlighting the difficulties of separating qualitative and quantitative methods in applied, problem-focused research, they argue in favour of a return to a strongly topic-oriented approach, where academic research remains a 'craft', where prevailing trends do not dictate which methods will be used, and where the selection of the most appropriate approach to a given research topic is paramount.

6.3 Designs for combining methods

A large number of categorisation proposals – each describing different designs, perspectives and approaches – have been put forward at one time or another (cf. for example Creswell/Plano Clark 2007; Greene 2007; Tashakkori/Teddlie 2003). We will focus on two proposals that are particularly suited to the projects in the focal topic.

Flick (2004) classified a number of approaches to mixing methods, in terms of the levels at which the methods were interlinked and in terms of how much weight was given to each method in the study design and the discussion of the results. From this, Flick derives a number of method combinations. In *mixed methods approaches* (such as proposed by Tahakkori/Teddlie 2003), qualitative and quantitative approaches are loosely combined, without necessarily being accorded equal weight. In models of the *integration of qualitative and quantitative methods* (Prein et al. 1993) the emphasis lies on equal integration. Flick (2004) remarks, however, that this emphasis is largely restricted to the research findings. By contrast, the *triangulation* approach (Flick 2004) integrates qualitative and quantitative methods, data and findings in equal measure during all phases of the research process. On that basis, findings are jointly formulated.

Mayring (2001) distinguishes research designs in terms of how the two methods are sequenced. This allows for a more precise classification of those approaches that Flick broadly labels "mixed methods". Mayring's *preliminary study model* largely applies qualitative methods in a preliminary explorative phase in order to subsequently develop, fine-tune and quantitatively assess the hypotheses and models. The qualitative phase thus serves to build up knowledge in the given field and comes to an end once sufficiently defined hypotheses have been formulated. In the *generalisation model*, the qualitative study is also conducted first, but it assumes a more important role. Once completed, its findings (or parts thereof) will be assessed against a large sample (i.e. a quantitative analysis). The *elaboration model* represents the opposite procedure: an initial quantitative study generates a general picture, whereupon the findings are qualitatively (e.g. case studies) assessed and fine-tuned.

Below we will take a closer look at how the above research designs have been applied by the various projects in the focal topic. It should be noted that these designs represent abstract 'textbook' categories, intended as guidelines rather than to be imported wholesale into projects. The projects of the focal topic, most of which are large and complex, tend to use a range of concurrent and consecutive strategies. For practical reasons, our account will focus on either certain steps within the research process or particular project characteristics. The intention is to show how particular combinations can serve particular aims. The following points will be addressed for each project (see also summary in Table 1):

- What type of method combination was selected?
- What were the associated objectives?
- What were the advantages of the combination of methods?
- What were the problems and difficulties?

The account below is not intended as a detailed content description of each project, but will focus on those aspects relevant in the context of mixing methods.

6.4 Mixed Methods approaches in the focal topic 'sustainable consumption'

A key objective of the "ENEF-Haus" project was to identify incentives and barriers for homeowners to carry out (energy-efficient) refurbishments. In a methodical multi-stage approach, the theory of planned behaviour (Ajzen 1991) as well as other literature was drawn on, and an initial model of decision-making behaviour was drafted. In a subsequent phase, qualitative interviews with homeowners served to modify and extend the original model. A third phase involved an extensive quantitative survey, in which homeowners who had recently invested in substantial refurbishments were questioned in standardised telephone interviews. On the basis of a statistical analysis of the survey, the decision-making model was then validated. Such an approach probably comes closest to a *preliminary study model*. The main objective was to match a suitable model to the field under investigation in order to subsequently extend and validate the model. The qualitative preliminary stage made it possible to complement the original model. The interviews conducted at that stage showed that opportunities for refurbishment not only presented themselves when certain components of the house needed to be replaced (e.g. facade or heating system), but also when certain 'social' changes took place, such as when a new house was bought, a child was born, or a building savings account had matured. The interviews also pointed to a discrepancy between how homeowners and experts perceived issues around refurbishment. On the basis of these qualitative interviews it was possible to restrict the quantitative analysis to those aspects that had been identified as relevant in the interviews. The quantitative surveys and analyses made it possible to test the hypotheses of the theoretical model and form target groups for the interventions. Once target groups were matched with data relating to buildings and refurbishments (these had also been generated by the quantitative analysis), target group specific energy savings were predicted. The both field-specific and representative results, together with the subsequent implementation of an appropriate intervention, were thus a direct consequence of combining several methods. No particular difficulties arose from combining methods in this particular project, but it was noted that the original theoretical model had to be considerably modified in the course of the empirical investigations.

The overall objective of the **"Intelliekon"** project was to evaluate user feedback in the context of domestic energy consumption, and in particular to assess how much energy was consumed and/or saved as a result of a smart metering technology. Here too, qualitative methods (problem-focused, semi-structured interviews, where customers of the participating energy companies were given the opportunity to tell their own story) were followed by a standardised study. In contrast to the approach taken in "ENEF-Haus", the theoretical models and typologies were formulated on the basis of the qualitative data and subsequently validated quantitatively. The findings of the qualitative project phase were further evaluated by means of a comprehensive panel survey: households in Germany and Austria were questioned at two points in time – shortly after a smart metering system had been installed in their home and six months after it had been installed. That group was additionally compared to a control group. The research design applied probably comes closest to a *generalisation model*. The aim was a) to develop models and intervention measures that were capable of matching the life context of the target groups and b) to be able to closely observe the target groups. The qualitative preliminary phase allowed researchers to understand the respondents' attitudes towards electricity use, electricity saving and their preferences for the design and graphics of the feedback instruments. The panel survey revealed that the preliminary qualitative research had led to high user acceptance of the feedback instruments. Additionally, on the basis of the motives for using the feedback (elicited in the interviews), several target groups were identified and categorised. The publication of this categorisation met with keen interest on the part of the field partners and other parties who had shown an interest in the study.[3] The study was able to demonstrate to field partners how qualitative research findings could lead to real, practical products. As a further step, the attitudes and preferences established on the basis of the quantitative survey were linked to a) the particular nature and the energy-efficiency of the electrical equipment in individual households, b) socio-demographic data, c) current user attitudes and evaluations of the feedback instrument, and d) the actual amount of saved energy. A quasi-experimental design further made it possible to conduct causal analyses on the basis of regression analyses. Systematic relationships could thereby be revealed between subjective, socio-demographic and technical factors (amount and energy-efficiency of electrical equipment in households) on the one

---

3 The paper by Birzle-Harder et al. (2008) has since been ordered over 250 times. On the basis of these results, a company consultant developed his own feedback products (cf. Gnilka/Meyer-Spasche 2010).

hand and electricity consumption and savings on the other. The standardised findings of this project phase were of key importance for research and for stakeholders (Federal Network Agency, energy suppliers), who were able to use them in their strategic planning. However, close collaboration with field partners sometimes hindered the combination of methods. Their high expectations occasionally impeded researchers in their attempt to adhere to a strictly methodical approach. Furthermore, it was not always possible to convey to field partners that the qualitative findings were only preliminary and would therefore be unsuitable for immediate implementation in their businesses. The researchers were thus confronted with two conflicting sentiments: on the one hand, they were pleased with the enthusiastic response of the field partners; on the other hand, they were not always able to proceed methodically through the planned stages of their research design.

The objective of the "**Consumer/Prosumer**" project was to investigate user patterns in the online trading of second-hand products with a view to gaining an insight into people's motives, barriers and general trading activity and identifying distinct user types. The project design was representative of the *elaboration model*: an initial, comprehensive quantitative online survey of second-hand trading was followed by a qualitative phase of semi-structured interviews. The qualitative study had been planned from the very beginning. Its purpose had been to observe in more detail the user types previously identified by the quantitative study. However, in view of the quantitative findings, the initial plan had to be abandoned. The expectation of the quantitative analysis had been to find socio-demographic aspects (such as gender, education, income) responsible for people's online trading motives. Instead, it was found that life circumstances (e.g. family life, retirement) were more influential in this respect. Moreover, the quantitative analysis had not been able to characterise the target groups accurately. With the focus of the qualitative elaboration study shifted, it emerged that a) retired people often used online trading in relation to their hobbies or leisure time, b) people on low incomes were able to partake of consumer-oriented activities thanks to online second-hand trade and c) parents used online trading to ensure a steady supply of clothes and equipment for their growing children. On that basis, target groups were classified in terms of life phases, and the findings, together with further possible explanations for online trading patterns, were presented to the field partners. The project did encounter one difficulty: because the quantitative survey had not investigated the life phases, the data required extensive modelling before it could be used.

The "**Transpose**" project also employed *elaborative* qualitative methods on the strength of a quantitative analysis. Proceeding from known domestic electricity saving potentials, the aim was to identify policy instruments for overcoming barriers to saving electricity – both at a consumer and structural level (trade, electricity suppliers) – that could be introduced across Germany. Methods were selected in accordance with the basic premise of the project that both levels – agents (consumers) and structures – should be taken into consideration. In the first stage of the project consumers' electricity savings patterns were captured by means of a standardised questionnaire. A further quantitative policy analysis served to identify policy instruments that had been successfully used in other countries. The findings of both studies were combined and used in the selection and design of the subsequent, qualitative case studies, involving expert interviews and focus group discussions. It was now possible to consider how intervention measures could best be adapted for Germany. The consistent combination of qualitative and quantitative methods throughout the process allowed for both a broad and an in-depth analysis. The variety of methods additionally also generated a large number of findings, which proved of interest to a range of field partners (e.g. political decision makers, consumer advice centres). No specific difficulties arose as a result of combining methods. While a variety of explanatory models were used, the different methods were implemented on the basis of a common understanding of the relevance of both agents and structures.

6.5 Integration of qualitative and quantitative findings into the focal topic of sustainable consumption

In the "**LifeEvents**" project, qualitative and quantitative methods were applied concurrently. The objective was to initially observe divergent and similar aspects of a phenomenon – such as changing consumption patterns after the birth of a child or after relocation – and the effect of a sustainability campaign in such a situation. As a next step, these aspects were integrated into an overall picture (cf. Jaeger-Erben et al. 2011). The quantitative phase of the project consisted of a theory-based intervention aimed at changing people's routines (in nutrition, energy and transport). Behavioural changes in people who had recently had a major life event were measured and compared to a group of people in stable life situations. A quasi-experimental control group design was used. This approach enabled researchers to reach reliable conclusions about the impact of the intervention on the sustainable behaviour of a range of target groups. The issue of how and why consumption patterns might change after life events was addressed in a qualitative phase of the project by means of problem-focused interviews and a procedure based on Grounded Theory. The aim was to investigate in detail how different individuals in their cultural contexts adopted new routines. A further aim was to investigate the role that participation in the campaign had played in the context of people's life course changes. A qualitative reconstruction of participants' perspectives made it possible to further investigate the rather surprising finding that the campaign had had no statistically significant impact on people in life changing situations, but had had an effect on people in stable circumstances. A number of possible explanations for these findings were generated. Combining results also allowed researchers to consider differences (quantitatively established beforehand) between the target groups, in terms of their life course changes and the resulting changes in their everyday needs and requirements. These differences might relate to, for example, increased use of local public transport by people who had relocated compared to people who had had a child. A particular challenge in this project phase emerged from some divergent findings. These sparked detailed discussions and obliged researchers to find methodological and theoretical explanations. Although there was a close communication between the two studies, the choice of research design (*integration of findings*) initially dictated the decision to conduct two concurrent, relatively independent studies with their own methodological and theoretical presuppositions. The final challenge was to come to a common appreciation of the methods used in each study and to integrate the findings into an overall theoretical framework. In the course of the project it had become clear that the definition of a common overall framework should have been prioritised at the beginning and throughout the project.

6.6 Triangulation of qualitative and quantitative methods in the focal topic of sustainable consumption

The procedure applied in the "BINK" project most closely matches the *triangulation design*. Here, a key objective was to generate a framework that could be applied for investigating the formal and informal features of educational institutions that influence young people's consumer learning. Strategies and instruments for promoting sustainable consumer behaviour could then be developed and evaluated on the basis of the framework. The following example illustrates how these research objectives were addressed, combining qualitative and quantitative methods at different stages and in different phases of the research process: in the development phase of the framework, group discussions involving pupils, students, teachers, lecturers and administrative staff served to identify consumption-related features of the organisational culture of their educational institutions. An accompanying qualitative study with open and semi-structured interviews was able to reconstruct the perceptions of the students involved in the "BINK" project regarding the changes that had resulted from the interventions. A further, large-scale quantitative survey, involving students and teachers from participating and external institutions generated additional information about how the culture of consumption in schools was perceived (cf. Barth et al. in this volume). Each qualitative and quantitative study contributed to an ever clearer picture of a 'sustainable culture of consumption'. By combining elaborative and generalising studies, both general and case-specific conclusions could be made. Workshops held in educational institutions facilitated the involvement of local stakeholders, exploiting, for instance, their knowledge of previous interventions. The findings of an accompanying qualitative study on successful changes in consumer behaviour in external educational institutions enabled researchers to feed back to both field partners and the project itself empirical knowledge on how to implement a sustainable culture of consumption in educational institutions. The result was close transdisciplinary collaboration between university researchers and field partners. A notable challenge in this context was the considerable overhead represented by the tasks of coordination and communication, as well as the required commitment from all parties involved. However, these overheads were not intrinsically linked to the fact that research methods had been combined, but represent a familiar challenge in transdisciplinary research projects.

An example of a *triangulation of methods and perspectives* that corresponds to the combination of different disciplines – such as social sciences, economics and engineering – in this project can be found in the **"Heat Energy"** project. The objective of the project was twofold: a comprehensive investigation of heat energy consumption from a social, economic and individual perspective and the development of strategies for the promotion of sustainable heat consumption. Similar to "BINK", the project consisted of various strands and project phases, each involving both qualitative and quantitative research methods. One strand consisted of a standardised, postal survey of tenants and homeowners, whose attitudes, behaviours, barriers and incentives in relation to pro-environmental heat energy consumption were investigated, on the basis of which a number of 'user types' were identified. In an elaborative second phase, focus groups were requested to discuss and evaluate possible instruments for the promotion of sustainable heat consumption. The findings from both project phases fed into recommendations for interventions, which were submitted to a panel of multi-disciplinary researchers. Using the Delphi technique they applied both quantitative and open methods to describe, assess and prioritise recommendations. Whereas the project phases described above largely involved sociological expertise (apart from the multi-disciplinary Delphi group), a complementary qualitative analysis – based on economic theories and models – used semi-structured experts interviews (with tradespeople, employees of energy companies) to identify barriers to and opportunities for sustainable consumption at a meso level. A subsequent 'cross impact analysis' focused on conceivable future impact networks. In semi-structured interviews, engineers commented on possible relationships between potential future incidents. The combination of methods in this project was, in a sense, a consequence of the collaboration of different disciplines, namely because each discipline brought into the project its own methodological and theoretical perspective. The triangulation was thus closely linked to the triangulation of perspectives and was partly a consequence of the general aim to create a comprehensive picture of sustainability in one area of consumption. In terms of content, combining methods meant that the eventual recommendations could be formulated on the basis of both a broad standardised survey and in-depth discussions of target groups (focus groups). The original intention of consistently addressing both technical and social aspects proved somewhat problematic: whilst the questionnaires failed to generate sufficient technical information, the technically-oriented 'cross impact analyses' did not adequately capture the various social dimensions.

6.7 Summary: goals, benefits and challenges of combined methods

The majority of projects presented above reflect what has often been referred to in the literature as the general objectives of combined methods and their particular advantages for complex research projects. Quantitative methods were primarily applied for gaining a comprehensive overview (where many participants were involved), comparing groups or establishing causal relationships on a large scale. They proved particularly useful for validating the impact and relative effectiveness of interventions and for calculating the ecological effects of everyday behaviour. A further major goal of most projects was to explore the lifeworlds of the target groups, i.e. the contexts for their everyday behaviours and challenges. This was particularly important where concrete strategies (relating to the environment, education, health, energy, policy) had to fit in with people's everyday routines, decision-making and learning. In contrast to quantitative methods, which inevitably focus on pre-selected variables, qualitative methods are useful for illuminating those dark corners of the 'black box'. They were therefore applied for discovering new facets (as in "ENEF-Haus" and "Intelliekon"), or elaborating on discoveries made in preliminary quantitative studies (as in "Consumer/Prosumer" and "Transpose"). In cases where quantitative findings did not answer the original research questions, but instead generated new ones, additional, qualitative findings were – at least in part – capable of contributing to completing the picture (as in "LifeEvents").

Table 1: Most important features of mixed method combinations in the seven project groups

Project name, research topic	Type of method combination	Goals	Advantages	Problems, difficulties
"ENEF-Haus" Driving and restricting factors for (energy-efficient) refurbishment by homeowners	1. Development of a basic theoretical model 2. Qualitative interviews which can be further developed 3. Quantitative survey for validation purposes (*preliminary study model*)	Situation-specific modification of a theoretical decision model and interventions based on the model	Added value emerging from preliminary qualitative studies (discovery of new aspects, more precision for the quantitative survey); model validation by quantitative analyses	No difficulties in the actual process, but excessive modification of theoretical model as a result of new findings
"Intelliekon" Impact of feedback on motivation to save energy in households and actual electricity savings	1. Qualitative development of models and types on the basis of interviews 2. Validation by panel survey (*generalisation model*)	Development of models and intervention measures meaningful in terms of the lifeworlds of the target groups	Added value in terms of content as a result of qualitative preliminary study (contextualising of attitudes and behaviour in relation to energy saving, survey of preferences for design of feedback instruments, development of hypotheses for standardised survey); divergent interests of stakeholders and general public could be taken into account	High expectations by stakeholders impeded systematic method combination
"Consumer/ Prosumer" Usage patterns and user types in online second-hand trading	1. Standardised survey of users of online trading platforms 2. Elaboration of individual findings from interviews (*elaboration design*)	Analysing in more detail a 'surprising' finding and extending knowledge about how the motivation to buy and sell can be influenced	Added value emerging from qualitative elaboration (formation of life phase target groups); exploration of other explanations, also for the benefit of stakeholders and research community	Direct linking of quantitative and qualitative research is difficult due to inadequacy of standardised survey concerning life phases
"Transpose" Development of policy incentives for energy saving in households	1. Standardised survey and quantitative, cross-national policy analysis 2. Case analyses (interviews, focus groups) (*elaboration design*)	Full consideration of obstacles at consumer and structural level (consumer's immediate context)	Case selection on the basis of quantitative surveys (behaviour in the context of saving electricity and impact of policy instruments); variety of methods generating large quantity of results, pertaining to a range of stakeholders	No difficulties in combining methods, since integration of various theoretical backgrounds was not intended

Project name, research topic	Type of method combination	Goals	Advantages	Problems, difficulties
"LifeEvents" Changing everyday consumption patterns after life events by means of targeted campaign at the time of life course transitions	1. Consecutive: qualitative interview study as well as standardised intervention study 2. Joint discussion and integration of findings (*integration of qualitative and quantitative findings*)	Naturalistic reconstruction of changing process in everyday life and measurement of behavioural effects of a campaign	Better understanding of the role of campaign in life course transitions, both in terms of everyday routines and measurable behavioural effects; qualitative results allowed to explain quantitatively insignificant effects on groups in life course transitions to field partners	Sometimes divergent findings; integrative framework was required as a result of divergent explanatory models
"BINK" Development of a framework of a 'sustainable culture of consumption' in educational institutions and development of an intervention strategy	Qualitative and quantitative methods in different phases of project, applied concurrently and consecutively (*triangulation of methods*)	Development of practically relevant framework for identification of starting points for intervention strategies and evaluation of intervention strategies	Cumulative development of understanding of a 'sustainable culture of consumption' in educational institutions in terms that are both specific and generalisable; qualitative methods allowing integration of the expertise of stakeholders; quantitative results allowing feedback from research; overall, good division of labour between research and practice	Need for extensive communication in the course of collaboration required strong commitment from all participants
"Heat Energy" Investigation of heat energy consumption under consideration of social, economic and individual attitudes, and development of interventions	Qualitative and quantitative strategies (e.g. standardised surveys and elaborative focus groups), applied consecutively and concurrently during different project phases and from different disciplinary perspectives (*triangulation of methods and perspectives*)	Development of broadly based recommendations for action etc., whereby barriers to and incentives for sustainable heat consumption were considered	Comprehensive picture of sustainability of different practices; identification of requirements and starting points for action; integration of different and discipline-specific knowledge and skills	Combination of social and technical aspects in surveys and cross impact analysis was difficult

The combination of qualitative and quantitative methods proved to be a strategic added value – whether intended or accidental – for a number of the projects, mainly because qualitative methods (such as focus groups) enabled a vivid exchange with target groups, stakeholders and field partners. Being aware of the concerns of external stakeholders in a project is crucial when it comes to developing recommendations, intervention instruments and products (e.g. "BINK"). Additionally, combining methods gave researchers the opportunity to present qualitatively diverse findings to fit the diverse needs of field partner e.g. for individual case histories and type descriptions or for larger scale and more representative results (e.g. "Intelliekon").

In principle, a combination of methods is appropriate for (content-related or strategic) complex research fields or to applied research, where both lifeworld-related and aggregated/standardised information is sought. The experience within this focal topic has shown that it can also be of benefit in the investigation of the specific field of sustainable consumption with its affiliation to sustainability research:

In almost all projects in the focal topic researchers felt the need to either develop field-specific and object-specific theoretical approaches and models or to adapt existing theories to their research topic. In a plenary of members of the projects described above, it was stressed that 'sustainable consumption' is a relatively young, constantly evolving field where new research questions cannot always be answered with existing theories. A combination of qualitative and quantitative approaches can thus contribute towards the development of new, more suitable theories. The "Intelliekon" project, for instance, applied a generalisation model to investigate the impact and use of a new technology in the field. After relying on a qualitative study for a categorisation of how the technology was used, a quantitative analysis was carried out to validate the findings. In the "Transpose" and "Consumer/Prosumer" projects quantitative methods applied to explore a new field were followed by a more in-depth qualitative analysis.

The collaboration between different disciplines in the "Heat Energy" project meant that the models and methods of each discipline were included and that, as a consequence, qualitative as well as quantitative data flowed into the project. A methodological research design that allowed for a concurrent, and sometimes consecutive, deployment of qualitative and quantitative methods thus enabled a range of disciplines to work together.

Combining methods also proved advantageous in both transdisciplinary collaboration and the exchange with field partners. It meant that requests for representative results made by some field partners, and requests for case descriptions made by others could be met. In the "Intelliekon" project, for instance, it was found that the information gleaned from qualitative procedures alone was highly useful and sufficient in itself as a basis for designing materials and tools. Closely linked to transdisciplinary research are participatory, communicative approaches. Qualitative methods such as focus groups are an important approach in this area of research (compare "BINK", "Heat Energy" and "Transpose", see also Henseling et al 2006; Dangschat 2005). In view of the fact that policy decisions were at stake in these projects, a quantitative assessment of efficiency and future impacts was equally important, and a combination of methods was therefore indispensable.

The challenges that the projects faced were mostly related to their inter- and transdisciplinary character. With regard to the combination of methods, in each of the projects it emerged that the layering of project elements and the systematic integration of appropriate methods not only required constant reflection throughout the project, but a considerable amount of advance planning.

6.8 Conclusion: mixed methods in sustainable consumption research

From the above considerations it can be concluded that a combination of qualitative and quantitative research methods is an appropriate and frequently applied approach in practice-oriented research into sustainable consumption. This is partly due to the fact that in such a complex field (similar to other socially important fields such as health and education) a range of levels (actors, institutions, structures) as well as the interactions between these levels have to be considered. A multi-method and multi-perspective approach is therefore generally recommended. As shown in the "Transpose" project, qualitative and quantitative methods are suitable at both a structural and agency level.

As the name implies, the research field of sustainable consumption consists of two strands, i.e. of consumption and sustainability. Both are investigated at interdisciplinary and transdisciplinary levels and by means of multi-methods approaches. In the case of consumption, it has been suggested that there is hardly another field with "more interdisciplinary research activity" (Solomon 2008, p. 39). Furthermore, research into sustainable consumption is part of research into sustainability. This in turn is concerned with the generation of system, target and transformational knowledge across disciplinary boundaries (cf. CASS 1997; Brand 2000), thereby integrating knowledge from a variety of sources.

A multi-methods approach is therefore more a natural orientation than an 'add-on' feature of research on sustainable consumption. The projects described above confirm that the high incidence of mixed method project designs within the focal topic is not due to a mere fashion, but is particularly suited to the topic of sustainability. It is thus not surprising that the difficulties and challenges that arose during the projects (if they were at all perceived as such) were not regarded as major obstacles, but instead as by-products of interdisciplinary and transdisciplinary research. The relative ease and naturalness with which the standardised and non-standardised procedures were integrated into the projects may well contribute to the future development of mixed methods research and to overcoming perceived dichotomies.

References

Ajzen I. (1991): The Theory of Planned Behavior. In: Organizational Behavior and Human Decision Processes 28: 179–211.

Barth M., Fischer D., Michelsen G., Rode H. (in this volume): Schools and their 'culture of consumption': a context for consumer learning.

Birzle-Harder B., Deffner J., Götz K. (2008): Lust am Sparen oder totale Kontrolle? Akzeptanz von Stromverbrauchs-Feedback. Ergebnisse einer explorativen Studie zu Feedback-Systemen in vier Pilotgebieten im Rahmen des Projekts Intelliekon. Frankfurt am Main. [https://shop.isoe.de/literatur/sonstige-materialien/lust-am-sparen-oder-totale-kontrolle-akzeptanz-2.htm; 18.06.2012]

Bohnsack R. (2005): Standards nicht-standardisierter Forschung in den Erziehungs- und Sozialwissenschaften. In: Zeitschrift für Erziehungswissenschaft 8 (4): 63–81.

Brand K.-W. (ed.) (2000): Nachhaltige Entwicklung und Transdisziplinarität. Besonderheiten, Probleme und Erfordernisse der Nachhaltigkeitsforschung. Berlin: Analytica-Verlag.

Bryman A. (2007): Barriers to integrating qualitative and quantitative research. In: Journal of Mixed Methods Research 1 (1): 8–22.

Bryman A. (2006): Integrating quantitative and qualitative research: How is it done? In: Qualitative Research 6: 97–113.

CASS (1997): Conference of the Swiss Scientific Academies (CASS), Forum for Climate and Global Change (ProClim-): Visions by Swiss Researchers. Research on Sustainability and Global Change – Visions in Science Policy by Swiss Researchers. Bern: ProClim-/SANW. [http://www.proclim.ch/4dcgi/proclim/en/media?1122; 03.05.2012]

Creswell J. W. (2009): Mapping the Field of Mixed Methods Research. In: Journal of Mixed Methods Research 3 (2): 95–108.

Creswell J. W., Plano Clark V. L. (2007): Designing and conducting mixed methods research. Thousand Oaks, CA: SAGE.

Dangschat J. S. (2005): Qualitative Sozialforschung und Partizipation. In: Forum Wohnen und Stadtentwicklung/vhw 6: 302–306.

Denzin N. K. (2010): Moments, Mixed Methods, and Paradigm Dialogs. In: Qualitative Inquiry 16 (6): 419–427.

Denzin N. K. (1970): Triangulation: A Case For Methodological and Combination Evaluation. In: Denzin N. K. (ed.): Sociological Methods: A Sourcebook. London: Butterworths. 471–475.

Giddings L. S. (2006): Mixed-methods research, positivism dressed in drag? In: Journal of Research in Nursing 11 (3): 195–203.

Gnilka A., Meyer-Spasche J. (2010): Umsetzbare Smart-Metering-Produkte: eine Handreichung für Energielieferanten. Berlin: LBD-Beratungsgesellschaft GmbH.

Gorard S. (2007): Mixing methods is wrong: An everyday approach to educational justice. London: British Educational Research Association.

Greene J. C. (2007): Mixed Methods in Social Inquiry. San Francisco: Jossey-Bass.

Greene J. C., Caracelli V. J. (eds.) (1997): Advances in mixed-method evaluation: The challenges and benefits of integrating diverse paradigms. San Francisco: Jossey-Bass.

Henseling C., Hahn T., Nolting K. (2006): Die Fokusgruppen-Methode als Instrument in der Umwelt- und Nachhaltigkeitsforschung. In: WerkstattBericht Nr. 82. Berlin: IZT – Institut für Zukunftsstudien und Technologiebewertung.

Jaeger-Erben M., Schäfer M., Bamberg S. (2011): Forschung zu nachhaltigem Konsum. Herausforderungen und Chancen der Methoden- und Perspektiventriangulation. In: Umweltpsychologie 15 (1): 7–29.

Mayring P. (2001): Kombination und Integration qualitativer und quantitativer Analyse. In: Forum Qualitative Sozialforschung 2 (1): Art. 6. [http://nbn-resolving.de/urn:nbn:de:0114-fqs010162; 14.06.2012]

Nastasi B., Hitchcock J., Sarkar S., Burkholder G., Varjas K., Jayasena A. (2007): Mixed Method in Intervention Research: Theory to Adaptation. In: Journal of Mixed Methods Research 1 (2): 164–182.

O'Cathain A. (2009): Mixed Method in Health Research – A quiet Revolution. In: Journal of Mixed Method Research 3(1): 3–6.

O'Cathain A., Murphy E., Nicholl J. (2007): Why, and how, mixed methods research is undertaken in health services research: A mixed methods study. In: BMC Health Services Research 7: 85.

Prein G., Erzberger C. (2000): Integration statt Konfrontation! Ein Beitrag zur methodologischen Diskussion um den Stellenwert quantitativen und qualitativen Forschungshandelns. In: Zeitschrift für Erziehungswissenschaft 3: 343–357.

Prein G., Kelle G., Kluge S. (1993): Strategien zur Integration quantitativer und qualitativer Auswertungsverfahren. Arbeitspapier Nr. 19. Bremen: Universität Bremen.

Solomon M. (2008): Consumer Behavior. Upper Saddle River: Prentice Hall.

Symonds J. E., Gorard S. (2010): Death of mixed methods? Or the rebirth of research as a craft. In: Evaluation/Research in Education 23 (2): 121–136.

Tashakkori A., Teddlie C. (2003): Mixed methodology. Combining qualitative and quantitative approaches. London: Sage.

Woolley C. (2009): Meeting the Mixed Methods Challenge of Integration in a Sociological Study of Structure and Agency. In: Journal of Mixed Methods Research 3 (1): 7–25.

Part 2

Findings from the project groups

Section A

Status of sustainability in investment decisions

Julika Weiß, Immanuel Stieß, Stefan Zundel

1 Motives for and barriers to energy-efficient refurbishment of residential dwellings

1.1 Introduction

Given that private consumption of heat energy represents one of the largest energy consumption sectors, it follows that an energy-efficient refurbishment of private dwellings would significantly reduce overall energy consumption and its associated CO_2 emissions. Particularly high potentials for energy and CO_2 savings are attributed to the refurbishment of single and two-family houses in Germany (e.g. BMVBS 2007; dena 2007; Schlomann et al. 2004). According to the German Energy Agency (dena 2007), comprehensive energy-efficient refurbishments of old buildings could reduce the consumption of fossil energy by an average of 80 percent. Currently, about 1 percent of the existing housing stock undergoes energy-efficient refurbishment every year; in terms of technical and economical feasibility, this figure could rise to about 3 percent.

Among the possible public policy measures for reducing climate change, the promotion of energy-efficient refurbishments of residential dwellings features right at the top in terms of cost-benefit assessment. The exact CO_2 avoidance costs of such measures depend on the measure in question, the condition of the building and further regional circumstances. However, according to the literature, CO_2 avoidance costs range from negative figures for some specific projects to around 60 euros per ton of CO_2-equivalent (Kuckshinrichs et al. 2009). In addition, many programmes create multiplier effects and thus represent highly effective stimuli for the economy at large. It is thus all the more regrettable if the estimated potential is not realised.

Our starting premise in this article is the claim that the main reason for unrealised refurbishment potentials is the inadequacy of information materials, communicative strategies and instruments on offer, and above all, the fact that these are insufficiently tailored to the complex decision-making processes involved in energy-efficient refurbishments. It is proposed here that it does not suffice to appeal to economic rationality alone, according to which energy-saving refurbishments pay off within a justifiable period of time. Instead, in addition to favourable technical and economic conditions, a number of other factors have to combine to entice homeowners to undertake extensive refurbishments. Furthermore, the contexts and combinations of motives tend to be as diverse as homeowners themselves.

Our argument here is as follows: in the second section, the above-mentioned thesis of untapped potential will be underpinned with our own considerations and calculations. In the third section, we will examine homeowners' possible motives or combinations of motives for energy-efficient refurbishment, as well as the role of sustainability in the decision-making process. It will be demonstrated that economic motives are important, but not always sufficient for energy-efficient refurbishments, and that the interpretation of the cost-effectiveness of these measures differ significantly between private homeowners and experts. In the fourth section, we will propose a typology of refurbishers. Given that each refurbisher type has his or her own motivations, it follows that each type should be considered as a specific target group as regards the promotion of energy-efficient refurbishments. The article concludes with a summary and a number of closing comments.

1.2   Analysis of potential

To date, there has been a dearth of representative data on the extent of the energy efficiency of existing homes. Hence, current estimates of potentials for energy-efficient improvements tend to disregard the actual state of the house in question (including insulation that might have been added), but instead take as a reference value the condition of an old house in its original state. In order to estimate the actual untapped potential, the "ENEF-Haus" project group carried out an analysis of potential (cf. Weiß/Dunkelberg 2010) on the basis of a sizeable data set of houses (n = 2,000). Although the data set is not universally representative, it does correspond to currently available statistical data in terms of the age distribution of the houses, size of living space and type of building. Estimates of potential savings were produced for the following energy-saving refurbishment measures: insulation of the facade, the roof, the attic floor, the basement ceiling, and the installation of condensing boilers and heating systems based on renewable energies.

In order to calculate the energy requirements of the houses before and after refurbishment, the calculation tool KVEP (Kurzverfahren-Energieprofil), developed by the German Institute for Housing and the Environment, was employed (IWU 2006).

A complete refurbishment of the building envelope of all single family and two-family houses in accordance with the standards set by the German Energy Conservation Act (EnEV: Energieeinsparverordnung) 2009 for existing buildings would, according to our calculations, result in a theoretical reduction of primary energy requirements of around 174 TWh per annum (Weiß/Dunkelberg 2010). Thus, by refurbishing the building envelope of single-family and two-family houses alone, the primary energy requirement for space heating of all private households could be reduced by about a third (according to destatis (2008) this amounted to 523 TWh in 2006). Savings are particularly high for buildings constructed before 1968; their refurbishment alone would reduce the energy requirement by 110 TWh. Including potential savings from the use of renewable energies, the refurbishment of single family and two-family houses could achieve a reduction of over 30 percent of the overall CO_2 emissions by private households.

In practice, this potential is limited by *direct* rebound effects (the lower the cost, the more heat is consumed) and *indirect* rebound effects (as a consequence of saving on heating, consumption rises in other areas, which in turn uses energy) (cf. de Haan 2009). In the context of space heating, a number of international empirical studies have shown that energy savings are on average considerably lower than theoretical forecast (cf. Biermayr et al. 2004). The Austrian MARESI project concludes that for the refurbishments of single family and two-family houses the rebound effects are not cost-related, but structural, namely that additional living space is often created as part of refurbishments.

However, this is only a partial rebound effect: as was shown in the "ENEF-Haus" project, living space is often extended in private homes without the intention of taking energy-conserving measures. Nonetheless, the increase in living space per person as recorded in Germany is such that overall energy savings are lower than the decrease in heat consumption (cf. Fischedick et al. 2006).

Despite the abovementioned limitations, it can be assumed that energy-efficient refurbishments lead to considerable energy savings. One requirement for realising the potential savings is the cost-effectiveness of the measures. This was assessed by the "ENEF-Haus" project in a profitability analysis using the annuity method (cf. Weiß/Dunkelberg 2010). According to the calculations, improvements to the facade of buildings (according to EnEV standards) are cost-effective over a period of 25 years – that is if they are taken during refurbishments that were independently due. In these cases, the energy savings compare only to the *added* costs incurred for the energy efficiency measures. The most advantageous measures (insulation of facade and roof) pay for themselves after 10 to 12 years, which represents an acceptable time frame for many homeowners. By contrast, compared to gas condensing boilers, heating systems based on renewable energies represent good financial value only in poorly insulated buildings with high energy consumption. However, in view of rising energy prices and decreasing costs for heating systems based on renewable energies, the installation of such heating systems will become cost-effective even in well insulated buildings in the coming years.

Even though the highly promising measures have been shown to be economically viable, given that they require considerable financial investment, many homeowners may struggle to meet the costs.

We are thus faced with a dichotomy: the theoretical possibility of achieving energy and cost savings by energy-efficient refurbishing vs. the reality of only about half of households taking up the opportunity to do so as part of their normal refurbishments cycles. If energy-efficient refurbishment does not become more widespread, it will not be before 2080 that the facades of all single family and two-family houses are insulated. Given the current slow pace of refurbishments, the "ENEF-Haus" project has come to the conclusion that it is crucial to promote energy-efficient measures as part of routine home refurbishments – even if they do not exceed the current statutory energy-efficiency levels. There is, of course, the risk of potential lock-in effects, as inadequate refurbishments undertaken now will still 'do the job' in 40 years' time (see Müller et al. 2010) – but opportunities (e.g. home purchase or renovation of the facade) that are open now may not present themselves again for years, or more likely for tens of years. As a consequence, decades will pass without comprehensive energy-efficient upgrading of our housing stock.

1.3   The significance of cost-effectiveness when refurbishing

In order to identify homeowners' motives for including energy-efficient measures in their refurbishment plans, the "ENEF-Haus" project conducted a survey of 2008 homeowners[1] who had spent a minimum of 4,000 euros on a refurbishment project between 2005 and 2008. The respondents were categorised into *progressive* homeowners who had carried out energy-efficient refurbishment measures with high energy efficiency potential (n = 541) and more *conventional* homeowners who had carried out standard energy-efficient refurbishments[2] (n = 467). The results of the survey that relate to the cost-effectiveness of the measures are presented below.

---

1   The survey took place between January and March 2009 and targeted homeowners of single family and two-family houses who had undertaken major refurbishments relating to the insulation of the facade and/or the upgrading of the heating system between 2005 and 2008. The survey consisted of a telephone interview of approx. 30 minutes. A detailed discussion of the methodology used for the survey and for the analysis of the results can be found in Stieß et al. 2010, pp. 10 ff.

2   The assignment was based on a list of measures that had previously been defined by the research project.

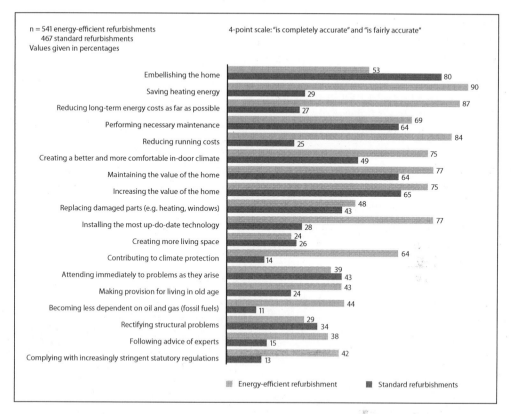

Figure 1: Reasons and objectives for refurbishments (source: Stieß et al. 2010, pp. 35 f.)[3]

When renovating a home with the aim of increasing its energy efficiency, large sums of money are generally needed. Investment costs for energy-efficient heating systems vary from 5,000 euros for modern condensing boilers up to 25,000 euros for ground source heat pumps. Insulating the exterior walls of an unrenovated house costs 30,000 euros (EnEV 2009 standard). Insulating the roof and replacing the windows also involve five-figure sums. Given these costs and the risk of financial losses if an incorrect decision is taken, homeowners could be expected to make more or less economically rational decisions.

To judge by motivations and targets for refurbishments cited by homeowners (see Figure 1), it appears that homeowners who have updated their properties to be more

---

3 The respondents were asked to comment on the statements in terms of "is completely accurate", "is fairly accurate", "is somewhat inaccurate" and "is completely inaccurate". The first two answers are reflected in the percentage indications in the table.

energy-efficient, have been indeed been influenced by economic factors. Yet, economic considerations link in with other factors, so that a range of aspects – including sustainability – flow into energy-efficient refurbishment decision-making. This combination of motivations will henceforth be referred to as 'alliances of motives'.

The most important motives stated by homeowners who had undertaken energy-efficient refurbishments were "saving heating energy" (90 percent), "reducing long-term energy costs as far as possible" (87 percent) and "reducing running cost" (84 percent), closely followed by "installing the most up-to-date technology" (77 percent). Next featured "maintaining and increasing the value of the property", followed by "making a contribution to climate protection" with a respectable 64 percent.

The significance of economic factors in the alliances of motives is underlined by the fact that as many as 80 percent of homeowners who had undertaken energy-efficient refurbishments, endorsed the questionnaire statements "I have examined thoroughly if a measure is cost-effective" with either "is completely accurate" or "is fairly accurate". These statements indicate that in the face of important financial outlay, homeowners carefully calculate and research their investments. In this sense, we can term these actions as reflected behaviours (see Kaufmann-Hayoz/Bamberg et al. in this volume).

When a homeowner is faced with practical decision-making, a whole range of issues relating to the cost-effectiveness of a refurbishment arise, and it appears that while homeowners rank cost-effectiveness highly in theory, they do not always strictly follow their investment plan in practice:

- Decisions on refurbishments are complex. Given that a decision on a major refurbishment might come up just once in the lifetime, homeowners often lack previous experience on which to base judgment.
- A large number of possible refurbishment measures are currently available. Some techniques are new or have not been in use for a long time. This is especially true in the field of energy-efficient refurbishments.
- Economic data, i.e. estimates of investment costs and potential savings from avoiding energy use are not very reliable. In particular, the price of energy is so volatile that even professionals experience great difficulties in forecasting energy costs. Additionally, in the light of constantly changing statutory regulations it remains unclear to what extent energy-efficient measures add to the long-term value of a property.
- Most homeowners find it difficult to calculate the return on their investment.

In essence, a decision to undertake energy-efficient refurbishments in a home is always taken in a context of considerable uncertainties, for which there is no simple economic solution. It is therefore worth investigating homeowners' perspectives of the decision-making process and which simplified criteria they resort to when it comes to deciding.

Our survey showed that respondents did regard amortisation periods (the period over which the improvement pays for itself) as an indicator of cost-effectiveness: those who had undertaken an energy-efficient refurbishment measure regarded an amortisation period of just under 10 years as acceptable (this figure is an average and varied widely). 39 percent felt that a 10–15 year period was acceptable. The willingness to accept fairly extended amortisation periods for the initial outlay of capital stands in stark contrast to the financial calculations made by professional investors.

It is somewhat surprising, however, that as many as 60 percent of respondents agreed with the statement "I am satisfied with the knowledge that I save energy, and I'm not that bothered about the amortisation period in a precise number of years" (this statement was rated as either "is completely accurate" or "is fairly accurate"). A considerable 33 percent of homeowners who had carried out refurbishments with a high energy efficiency potential endorsed the following two statements: "I meticulously check the cost-effectiveness of energy-efficient refurbishment measures" and "I am satisfied with the knowledge that I save energy, and I'm not that bothered about the amortisation period in a precise number of years". These responses indicate that the cost-effectiveness criterion is often not as important to homeowners as experts have hitherto assumed. "Cost-effectiveness" appears to be interpreted by homeowners in the broader sense of achieving a noticeable overall saving.

These observations lead to the following overall picture: homeowners are largely "liberal" in their interpretation of cost-effectiveness. In contrast to experts, private homeowners are satisfied if the ongoing costs decrease noticeably. The exact time frame for the amortisation of a refurbishment measure is less important. Accordingly, advice services need not supply potential refurbishers with the potentially highest and most precise figures on savings, but simply need to reassure homeowners that overall savings will be achieved (for a detailed description of assessments of cost-effectiveness see Albrecht/Zundel 2010).

The reality that the actual refurbishments undertaken lag behind the potentially most cost-effective measures is probably due to other factors – including financial constraints. This became apparent when barriers to energy-efficient refurbishments were considered in more detail as part of the survey. All refurbishers were asked to state possible objections to an energy-efficient refurbishment of their home. The response categories comprised structural, technical and financial aspects, as well as attitudes towards the execution and the outcome of the refurbishment.

Figure 2 shows the number of barriers relative to the extent to which they can be overcome. The x-axis shows the number of the barriers in terms of how many times they were mentioned. The y-axis shows the extent to which the barriers can be dismantled – ranging from "easy to overcome" to "difficult to overcome". Thus, the upper

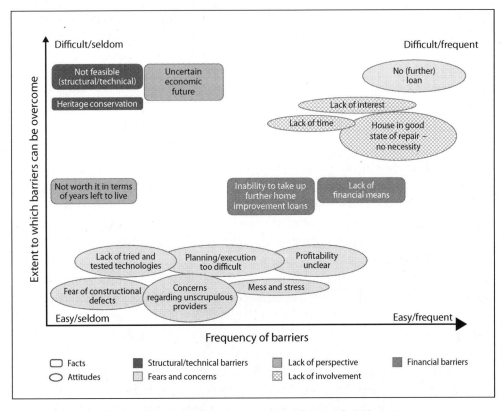

Figure 2: Frequency of barriers to energy-efficient refurbishments and extent to which they can be overcome (source: ENEF-Haus 2010, p. 10)

right quadrant features frequent barriers and those that are difficult to break down. Attitudes such as aversion to loans are represented in elliptical form. "Objective" barriers such as heritage conservation or a lack of finances are represented as rectangles.

The most frequently cited argument against undertaking comprehensive energy-efficient refurbishment is the unwillingness to take on an additional loan in the face of the considerable financial cost of such a measure. The second most frequent reason is the assertion that the house is sufficiently energy-efficient and therefore requires no improvements. More tangible financial reasons such as a lack of funds or an inability to take up a loan also play a role, but they are far less regarded as an obstacle to comprehensive energy-efficient refurbishment. Further barriers to energy-efficient refurbishment are a lack of interest in any refurbishing beyond essential repairs, uncertainty about the cost-effectiveness of energy-efficient measures and a lack of time to engage intensively with energy-efficient refurbishing. These reasons are given primarily by those with no previous experience of energy-efficient refurbishment. In comparison, "hard" technical or structural constraints, such as heritage conservation, and concerns about the uncertain economic climate (e.g. unemployment and uncertain work prospects) are rarely cited.

1.4 Typology of refurbishers

The data obtained from the standardised questionnaire was analysed by way of multivariate statistical methods, on the basis of which a typology for refurbishers was drawn up. The statistical analysis comprised several steps: first, a factor analysis captured the following: contexts and motives for refurbishments; attitudes towards professional advice and financing the refurbishments; barriers to energy-efficient refurbishing. Second, a logistic regression analysis was used to assess the influence of the previously identified factors on the dependent variable "implementation of energy-efficient refurbishment measures". Additionally, further regression analyses with socio-demographic variables and building-related features as well as an overall model with all three groups of features were employed. The results indicated that socio-demographic and building-related factors – with the exception of "previously existing insulation" – influenced the quality of both standard and extended refurbishments only very slightly (Zundel/Stieß 2011).

A cluster analysis was used to identify five types of refurbishers (cf. Figure 3). Those factors that had a particularly strong influence on the type of refurbishment were

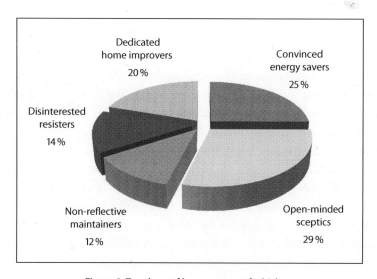

Figure 3: Typology of homeowner refurbishers

included as classification variables (the cluster analysis was performed using a non-hierarchical, disjoint partitioning method, i.e. k-means clustering). The types were characterised by attitudes, alliances of motives and barriers to energy-efficient refurbishment. They diverge significantly at the level of the dependent variable "type of refurbishment" ($Chi^2 = 321.56$; $df = 4$; $p < 0.001$). The individual types of refurbishers are described below:

"Convinced energy savers"

A quarter of all respondents fall into the group of "convinced energy savers". They strongly identify with their house, they show above-average sustainability orientation and they are intrinsically motivated to make their house highly energy-efficient. They are well-versed in sustainability issues as they constantly keep abreast of the latest developments.

When it comes to refurbishing, the "convinced energy savers" are motivated by a whole range of factors: apart from saving heating energy and reducing energy costs in the long term, reversing climate change and not depending on fossil fuels are key motivators for this group. They are tech-savvy and see themselves as pioneers in innovative building techniques. Other motives include the desire for a comfortable and cosy living environment and the retention of, or an increase in, the value of their house. The primary constraint on any additional energy-efficient refurbishment measures tends to be a lack of financial resources.

"Open-minded sceptics"

With a share of 29 percent the "open-minded sceptics" are the largest group. Their refurbishment measures tend to be triggered by the need for routine maintenance of the house. The "open-minded sceptics" are motivated by a wide range of factors: a refurbishment should ensure and raise the value of the house and bring it up to the latest technological standard – not least in order to save on heating and reduce costs in the long run. While living comfort also ranks high, climate protection tends to be of less concern.

Despite their openness to energy-efficient refurbishments, "open-minded sceptics" have some concerns and reservations. Many are uncertain as to whether technologies are sufficiently mature – or they fear structural damage to their house as a result of energy-efficient refurbishments. Additionally, "open-minded sceptics" are often doubtful as to the usefulness of energy-efficient improvements. A further common concern is the excessive demands and stress that the planning and execution of a refurbishment might cause, as well as a fear of unscrupulous tradespeople.

"Non-reflective maintainers"

With 12 percent the "non-reflective maintainers" are the smallest group among the respondents. This group neither identifies intensively with comprehensive energy-efficient refurbishments, nor does it have fundamental reservations in that respect. Refurbishment decisions, such as replacing a defective part (e.g. window) or maintenance work, are taken spontaneously in response to immediate problems. "Non-reflective maintainers" tend to take refurbishment measures one at a time, in response to expert advice. However, this does not mean that their sole aim is to maintain the existing state of their house. Motives include savings on heating and reducing energy costs – even though the percentage of "non-reflective maintainers" stating this motivation was below the average of the whole sample.

Any refurbishment that exceeds the strictly necessary is of little interest, and a principal barrier for the "non-reflective maintainers" is their limited view of energy conservation: energy saving is primarily equated with improving the heating system of the home. Additionally, an aversion to taking on an (additional) loan for the purpose of refurbishing effectively limits the extent of any potential work.

"Disinterested resisters"

"Disinterested resisters" make up 14 percent of respondents. This type does not go beyond what is strictly necessary. In order to save money, the refurbishment work is carried out by the homeowner. Because refurbishment is felt to be a burden, anything beyond the most pressing repairs is avoided. Furthermore, "disinterested resisters" show a distinct lack of curiosity about innovative technologies and are extremely averse to risk taking. Many dread the mess and stress caused by refurbishments and feel overwhelmed by the thought of planning and executing energy-efficient refurbishments. The motivation to take up a loan for comprehensive energy-efficient refurbishments is low – not least because the economic benefits of such measures are unclear.

"Dedicated home improvers"

"Dedicated home improvers" make up 20 percent of respondents. They live in relatively new and energy-efficient houses. Their refurbishment ambitions tend to be limited to the embellishment and extension of their living space. They are not interested in energy-efficient refurbishments, because their houses are already energy-efficient. Nonetheless, this group could represent a potential target group for energy-efficient refurbishments at a point in time when the home is due for a routine overhaul.

An overview of refurbisher types and respective energy savings achieved

The above five types of refurbishers not only have dissimilar alliances of motives and barriers to energy-efficient refurbishment, but they also differ in the way they go about refurbishments and in relation to how much energy they save (cf. Figure 4). A comparison of the energy requirements of the houses before and after the refurbishments shows this clearly.[4] The "convinced energy savers" achieved the highest savings in heating and primary energy. The "open-minded sceptics" and to some extent the "non-reflective maintainers" realised significant savings. The other two types achieved savings of barely 5 percent.

Overall, the "convinced energy savers" and the "open-minded sceptics", representing about half the respondents, can be considered the primary target groups for energy-efficient refurbishments. Additionally, the "non-reflective maintainers" may be a potential target group. The other groups would probably not count as target groups: the "disinterested resisters" have strong reservations about energy-efficient refurbishments, and the "dedicated home improvers" already live in relatively energy-efficient houses.

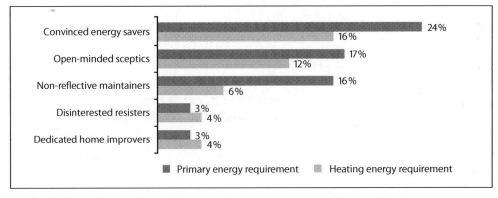

Figure 4: Decrease in heating and primary energy requirement after energy-efficient refurbishment – comparison of refurbisher types (source: Weiß/Dunkelberg 2010)

---

4 As part of a standardised survey, the condition of the building was recorded before and after refurbishment. For the collection of the buildings data the calculation tool KVEP (Kurzverfahren-Energieprofil), developed by the IWU (2006), was adapted to the needs of the present investigation. On the basis of that data, the change in heating and primary energy requirement after refurbishment was recorded for each building (Weiß/Dunkelberg 2010).

1.5 Summary and conclusions

The analysis of potential revealed a number of cost-effective refurbishment measures that lead to energy and cost savings in the medium term, which – even under conservative estimates – pay off in a time period considered acceptable by many homeowners, i.e. these refurbishments pay off, if they are carried out as part of a routine major updating programme of a house or at the time when the house is purchased.

In order to optimally benefit from such windows of opportunity, it is important to make the right decisions at the right point in time. The most cost-effective measures often require considerable investment. Conversely, some initial low-cost partial refurbishments show relatively poor profitability. As a result of homeowners' lack of capital or inability to take up a loan, comprehensive, integrated solutions are forgone in favour of lower-cost partial refurbishments. Yet, these cheaper refurbishments with faster amortisation periods represent worse value for money in the long term. This is an area where information campaigns could prevent homeowners from taking the wrong decisions.

The fact that energy-efficient refurbishments also make economic sense speaks in favour of including them in refurbishment programmes. Routine refurbishments constitute ideal contexts in which to familiarise homeowners with the potentials of energy-efficient measures. Targeted communication campaigns should therefore use these opportunities to approach a range of stakeholders (e.g. tradespeople, energy consultants, home improvement stores) as well as information channels, who can in turn present to homeowners energy-efficient home improvement solutions (cf. ENEF-Haus 2010).

Given the diverse alliances of motives and barriers to the implementation of energy-efficient refurbishments, wide-ranging refurbishment of Germany's housing stock is only possible if communication campaigns make use of opportunities for promoting energy-efficient refurbishment and take account of homeowners' specific alliances of motives. The typology presented here provides an important starting point for a differentiated approach in promoting energy-efficient measures. The most important target groups are the "convinced energy savers", the "open-minded sceptics" and the "non-reflective maintainers" (Stieß et al. 2011).

Even though an information campaign as outlined in this article may be effective in its own terms, it does not necessarily reach everyone. Moreover, it largely fails to motivate certain types of refurbisher. Seen in this light and with long-term goals in mind, there seems to be a need for promoting energy-efficient refurbishment measures as part of a regulatory framework (cf. Weiß et al. 2012).

Among the motives for energy-efficient refurbishment, cost-effectiveness undoubtedly plays a major role – although the relative importance of this motive varies among the target groups. Moreover, homeowners and experts have different understandings of cost-effectiveness. Supposedly meticulous expert profitability analyses do not always have the desired impact. Most homeowners are content to know simply that they are going to make significant overall savings. The motive of "providing for the future", for instance, shared by a large number of homeowners, can be fostered by stressing that savings will increase in line with energy price hikes. This motive would undoubtedly be further strengthened if the energy-efficiency of each house were reflected in property prices.

Non-economic barriers to energy-efficient refurbishment should not be overlooked: indifference, prejudice, lack of knowledge, as much as uncertainties, fear of being overwhelmed and misleading advice can prevent homeowners from undertaking comprehensive energy-efficient refurbishments. This is where targeted communication campaigns and advice services can make a significant contribution to reducing such barriers.

References

Ajzen I., Fishbein M. (1980): Understanding attitudes and predicting social behaviour. Englewood Cliffs: Prentice Hall.
Albrecht T., Zundel S. (2010): Gefühlte Wirtschaftlichkeit – Wie Eigenheimbesitzer energetische Sanierungsmaßnahmen ökonomisch bewerten. Senftenberg: Hochschule Lausitz.
Biermayr P., Ernst Schriefl E., Baumann B., Sturm A. (2004): Maßnahmen zur Minimierung von Reboundeffekten bei der Sanierung von Wohngebäuden (MARESI); final report. Vienna.
BMVBS [Bundesministerium für Verkehr, Bau und Stadtentwicklung] (2007): CO_2-Gebäudereport 2007. [http://www.ibp.fraunhofer.de/Images/2009_64-CO_2-Gebaeudereport_tcm45-93208.pdf; 22.06.2012]
De Haan P. (2009): Energie-Effizienz und Reboundeffekte: Entstehung, Ausmass, Eindämmung – final report. Zürich.
dena [Deutsche Energie-Agentur] (2007): Hauseigentümer können 80 Prozent Heizenergie einsparen. [http://www.umweltjournal.de/AfA_bauenwohnengarten/12004.php; 22.06.2012]
destatis (2008): Energieverbrauch der privaten Haushalte. Wohnen, Mobilität, Konsum und Umwelt. Berlin: destatis.
ENEF-Haus (ed.) (2010): Zum Sanieren motivieren: Eigenheimbesitzer zielgerichtet für eine energetische Sanierung gewinnen. Bearbeitet von Tanja Albrecht, Jutta Deffner, Elisa Dunkelberg, Bernd Hirschl, Victoria van der Land, Immanuel Stieß, Thomas Vogelpohl, Julika Weiß, Stefan Zundel. Frankfurt am Main: ISOE.
Fischedick M., Schüwer D., Venjakob J., Merten F., Mitze D., Krewitt W., Nast M., Schillings Ch., Bohnenschäfer W., Lindner K. (2006): Anforderungen an Nah- und Fernwärmenetze sowie Strategien für Marktakteure in Hinblick auf die Erreichung der Klimaschutzziele der Bundesregierung bis zum Jahr 2020. Endbericht, Wuppertal, Stuttgart, Leipzig.
IWU [Institut für Wohnen und Umwelt] (2006): Kurzverfahren Energieprofil. Exceltool. [http://www.iwu.de; 22.06.2012]

Kaufmann-Hayoz R., Bamberg S., Defila R., Dehmel C., Di Giulio A., Jaeger-Erben M., Matthies E., Sunderer G., Zundel S. (in this volume): Theoretical perspectives on consumer behaviour – attempt at establishing an order to the theories.

Kuckshinrichs W., Hansen P., Kronenberg T. (2009): Gesamtwirtschaftliche CO_2-Vermeidungskosten der energetischen Gebäudesanierung und Kosten der Förderung für den Bundeshaushalt im Rahmen des CO_2-Gebäudesanierungsprogramms. STE Research Report [http://www.kfw.de/kfw/de/I/II/Download_Center/Fachthemen/Research/PDF-Dokumente_Evaluationen/33807_p_0.pdf; 26.06.2012]

Müller A., Biermayr P., Kranzl L., Haas R., Altenburger F., Bergmann I., Friedl G., Haslinger W., Heimrath R., Ohnmacht R., Weiss W. (2010): Heizen 2050 – Systeme zur Wärmebereitstellung und Raumklimatisierung im österreichischen Gebäudebestand: Technologische Anforderungen bis zum Jahr 2050. Final report on research project nr. 814008. Vienna.

Schlomann B., Gruber E., Eichhammer W., Kling N., Diekmann J., Ziesing H.-J., Rieke H., Wittke F., Herzog T., Barbosa M., Lutz S., Broeske U., Merten D., Falkenberg D., Nill M., Kaltschmitt M., Geiger B., Kleeberger H., Eckl R. (2004): Energieverbrauch der privaten Haushalte und des Sektors Gewerbe, Handel und Dienstleistungen (GHD); [Fraunhofer-Institut für Systemtechnik und Innovationsforschung (ISI), Deutsches Institut für Wirtschaftsforschung (DIW) et. al.]. Karlsruhe, Berlin.

Stieß I., van der Land V., Birzle-Harder B., Deffner J. (2010): Handlungsmotive, -hemmnisse und Zielgruppen für eine energetische Gebäudesanierung. Ergebnisse einer standardisierten Befragung von Eigenheimsanierern. Frankfurt am Main: ISOE.

Stieß I., van der Land V., Deffner J., Birzle-Harder B. (2011): Eigenheimbesitzer zielgruppengerecht zur energetischen Sanierung bewegen – Ansatzpunkte, Barrieren und Überblick zu geeigneten Instrumenten zur Förderung energetischer Sanierungen bei Eigenheimsanierern. Frankfurt am Main: ISOE.

Weiß J., Dunkelberg E. (2010): Erschließbare Energieeinsparpotenziale im Ein- und Zweifamilienhausbestand. Berlin: IÖW.

Weiß J., Vogelpohl T., Dunkelberg E. (2012): Improving policy instruments to better tap into homeowner refurbishment potential: Lessons learned from a case study in Germany. In: Energy Policy 44: 406–415.

Zundel S., Stieß I. (2011): Beyond Profitability of Energy Saving Measures – Attitudes towards Energy Saving. Journal of Consumer Policy. In: Journal of Consumer Policy 34 (1): 91–105.

Joachim Schleich, Bradford F. Mills

2 Determinants and distributional implications in the purchase of energy-efficient household appliances

2.1 Introduction

Sustainable consumption decisions are not only characterised by their ecological and economic effects, but also by their social impact. Measures intended to promote sustainable consumption can, for instance, have distributional impacts, which in turn contribute to increased social cohesion. This article attempts to quantify the distributional impacts that energy efficiency measures taken by households can have. The example used here for an illustration of such measures is the promotion of the diffusion of highly energy-efficient cooling appliances.

Major household appliances account for around 15 percent of energy consumption and 35 percent of electricity consumption in private households in the European Union (European Commission 2011a; Bertoldi/Atanasiu 2009). Cooling appliances alone account for 15 percent of electricity consumption in private households, washing machines for 4 percent, and dishwashers, electric ovens and dryers for approximately 2 percent each. Improving energy efficiency – for instance by increasing the diffusion of energy-efficient appliances – therefore represents an important opportunity for achieving policy targets for climate protection and energy efficiency. According to the "Low carbon roadmap" set out by the European Commission, by the year 2020, 25 percent of greenhouse gas emissions should be saved compared to 1990; by 2050, these savings should rise to between 80 and 90 percent (European Commission 2011a). Compared to projections, energy efficiency in the EU should improve by 20 percent by the year 2020 (European Council 2007; European Commission 2008). If all the measures of the new energy efficiency plan are implemented, each household could save up to an estimated 1000 euros annually (European Commission 2008, 2011b). This shows that measures designed to enhance the diffusion of energy-efficient technologies do not only contribute to achieving environmental goals, but can also lead to a fairer distribution of income.

Since its inception and implementation in the 1990s, the EU regulation for energy consumption labelling, stipulating the labelling of certain household appliances, has been the most important policy instrument for an accelerated diffusion of energy-efficient appliances (Bertoldi/Atanasiu 2009). It is envisaged that more widespread labelling of household appliances and other energy-consuming equipment will particularly contribute to achieving future energy and climate protection targets (European Commission 2011b).

Generally, labelling is considered a cost-effective measure to overcome barriers such as information gathering and search costs, or a lack of rational thinking on the part of the buyers (Howarth et al. 2000; Sutherland 1991; Sorrell et al. 2004). The effectiveness of energy labelling and thus the targeted achievement of sustainability goals, however, depends on whether the buyers of appliances are familiar with the labels in the first place and whether labelling will affect buyers' decisions. With the help of a large data set drawn from German households, Mills and Schleich (2010a) recently studied the influence of socio-economic factors and residence characteristics on a) individuals' knowledge of the energy ratings of their major household appliances, and b) the likelihood of people purchasing highly energy-efficient appliances. The work undertaken by Mills and Schleich (2010a), which formed part of the "Seco@home" project group, thus constitutes an empirical basis for the study of energy consumption behaviour among households and contributes to a positive theory of sustainable consumption (Fischer 2002). Based on the results of Mills and Schleich (2010a), this paper considers the distributional implications of the purchase of energy-efficient cooling appliances. As part of the "Seco@home" project group it contributes to an analysis of the effects of sustainable consumption patterns and of policy measures designed to promote such behaviour on a larger scale.

This paper first presents the methodology employed and the results reached by Mills and Schleich (2010a), in terms of the determinants of a) the purchase of energy-efficient household appliances and b) people's awareness of the actual energy efficiency rating of their appliances. Then follows an analysis of the distributional effects of the adoption of energy-efficient household appliances (compared to average energy-consuming models), i.e. which income brackets benefit the most from the electricity cost savings involved in the acquisition of energy-efficient appliances. Finally, the results are summarised and discussed in relation to more general statements and to implications for policy interventions.

2.2 Methodology and results

2.2.1 Determinants of knowledge about energy labels and purchase decisions

Recent observations about the factors determining the knowledge about energy labels and the decisions to buy energy-efficient major household appliances draw on Mills and Schleich (2020a), whose econometric analysis is based on data from a survey of 20,000 German households in 2002 (Schlomann et al. 2004).[1] The data contains socio-economic information as well as residence characteristics of households. The questionnaire does not include questions on attitudes, values and purchasing motives and therefore does not allow for an analysis of the influence of people's attitudes or of social and individual norms. Households were also asked to state the energy efficiency class of several of their major household appliances. Other characteristics of these appliances, such as quality, purchasing price or operating costs are not known.

Separate equations were estimated econometrically to investigate people's knowledge of the energy efficiency class of their personal appliances on the one hand, and to evaluate their personal choice of appliance on the other. Socio-economic factors and residence characteristics featured as explanatory variables. The following socio-economic explanatory variables were taken into account: number of household members, number of children under six, age of the head of household, income bracket (16 brackets in total), level of education, regional electricity prices, as well as indicator variables for professional people's management functions, pensioners, households in the former East German states, ownership of a PC and ownership of another household appliance with the highest energy-efficiency rating. Residence characteristics were captured by the following variables: dwelling size and age of building, as well as variables indicating if a residence was owned or rented and if it was a detached building. Furthermore, consideration was given to the fact that consumers who were familiar with the labelling system might act differently to those who had not come across it.

The results of the econometric estimates of people's knowledge of their appliances' energy efficiency ratings suggest that there is a higher degree of awareness among people who live in newer buildings or in rented apartments, who have a higher level of education, who are comparatively high earners, who know how much energy they consume, who have a younger head of household, who live in a household with a

---

[1] The fact cannot be ignored that a different data set may lead to different findings. The current data, however, is unique in that it allows for a methodologically consistent connection between the analyses of the determinants of knowing the energy ratings of one's own household appliances (or for the purchase of energy-efficient appliances) on the one hand, and the distributional impacts on the other. Additionally, this article aims to generate knowledge of a general nature. In this sense, the results gleaned from the 2002 data are exemplary in character.

retired person as head of household, who pay higher electricity prices and who own a PC or another appliance within the highest energy efficiency class (i.e. Class-A).

The decision to purchase energy-efficient household appliances, however, seems to depend less on socio-economic factors. Residence characteristics seem to be more important here. Efficient appliances are more likely to be purchased by households who live in more modern buildings and by those who rent. Equally, the level of income does not seem to be statistically significant, according to Mills and Schleich (2010a).[2] Other empirical studies have generally found that higher incomes lead to an accelerated diffusion of energy efficiency. Dillman et al. (1983) and Long (1993) have established such a relationship for the USA, Walsh (1989) and Ferguson (1993) for Canada, Sardianou (2007) for Greece and Mills and Schleich (2010b) for Germany. The study by Young (2008) for Canada also indicates that higher-income households replace their appliances more often, so that old equipment tends to be replaced by newer, usually more energy-efficient appliances. This means that higher-income households are less likely to be restricted by income or credit considerations, which would prevent them from buying new appliances in general and energy-efficient appliances in particular. Interestingly, the price of electricity does not seem to affect the choice of energy efficiency class in a household appliance (Mills and Schleich 2010a).

2.2.2 Distributional effects of the purchase of energy-efficient household appliances

The distributional effects of the adoption of energy-efficient household appliances can be represented as a function of two linked components: the effect of income on the likely purchase of energy-efficient appliances on the one hand, and the electricity cost savings in relation to income on the other.[3] For the household data referred to above, no statistically significant relationship between income and adoption of energy-efficient appliances can be determined (Mills/Schleich 2010a). In other words, households of all income brackets are equally likely to buy an energy-efficient household appliance – so long as the influence of other socio-economic factors and residence characteristics are also factored in. Therefore, the subsequent analysis of the distributional effects (based on the same data as the study by Mills and Schleich 2010a) assumes

---

2 Mills and Schleich's data (2010a) contains, apart from income, numerous other explanatory variables that correlate with income, such as flat size, age of building or level of education. Studies that only comprise income as an explanatory variable can overestimate the influence of the income factor, because the estimated parameter value also reflects the influence of the other (omitted) variables.

3 The investigation abstracts from potential purchase price or other cost differentials between an energy-efficient device and the average appliance considered in the survey. More efficient appliances need not be more expensive to buy.

that there are no differences in adoption behaviour between different income brackets. Thus, only the potential electricity cost savings relative to the income of each of the sixteen income brackets are taken into account. This analysis is based on the example of household cooling appliances. The calculated Class-A adoption frequencies of German households are 40.6 percent for refrigerators, 48.1 percent for refrigerator-freezer combination units and 44.6 percent for freezers.

The data in Table 1 are based on a comparison between the mean energy rating of the appliances captured in the survey (Schlomann et al. 2004) and the most efficient appliance available on the market in 2003.[4] Thus, compared to the average, the current electricity savings achieved by a Class-A refrigerator stands at 31.6 percent. The corresponding values for refrigerator-freezer combination units and freezers stand at 15.4 percent and 40.4 percent respectively.[5] The associated annual savings using a electricity price of 0.156 euro per kWh (average price in survey) are around 13 euros for refrigerators, 8 euros for refrigerator-freezer combination units and 20 euros for freezers. Since all of these appliances typically run on a continuous basis, there is no reason to expect that annual savings differ across the different households.

Table 1: Efficiency gains and electricity cost savings from Class-A appliances (reference year 2003)

	Electricity use of stock average	Electricity consumption of most efficient appliance	Reduction in electricity consumption	Electricity cost savings
	kWh/year	kWh/year	%	Euro
Refrigerators	256	175	31.6	12.64
Refrigerator-freezers	344	291	15.4	8.27
Freezers	319	190	40.4	20.12

---

4 There were no data available for 2002, but it is estimated that these would not differ significantly from the data used in Table 1.

5 Here, as far as possible, the specifications for those A-Class appliances were used that matched the average appliance in the survey in terms of volume. It is self-evident that appliances can differ in other respects, but this was not taken into account in the current analysis.

Electricity expenditure by income brackets

Non-parametric estimates of the distribution (density) of annual household electricity expenditures for the sixteen household income brackets are presented three-dimensionally in Figure 1. It shows clearly that the mean value of household electricity expenditure is smaller for households in the lower income brackets. The bulk of households in the lowest income bracket (i.e. bracket 1) are, for instance, much closer to the starting value of the electricity expenditure axis (electricity expenditure proceeds from right to left) than households in higher income brackets. Equally, the variance of electricity costs is greater for higher income brackets than for the lower ones. In Figure 1, for instance, the observed values of electricity costs for the lowest income brackets are much closer to each other than is the case for the higher income brackets.

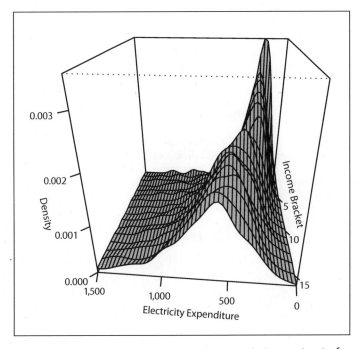

Figure 1: Electricity expenditure density estimates by income bracket[6]

---

6   The density function is, in principle, a three-dimensional histogram. Here, it shows the distribution of the probabilities that, in terms of the data set used, certain combinations of electricity costs (in euro per annum) occur in the different income brackets. Bracket 1 and bracket 16 represent the lowest and the highest income bracket respectively.

If electricity expenditure is viewed in absolute terms and in relation to income (Figure 2), the trends are as expected: in absolute terms, high earners have higher electricity outgoings (continuous line, Figure 2), but the share of electricity expenditure measured in terms of the overall disposable income is higher for lower income brackets (broken line, Figure 2).

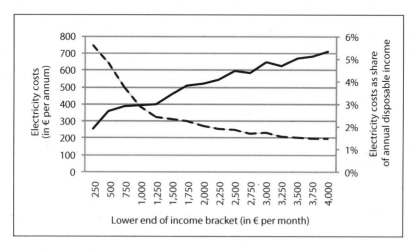

Figure 2: Absolute annual electricity expenditure and as share of income for different income brackets

Electricity cost savings by income brackets
Figure 3 shows that, by purchasing refrigerators, refrigerator-freezers and freezers, lower income households save more money on electricity than higher income households. A similar trend is observed in terms of electricity cost savings in relation to disposable income. For instance, a household from the lowest income bracket, with an income of less than 250 euros, who buys a Class-A refrigerator, saves 5 percent of its overall electricity expenditure, i.e. 0.07 percent of its disposable income. In comparison, the savings for a household in the highest income bracket with a monthly income of over 4,000 euros saves 1.8 percent of its overall electricity expenditure, which amounts to less than 0.007 percent of its disposable income. Similar trends can be observed for refrigerator-freezers and freezers.

It is true for all three cooling appliances that purchasing a Class-A appliance leads to electricity cost savings. Relative to overall disposable income, households in the lowest income bracket save around 10 times more than households in the highest income brackets.

Although dishwashers and washing machines have lower base levels of electricity consumption (owing to their intermittent use), the potential savings of energy-efficient models are comparable to those of refrigerators. In other words, the distribution impact of energy-efficient dishwashers and washing machines is qualitatively similar to that of refrigerators.

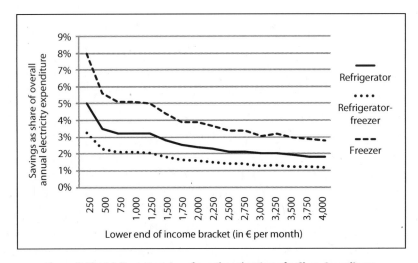

Figure 3: Electricity cost savings from the adoption of a Class-A appliance as share of the overall electricity expenditure (according to income bracket)

2.3 Conclusion

In summary, it can be noted that the adoption of energy-efficient household appliances not only has positive environmental effects – it contributes to a reduction in energy consumption and pollutant emissions caused by the generation of electricity – but that it can also lead to socially desirable distributional effects. On the basis of an extensive data set collected in Germany, this paper has attempted to quantify these distributional effects. It concludes that households in the lowest income bracket can save between 3 and 8 percent of their electricity bills by purchasing a highly energy-efficient refrigerator, while households in the highest income brackets can save between 1 and 3 percent. The fact that lower-income households benefit relatively more from the purchase of energy-efficient appliances is also reflected in the fact that, relative to disposable income, households in the lowest income brackets save about 10 times more when they purchase the most energy-efficient appliance than households in the highest income classes. If other electricity consuming household appliances were taken into account, this proportion would in fact rise considerably and would be significant not only in relative, but also in absolute terms. Therefore, at least for the major household appliances, it seems that policy interventions designed to support sustainable consumption also contribute to distributional equity.

Nonetheless, the investigations into the determinants of the purchase of energy-efficient household appliances, and into people's knowledge of the energy efficiency ratings of the appliances (these formed the basis of the analysis of the distributional effects), show that the income level is only significant in terms of people's knowledge of the energy efficiency class, but is insignificant when it comes to selecting appliances. That is, those households that on distributional criteria would benefit most from the purchase of energy-efficient equipment do not seem to pay closer attention to potential energy consumption than high-income households. In addition, lower-income households seem to be less familiar with the energy efficiency class of their household appliances than households with higher incomes. Equally, the results for other explanatory variables (dwelling size, level of education, computer ownership) lead to the general conclusion that wealthier households are more likely to know the energy efficiency ratings of their appliances. These results therefore warrant policy interventions (based on a relatively broad statistical basis) that are designed to distribute information about energy consumption and energy costs (as far as major household appliances are concerned) specifically to those households with lower levels of education and income. If income or credit restrictions prevent people from buying energy-efficient equipment, replacement bonuses, i.e. grants for replacing old appliances with new, more energy-efficient appliances, could help people overcome their reservations.[7] In order to prevent windfall effects, eligibility for such grants should be restricted to low-income households. In order to avoid "rebound effects", the payment of the replacement bonus should be subject to an obligation to dispose of old appliances permanently (e.g. by handing them in to a dealer or collection point).

The results for the explanatory variables age of the building and residence ownership or rental – in relation to how likely people were to buy energy-efficient household appliances and how likely they were to be familiar with energy efficiency ratings of appliances – suggest that people who move frequently (tenants) or have just moved into a building (more modern buildings) are not only more knowledgeable about the energy efficiency ratings of their appliances, but they also tend to buy more energy-efficient appliances. Thus, the results of the statistical and econometric analysis of house-

---

7 Cf. Dehmel (2010) and Brohmann/Dehmel et al. in this volume for an overview of experiences with replacement bonuses for refrigerators in Austria, Denmark and the Netherlands. These investigations were carried out as part of the "Transpose" project.

hold cooling appliances support the thesis by the "LifeEvents" project that changes in people's lives can represent opportunities for more sustainable consumption (cf. Schäfer/Jaeger-Erben in this volume; Jaeger-Erben 2010; Schäfer/Bamberg 2008).

The results also indicate that an increase in electricity prices (e.g. as a result of an energy tax) do not directly lead to an increased adoption of energy-efficient household appliances. As expected, households are more likely to be familiar with the energy efficiency class of their appliances in a climate of high electricity prices, as there is a stronger financial incentive to shoulder the increased information costs. The level of electricity prices, however, does not seem to affect the choice of energy efficiency class of the appliance. From a predominantly economic perspective, it could be expected that higher electricity prices would lead people to choose more energy-efficient appliances, as these reduce energy expenditure. The current result therefore calls into question the validity of models based on neoclassical economic theory, according to which households maximise their benefits by acting rationally and by consistently selecting the most efficient amongst several alternative behaviours (homo economicus). By contrast, the result points to people not always behaving rationally. In terms of policy interventions, it follows that price incentives will only have a limited effect, and that instead, other more effective measures should be employed. Such measures should include strategies based on providing information and communication (e.g. energy labelling).[8] In the context of distributional equity, increasing electricity prices may even prove counterproductive in that they disadvantage low income households disproportionately and do not directly increase the adoption of energy-efficient appliances.

In summary, it can be noted that socio-economic variables seem to have little influence on whether people buy energy-efficient household appliances. However, it was found that households are more likely to buy an energy-efficient appliance if they already own other such appliances. Further research is needed in this area in order to identify potential factors that have not been considered in this study, but may influence the purchase of energy-efficient appliances. These include attitudes towards the environment, other psychological factors, as well as social norms (Kahn 2007; Gilg/Barr 2006; Wilson/Dowlatabadi 2007). For instance, the study by Brandon and Lewis (1999) concludes that attitudes towards the environment and individual convictions can be just as relevant as financial considerations.

---

8 The possibility that households only act rationally to a limited degree should not only be taken into account in the policy mix, but also in the design of concrete measures. In the context of the revision of the EU energy labelling, Heinzle and Wüstenhagen 2010 (within the "Seco@home" project) conclude that the design of energy labels can systematically influence the choice of appliance and can lead to less than optimal results (improving the energy efficiency labels by rescaling their threshold value versus improving the labels by introducing additional energy efficiency classes).

References

Bertoldi P., Atanasiu B. (2009): Electricity Consumption and Efficiency Trends in European Union – Status Report 2009. European Commission Joint Research Centre, Institute for Energy. Ispra.

Brandon G., Lewis A. (1999): Reducing household energy consumption: a qualitative and quantitative field study. In: Journal of Environmental Psychology 19: 75–85.

Brohmann B., Dehmel C., Fuchs D., Mert W., Schreuer A., Tews K. (in this volume): Bonus schemes and progressive electricity tariffs as instruments to promote sustainable electricity consumption in private households.

Dehmel C. (2010): Austausch von Kühlgeräten durch effiziente Neugeräte in privaten Haushalten. Die Trennungsprämie in Österreich im Vergleich zu ähnlichen Programmen in Dänemark und den Niederlanden. TRANSPOSE Working Paper No. 9, Münster/Berlin: 83.

Dillman D., Rosa E., Dillman J. (1983): Lifestyle and home energy conservation in the United States: the poor accept lifestyle cutbacks while the wealthy invest in conservation. In: Journal of Economic Psychology 3: 299–315.

European Commission (2008): Communication from the Commission: Energy efficiency: delivering the 20 % target. COM(2008) 772. Brussels.

European Commission (2011a): Communication from the Commission to the European Parliament, the Council, the European Economic and Social Committee and the Committee of the Regions. A Roadmap for moving to a competitive low carbon economy in 2050. COM(2011) 109 final. Brussels.

European Commission (2011b): Communication from the Commission to the European Parliament, the Council, the European Economic and Social Committee and the Committee of the Regions. Energy Efficiency Plan 2011. COM(2011) 109 final. Brussels.

European Council (2007): Presidency Conclusions of the European Council of 8/9 March 2007, 7224/1/07 REV 1. Brussels.

Ferguson M. R. (1993): Energy-saving housing improvements in Canada (1979–82): a nested logit analysis. In: Environment and Planning A 25: 609–625.

Fischer C. (2002): Nachhaltiger Konsum – Zum Stand der Forschung, TIPS Discussion Paper 2.

Gilg A., Barr S. (2006): Behavioural attitudes towards water saving? Evidence from a study of environmental actions. In: Ecological Economics 57: 400–414.

Heinzle S., Wüstenhagen R. (2010): Disimproving the European Energy Label's value for consumers? – Results of a consumer survey, Working Paper No. 5 within the project Seco@home.

Howarth B., Haddad B. M., Paton B. (2000): The economics of energy efficiency: insights from voluntary participation programmes. In: Energy Policy 28: 477–486.

Jaeger-Erben M. (2010): Zwischen Routine, Reflektion und Transformation. Die Veränderung von alltäglichem Konsum durch Lebensereignisse und die Rolle von Nachhaltigkeit – eine empirische Untersuchung unter Berücksichtigung praxistheoretischer Konzepte. Dissertationsschrift. Berlin: Technische Universität Berlin.

Kahn M. E. (2007): Do greens drive Hummers or hybrids? Environmental ideology as a determinant of consumer choice. In: Journal of Environmental Economics and Management 54: 129–154.

Long J. (1993): An econometric analysis of residential expenditures on energy conservation and renewable energy sources. In: Energy Economics 15: 232–238.

Mills B., Schleich J. (2010a): What's Driving Energy Efficient Appliance Label Awareness and Purchase Propensity? In: Energy Policy 38: 814–825.

Mills B., Schleich J. (2010b): Why Don't Households See the Light? Explaining the Diffusion of Compact Fluorescent Lamps. Resource and Energy Economics 32 (3): 363–378.

Sardianou E. (2007): Estimating energy conservation patterns of Greek households. In: Energy Policy 35: 3778–3791.

Schäfer M., Bamberg S. (2008): Breaking Habits: Linking Sustainable Consumption Campaigns to Sensitive Life Events. In: Proceedings: Sustainable Consumption and Production: Framework for Action, 10–11 March 2008, Brussels, Belgium. Conference of the Sustainable Consumption Research Exchange (SCORE!) Network. [http://www.lifeevents.de/media/pdf/publik/Schaefer_Bamberg_SCORE.pdf; 26.06.2012]

Schäfer M., Jaeger-Erben M. (in this volume): Life events as windows of opportunity for changing towards sustainable consumption patterns? The change in everyday routines in life-course transitions.

Schlomann B., Gruber E., Eichhammer W., Kling N., Diekmann J., Ziesing H. J., Rieke H., Wittke F., Herzog T., Barbosa M., Lutz S., Broeske U., Merten D., Falkenberg D., Nill M., Kaltschmitt M., Geiger B., Kleeberger H., Eckl R. (2004): Energieverbrauch der privaten Haushalte und des Sektors Gewerbe, Handel, Dienstleistungen (GHD). In Zusammenarbeit mit dem DIW Berlin, TU München, GfK Nürnberg, IE Leipzig, im Auftrag des Bundesministeriums für Wirtschaft und Arbeit. Karlsruhe, Berlin, Nürnberg, Leipzig, München.

Sorrell S., O'Malley E., Schleich J., Scott S. (2004): The economics of energy efficiency: Barriers to cost-effective investment. Cheltenham, UK: Edward Elgar.

Sutherland R. J. (1991): Market barriers to energy efficiency investments. In: The Energy Journal 12: 15–34.

Walsh M. (1989): Energy tax credits and housing improvement. In: Energy Economics 11: 275–284.

Wilson C., Dowlatabadi H. (2007): Models of Decision Making and Residential Energy Use. In: Annual review of environment and resources 32: 169–203.

Young D. (2008): When do energy-efficient appliances generate energy savings? Some evidence from Canada. In: Energy Policy 36: 34–46.

# Section B

Changing everyday consumption patterns

Martina Schäfer, Melanie Jaeger-Erben

3 Life events as windows of opportunity for changing towards sustainable consumption patterns? The change in everyday routines in life-course transitions

3.1 Introduction: consumer behaviour as part of everyday routines

Despite extensive efforts to promote sustainable consumption by way of information campaigns, economic incentives, public awareness campaigns and other interventions, the mega trends of unsustainable consumer behaviour persist. Examples are people's aspirations to live in larger dwellings, to embrace a 'convenience' lifestyle and to be more mobile, especially as far as air travel and individual transport is concerned (European Environment Agency 2006; OECD 2002).

A major challenge for policy intervention measures and for interventions by environmental and consumer groups is the fact that unsustainable behaviour is deeply embedded in structural, sociocultural and interactional contexts (cf. Kaufmann-Hayoz/ Bamberg et al. in this volume). Brand's model (2010, 2006) distinguishes further contextual elements influencing individual consumer behaviour. These include macrostructural trends (e.g. globalisation, changing values, technological developments), living styles propagated in social discourse (e.g. in terms of living or mobility), and regulatory measures such as laws, subsidies and pricing. On a smaller scale, Brand sees consumption patterns embedded in everyday social practices, originating from the interaction between consumers' everyday routines and available 'systems of provision'.

This network of social practices consisting of interrelated everyday routines allows people to meet the diverse demands of everyday life. It has evolved over a long period of time through negotiation and cooperation between individuals and their social environment (at home, at work etc.) and has become part of people's identity. The routines have become established as 'the best solutions' under the given circumstances and are thus resistant to change.

Life events can be a point of departure for changing consumption patterns towards greater sustainability, because by their very nature they call for adaptations and changes in people's routines. This article first discusses theoretical approaches to changing everyday routines and life-course transitions (section 3.2), and to existing research on life events as windows of opportunity for sustainable consumption (section 3.3). Sections 3.4 and 3.5 present the approach and results of a qualitative study within the "LifeEvents" project group. The aim of the study was to trace the changes in people's everyday consumption routines in the context of major events in their personal lives. In the overall context of the project, the study sought to address the potential of two particular life events – the birth of a first child and relocation to a large city – for different kinds of changes in behaviour towards greater sustainability (eating habits, mobility and energy consumption).

3.2 Theoretical perspectives on everyday life and life-course transitions

3.2.1 Daily routines

Research approaches to everyday life can be divided into two categories: the first views structures and societal constellations as the major influence on individuals' everyday behaviour (structural-functional perspective as in Herkommer et al. 1984; Lefèbvre 1975); the second focuses on individual actors and their interaction with the social environment (subjective-interpretative approaches such as Schütz/Luckmann's (1979) or Hitzler/Honer's (1991) 'life-world' approach).

Voß views task-oriented approaches as integrative, which take as a point of departure "principally, the practical actions of people in everyday life" (as opposed to structures of knowledge and meaning or complementary practices of symbolisation and interpretation). He does, however, stress "the need to keep in mind 'objective' social constraints and, in particular, the surrounding social structures." (Voß 2000, p. 9; translated by C. Holzherr). He cites 'conduct of everyday life' as the most important concept relating to everyday practices (e.g. Jurczyk/Rerrich 1993; Voß 1991). 'Conduct of everyday life' is to be understood as a 'system of action' based on active 'construction' by each individual. The research interest here is to establish which (e.g. temporal, spatial, material, social) resources are exploited by the individual in performing the daily routines, where the resources originate and how they were acquired by the individual. Individual construction is surrounded by structural constraints and social circumstances. As a frequent consequence of situational or ad-hoc decisions, individual construction tends to be unreflective in nature.

Giddens (1984) makes a distinction along similar lines between 'structure' and 'agency'. At the core of his theory of structuration is the assumption that individual behaviour (i.e. actors) and structures (institutional, spatial, socio-technical structures as well as infrastructures) are part of a continuous structuration process, wherein the structures determine the conditions for action, and the actors reproduce and transform structures. Empirical social research should therefore focus on social practices[1] that emerge out of this process of structuration. According to Giddens (ibid.), people go about their daily tasks – which mostly consist of routines – with an implicit knowledge of social interconnections and the conditions for action. Practical consciousness consists of all things which actors know tacitly about how to 'go on' in the contexts of social life without being able to give them direct discursive expression. In line with conclusions which can be drawn from the concept of daily conduct of life, Giddens proposes that social research should be "sensitive to the complex skills which actors have in co-ordinating the contexts of their day-to-day behavior" (Giddens ibid., p. 285). These two approaches allow the conclusion that an empirical study should take as a point of departure everyday social practices and their contexts.

3.2.2 Life events as triggers for transition phases

Life events are ascribed different connotations by different academic disciplines, such as developmental psychology or biographical research, in which they may be viewed as triggers for crises, or processes of maturation and socialisation. A frequent distinction is made between life events that are expected, i.e. those that are part of a 'normal biography' (sexual maturity, starting school, coming of age), and those that are not expected or planned (chronic illness, unemployment) (cf. e.g. Filipp/Ferring 2002). Even life events that are the result of more or less conscious decisions, and can therefore be predicted and prepared (such as the birth of a child or relocation), represent a 'rupture' in a stable life situation. At such times, individuals may call into question their habitual self-understanding. Life events impact on everyday life in a variety of ways: apart from a small number of 'extraordinary events', numerous smaller incidences, new contexts and challenges arise, which have to be perceived and dealt with by the individual (Große 2008).

---

1 According to Schatzki (2002) and Reckwitz (2003), social practices are 'logically' organised clusters of activities (doings) and supporting matter (artefacts, technological and other media) that are spatially and temporally structured (see contribution by Kaufmann-Hayoz/Bamberg et al. in this volume, i.e. the portrayal of theories of social practices). Everyday consumer practices such as travelling to the shops, the shopping itself, the storage and preparation of food all form part of complex, tightly interconnected networks, which in turn are embedded in certain structures.

Hoerning (1987) distinguishes between 'conservative transformations' – where old behaviour patterns are reproduced – and 'evolutionary transformations' – where patterns are reassembled in a new way. Here, the individual interpretation of a situation is regarded as crucial: either the event is perceived as a beginning of a new life phase or merely as a trigger for a few changes that will not affect the 'big picture'. Hoerning (ibid.) further notes that the ease with which people adapt to a new life context depends on how well the event fits with their particular age range. If the event is 'normal', i.e. is part of a 'normal biography', institutionalised social structures tend to be in place to support individuals in their transition phase.

Thus, two important aspects can be derived from biographical research for the analysis of changes that occur within transition phases: firstly, people's own interpretation of a situation and their readiness to accept and embrace changes as part of the new situation; secondly, the degree of social embeddedness of a life event and the structures available to support and guide individuals in the new life context.

3.3   Empirical findings on life events and sustainable consumption

A number of research projects on sustainable consumption have found that individuals tend to change their usual consumption behaviour in certain life situations, and that these include potentials for sustainable consumption. Several studies have examined the impact of relocation on car ownership or car use (e.g. Bamberg 2006; Harms/Truffer 2005; Heine et al. 2001; Klöckner 2005; Prillwitz et al. 2006; Thøgerson 2009). In general, these studies provide empirical evidence about the impact of context changes on car ownership and choice of transport. For example, people who move from rural to urban areas tend to use their car significantly less often than those who move from one city to another (Bamberg 2006). What is more, compared to people who have lived in a city for some time, recent relocators are more likely to change to local public transport as a result of information campaigns or incentive schemes such as free tickets (see also Thøgersen 2009). By contrast, a study by Heine et al. (2001) has shown that life events such as the birth of the first child can lead to less sustainable transport arrangements, i.e. increased car use or car ownership. In these cases, the change tends to be linked to moving to the suburbs.

Similarly, surveys on eating habits (Brunner et al. 2006; Herde 2007; Schäfer et al. 2010) indicate that life events such as parenthood, illness or retirement offer potential for changes towards greater sustainability. However, this simultaneously depends on other factors such as socialisation, perception and awareness of one's own body and level of education. For instance, after the birth of the first child, parents often reassess their eating habits, opting for organic, fresh, seasonal and local produce (cf. Herde 2007) – primarily for health reasons. Other studies indicate that mothers do place value on a healthy diet for their offspring during the first months after birth, but often pay little attention to their own nutritional needs (Noble et al. 2007).

To date, there are no studies on the detailed processes involved in the destabilisation and reconstruction of everyday practices during major life events. The "LifeEvents" project thus fills a gap in recent research.

3.4 The "LifeEvents" project group

3.4.1 Research questions and theoretical perspective

This contribution presents the results from the qualitative investigations undertaken as part of the "LifeEvents" project group, which served to find out why and how everyday consumption patterns change during life-course transitions.[2] Whilst the emphasis was on understanding and theoretically describing the processes observed, a further intention was to outline some first assumptions as to possible interventions.

The approaches referred to above, together with the important aspects derived from research on life-course transitions, provided the theoretical framework. Consumption was understood as a component of social practices (Warde 2005). Selected life events (birth of a first child and relocation) were investigated to see what challenges they posed to everyday life, i.e. how the social network of interconnected practices was affected by them. Against this background, particular focus was placed on the ways in which the new life event impacts on everyday life, i.e. which new practices will be adopted and which old ones will change.

3.4.2 Methods and sample

Given that the qualitative sub-project emphasised a better understanding of the changes in everyday behaviour, the analysis was based on the coding system of the Grounded Theory approach (Strauss/Corbin 1988). The aim of this approach is the gradual development of theoretical assumptions and models on the basis of qualitative data. The basic data comprised problem-focused interviews (Witzel 2000) with participants who had recently had a first child, had recently moved house and who were in stable life situations. Observational reports and memos from the interview situations were used as complementary data. The interview began with a narrative sequence relating to the life

---

2 Further literature relating to the results of the entire project group can by requested from the authors and from http://www.lifeevents.de.

event itself; next, the participants were encouraged to talk about their everyday life before and after the event – in terms of their consumption practices. Attention was paid to obtaining a highly detailed picture of participants' everyday practices and their contexts. In order to better understand each individual's daily temporal and spatial organisation, the interviewees were asked to draw a map of their daily movements and whereabouts.

In accordance with Grounded Theory, the first step was to 'open code' the different phenomena thrown up by the data. Here, the focus was on consumption-related processes, contexts, actions, interactions and motives. The codes were further developed once all the cases had been considered; at this stage, the focus was on discovering interrelations between codes. In subsequent steps, the codes were consolidated and grouped into superordinate categories (axial and selective coding), and theoretical constructs capable of representing the areas under investigation were developed. Theoretical approaches relating to everyday routines, practical and biographical aspects were integrated into the analysis in the form of 'sensitising concepts'.

Between October 2008 and October 2009, 46 qualitative interviews were held. The sample comprised 23 parents, 17 relocators and 6 individuals in stable life situations. The life event in question took place two to six months before the interview of the participants. Ages ranged between 19 and 89 and the sample of people having experienced life events was considerably younger than the group in stable life situations (on average 34 versus 52 years). The majority of the interviewees were women (approx. three quarters) and individuals with academic backgrounds (approx. two thirds). Although some interviewees came from backgrounds with very low incomes or were in receipt of state benefits, most participants came from an 'above average' socio-economic background – a circumstance that needs to be taken into account when conclusions are drawn.

3.5 Results

In the following, the focus is on the key aspects of the qualitative investigations referred to in section 3.4.1.[3]

3.5.1 Changes in everyday life and consumption after life events

First of all, it should be noted that the two life events differ in respect of the extent of challenges and adjustments to everyday practices that are required. Table 1 compares the two life event groups in terms of the changes that are experienced.

---

[3] For a detailed description of the methodology and results of the qualitative investigations refer to Jaeger-Erben (2010).

The interviewees associated *the birth of the first child* with profound changes in daily routines. The changes were primarily characterised by the necessity to integrate the baby's diet, rest, safety and hygiene needs into everyday life. For parents this does not only present practical challenges, but it is a learning process in terms of identifying and dealing with the baby's needs. Knowledge, skills and new practices (e.g. in baby care) have to be acquired in order to carry out the new tasks. At the same time, parents are part of the social construction of hygiene, safety and health. With the birth of a child, social roles and allegiances also change: individuals are now 'fathers' and 'mothers', and as such they become part of a specific target group for promoters of parenting products, information materials, etc. The information they receive may, however, recommend contradictory practices, which, instead of providing guidance, may create confusion.

> *"Some say you should give cow's milk, others say 'no cow's milk whatsoever for the first year'. Some say you should take this formula or that powdered milk, and in my case – I suffer from hay fever – some say, those babies should be given hypoallergenic formula" (mother, 36 years).[i]*

Despite occasionally being given contradictory information in relation to the changes taking place, the parent group was far more homogeneous than the group of relocators. A move towards a healthier diet before the child was born was observed with most interviewees, and all interviewees stated that at least for their baby they primarily bought organic products – even if organic products had not been on their shopping list before. Furthermore, a warm, cosy and clean home became more of a priority. As a consequence, heating, washing and hot water consumption increased, which in turn raised the overall household energy expenditure. The rise in costs was accepted as a matter of course and was generally not even noticed. It was also found that parents' routines after the birth were similarly governed by new 'ordering principles'. Key concerns now seemed to be consistency in daily routines and child compatibility of everyday activities and frequented locations.

> *"I now always think in a two hour rhythm, because I don't want to find myself at Rossmann's or in the middle of Zehlendorf on a park bench when it comes to breastfeeding my baby" (mother, 34 years).[ii]*

Consumption-related infrastructures were assessed by parents in terms of their accessibility for people with young children. Many respondents relied less on the local public transport because it was inconvenient for people with prams. Some saw the car as a more comfortable and safer option, and others even bought a car for the purpose of overcoming transport problems. By contrast, some new parents avoided driving, because they were afraid of having an accident. All respondents indicated that they walked more frequently since the birth of their child and that they tried to include shopping in their daily walks. This is where local infrastructure became increasingly important when it came to their shopping needs.

Differences in attitudes within the parent group were particularly noticeable in terms of how different parents dealt with the new life event, i.e. whether they adopted a more proactive or more cautious approach. Proactive parents tended to intensively seek out information on a variety of topics; they made numerous new purchases or modified their living space to such an extent that some ended up feeling overwhelmed.

Compared to the parents, the relocators were highly heterogeneous. For this group, the deciding factor of whether their relocation was a trigger for major changes in their daily routines seemed to be the occurrence of parallel, interconnected events. Major routine changes were observed with those individuals who moved into a new multi-person household. A key challenge seemed to be adjusting their consumption requirements to the new environment. People who established a new multi-person household tended to focus on negotiating the consumption practices with the other household members (handling equipment, organising the shopping, preparation and design of meals). People who were living on their own had to find their own structures. When establishing new routines, the local infrastructure played an important role, as individuals wanted to incorporate their shopping into their daily comings-and-goings. The daily diet seemed to change most for those individuals whose local environment offered new ways of shopping (e.g. weekly market), or where the other members of the household had different consumption habits. This led some respondents to adopt healthier diets, but had the opposite effect on others:

> "Well, my eating habits have deteriorated somewhat, I go for fast food now and then, because I feel like it and because in Berlin… Curry-sausage stalls and the like…" (relocator, 28 years).[iii]

As regards mobility, good public transport, limited parking space and heavy traffic, all seemed contributory factors to a drop in car use. Local amenities were used similarly by relocators as by new parents. Even after months of moving into the new home, exploratory walks in the local area (referred to as 'Kiez' in Berlin) formed part of everyday life. Many respondents stated that since their move they tended to go out more often in order to make use of what was on offer and 'participate in street life'. The more the new environment differed from the old one, the greater was the need to adopt new practices (e.g. using local public transport).

In contrast to the group of parents, the relocators placed more value on saving energy. Long-term measures were often taken in equipping the new home: energy-efficient household appliances were bought, the lighting was optimised, power strips were added. Overall, the relocation was seen as an opportunity to render the new home more energy-efficient.

Table 1: Changes in everyday routines and consumption for each of the life events

	Birth of the first child	Relocation to Berlin
What are the challenges for everyday consumption?	Everyday consumption needs are redefined and adapted to the demands of the new situation Ensuring (healthy) diet, care and safety of the baby	Consumption needs are adapted to the new environment Negotiation between new household members of everyday consumption processes or establishment of a new daily structure
What are the main triggers for changing routines?	Fundamental change in social role and daily responsibilities Change in group identification and corresponding supply systems	Alternative supply systems/infrastructures New household situation
What are the new ordering principles in a new lifestyle?	Child-compatibility, establishment of routines, balance between comfort, reduction of effort, safety and variety	With expansion of household: communality (shared time and tasks)
What practices are reinforced and which ones are added?	New: breast feeding; preparation of baby food from scratch; going to parent-child cafes; family mealtimes; bathing the baby; buying children's clothes Reinforced: heating; cleaning; attending courses	New in multi-person households: e.g. communal meals Reinforced: eating/drinking out; outings into immediate surroundings; use of local public transport (some of these are also newly learnt)
What are the changes in terms of sustainability?	Healthier, more structured diets, more organic products Increased energy consumption Increased car use/purchase of car (sometimes); increased walking	Increased use of local public transport (change from car use) Energy-efficiency measures in equipping new home More frequent outings/more fast food

In summary, it can be concluded that the life event 'birth of the first child' is more likely to challenge 'practical consciousness' than moving to another city. In contrast to relocators, new parents focus much more on adopting new knowledge, skills and social practices. Their consumption needs are more likely to change in response to the actual life event, whereas relocators' consumption needs tend to change in reaction to different external circumstances, such as new infrastructures or new household situations. In both groups, the changes are somewhat indirect – and positive as well as negative in terms of sustainability. This is explicable by the fact that needs[4] and demands for health, safety, comfort, communality, etc. tend to change over time. An increase in the desire for sustainable consumption – where it had not existed before the life event – was noted with only a few parents, some of whom felt a greater responsibility towards a healthy and more equitable world after the birth of their child.

3.5.2 Life events as triggers for a behaviour adaptation process comprising several stages

Change in everyday routines and consumption patterns is not abrupt, but can best be understood as a continuous process. On the basis of the available data, three qualitatively different phases were identified: the preparation stage, the adaptation stage and the stage of re-establishing routines.

Prior to both life events, participants took measures that would influence their everyday life after the life event. The 'preparation stage' comprised interrelated activities at an organisational level (e.g. agreements about parental leave, search for new apartment) and a spatial-material level (e.g. rearranging the home, purchase of new appliances). Additionally, knowledge necessary for handling the new circumstances was acquired (e.g. nutrition, care of the baby, environment of the home). On that basis, parents and relocators developed ideas about how they imagined everyday life and consumption would be after the life event. The participants subsequently tried to implement these ideas in their new situation. The preparation activities are intended to engender a feeling of control over an uncertain future; as such they can be seen as an 'initiation' to the new life. Related to this process, the purchase of certain consumer goods during this stage (e.g. cot, pram) symbolises the impending transition.

In the 'adaptation stage' after the event, individuals assess how their preparations match the new life stage. The challenges of the transition now take on a more concrete shape and solutions have to be found fast. Whilst individuals considered all kinds of information materials as interesting in the preparation stage, now their focus was much more on concrete solutions for specific problems (e.g. close-by tube stations with elevators, closest weekly market). This stage is also characterised by negotiations within the household (e.g. clarification of responsibilities), and the new home or the

---

4 Needs are to be understood in this context as individual desire, or a perceived need, without further distinguishing between objective needs, subjective desires and ideas on degree and extent of need satisfaction, as discussed in the contribution by Di Giulio et al. in this volume.

new multi-person household is established by the institution of rituals (e.g. communal meals, rituals around going to bed and getting up in the morning with a child).

In the final 'stage of re-establishing routines', those that have proven practicable in terms of spatial and social arrangements and the needs and demands of individuals are further entrenched. The person now feels settled in the new environment or in the new life phase, a feeling that is confirmed above all by the existence of functioning routines. Unlike the group of relocators, the parents tend to re-assess their routines more quickly – in line with the fast-changing needs of a child in its first years of life.

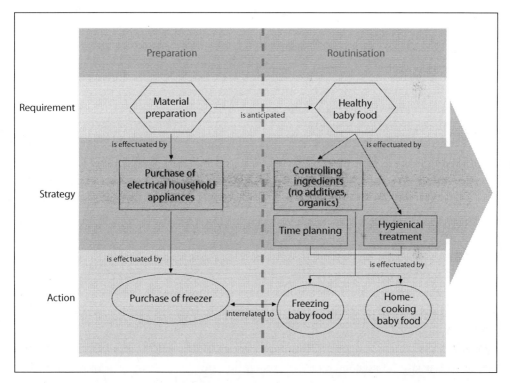

Figure 1: Prestructuring of consumption behaviour by activities in the preparation stage, illustrated by nutrition[5]

---

5   This example shows clearly how perceived models of baby feeding can influence the planning of future everyday practices and associated purchases of electrical appliances. The diagram shows how general requirements (hexagonal boxes) transform into strategies (rectangular boxes), which in turn lead to concrete actions (circles and ellipses), even in the preparation stage.

Figure 1 illustrates (with baby feeding) how preparatory activities can be influenced by anticipating everyday practices after a life event. It also shows how a change towards a more sustainable diet (home-cooking with fresh organic ingredients) can lead to a less sustainable use of energy (purchase of an additional cooling appliance). Key in this example is the intention to feed the baby home-cooked meals, which is related to the requirement of feeding the baby as healthily as possible. Many parent respondents specifically stated this intention. Parents learned in this process that a healthy diet was primarily achievable through a conscious choice of ingredients (mainly fresh organic vegetables and fruit) and by preparing them hygienically. Parenting guidelines often point out that it is not only more hygienic to freeze unused portions of baby food, but that cooking larger portions in order to freeze them can also save time. With this in mind, some respondents purchased suitable appliances during the material preparation stage (not only were freezers bought, but also blenders, microwaves and other devices).

3.6   Discussion and conclusions

The qualitative investigations have shown that, during life-course transitions, everyday routines are first destabilized and subsequently restructured in a multi-stage process. The extent of changes in everyday social practices depends firstly on the type of life event: where the birth of a first child is concerned, a great number of routines (diet, transport, leisure time, structuring of the day, task sharing by parents etc.) inevitably change, whereas relocations may only necessitate a small number of changes in everyday routines. Evolutionary transformations were mostly observed in the parent groups. Proactive parents underwent even greater changes, as they had acquired and integrated into their daily lives a wide range of new skills and knowledge prior to the birth (e.g. a healthier diet, more home-cooking). The extent to which relocators changed their behaviour tended to depend more on the extent to which their spatial and social situation had changed. The greater the contrast to the previous life situation, the greater the need for renegotiating old and acquiring new practices (e.g. if the previous location did not offer good local public transport).

With regard to sustainability, it is important to note that in both groups, sustainable and non-sustainable consumption patterns arose. Of particular interest are the interactions between the different categories of consumption. Hence, areas such as food and energy, for instance, should not be investigated in isolation – neither in academic research projects, nor in the actual development of strategies for changing everyday life.

The present study has shown how and why changes can arise prior to the event and how these are handled. The de-stabilization of consumption routines already starts in the preparation stage. The contributory factors at this stage (especially actors who provide information) deserve particular attention. During the stage of re-establishing routines, the knowledge acquired, the organisational activities and the spatial-material preparations that took place in the preparation stage provide guidance. These preparations do, however, also create their own realities. For instance, an appliance or a car that has been bought almost 'obliges' the owner to use it after the event – even if it turns out to be less practical than previously assumed.

The following aspects have been shown to be central when assessing life events as starting points for increased sustainable consumption (and interventions designed to promote it):

Recommendations or interventions designed to promote sustainability should take into consideration the emerging new needs and challenges that lead to changes in consumer practices in the wake of life events. For instance, parents may use more heating, do more washing and possibly use the car more frequently after a birth. From their perspective, these are meaningful and necessary changes in their everyday practices.

Furthermore, the type of life event determines which social structures and supply systems will be of significance (e.g. as a consequence of changes in social roles, locations or time availability). Supply systems, social discourse and assignment processes within society all play a part when it comes to which new social practices are adopted, which existing ones are reinforced and which old practices are abandoned. The analysis of changing everyday consumption patterns should not only take these factors into account, but should look at the norms and linked social practices at the level of discourse, together with the implications for sustainable consumption (e.g. How does a caring mother behave? How do joint households function?).

Sustainability interventions (e.g. consumer campaigns) should also be sensitive to the chronological sequences within life-course transitions. Thus, new parents and relocators should be addressed prior to the event, as it is in that period that the ground is prepared for new subsequent consumption practices. Given that a range of issues and challenges arise in the various stages of life-course transitions, an accompaniment throughout the adaptation process seems a promising strategy. In view of the fact that target groups are not easily identifiable before a life event (particularly in the case of parents), and that information and support are often at hand, it seems sensible to opt for combining existing services and collaborating with the entities behind these services (e.g. accommodation portals, online classified directories, health insurance services, course providers, parenting websites).

Life events provide opportunities for changes towards both more or less sustainable consumption patterns. Interventions designed to increase sustainability can therefore opt for a strategy that picks up on and reinforces existing positive tendencies, or a strategy that counteracts negative changes. For example, parents can be supported in their efforts to provide healthy and sustainable baby food (reinforcing positive trends) – or the needs for a child-compatible transport and warm, clean homes can be addressed in such a way that negative sustainability effects (e.g. higher energy consumption) can be avoided (counteracting negative trends).

Endnotes

i   "Manche sagen, man soll Kuhmilch nehmen, manche sagen 'Im ersten Jahr auf keinen Fall Kuhmilch'. Manche sagen, man soll diese Baby-, diese Pulvermilch nehmen und dann – da ich Heuschnupfen hab' – sagen auch manche, solche Babys sollen die hypoallergene Pulvermilch nehmen" (Mutter, 36 Jahre).

ii "Also denk ich ja immer im Zwei-Stunden-Rhythmus, weil ich auch nicht, wenn ich jetzt grade bei Rossmann bin, ihn dann stillen möchte oder in Zehlendorf Mitte auf der Parkbank" (Mutter, 34 Jahre).

iii "Also ich bin wieder ein bisschen schlechter geworden mit Essen, also auch schon mal so Fast-Food-Sachen, weil ich Lust drauf hatte, und weil Berlin… Curry-Wurst-Buden und so…" (Umzüglerin, 28 Jahre).

References

Bamberg S. (2006): Is a Residential Relocation a Good Opportunity to Change People's Travel Behavior? Results from a Theory-Driven Intervention Study. In: Environment & Behavior 38: 820–840.

Brand K.-W. (2006): Konsum im Kontext. Der "verantwortliche Konsument" – ein Motor nachhaltigen Konsums? In: Lange H. (ed.): Nachhaltigkeit als radikaler Wandel: Die Quadratur des Kreises? Wiesbaden: VS Verlag für Sozialwissenschaften. 71–93.

Brand K.-W. (2010): Social Practices and Sustainable Consumption: Benefits and Limitations of a New Theoretical Approach. In: Gross M., Heinrich H. (eds.): Environmental Sociology: European Perspectives and Interdisciplinary Challenges. Dordrecht: Springer. 217–235.

Brunner K.-M., Kropp C., Sehrer W. (2006): Wege zu nachhaltigen Ernährungsmustern: Zur Bedeutung von biographischen Umbruchsituationen und Lebensmittelskandalen für den Bio-Konsum. In: Brand K.-W. (ed.): Von der Agrarwende zur Konsumwende? Ergebnisband 1. München: oekom. 145–196.

Di Giulio A., Brohmann B., Clausen J., Defila R., Fuchs D., Kaufmann-Hayoz R., Koch A. (in this volume): Needs and consumption – a conceptual system and its meaning in the context of sustainability.

European Environment Agency (EEA) (2006): Household consumption and the environment, EEA Report No 11/2005. [http://www.eea.europa.eu/publications/eea_report_2005_11, 22.06.2012]

Filipp S.-H., Ferring D. (2002): Die Transformation des Selbst in der Auseinandersetzung mit kritischen Lebensereignissen. In: Jüttemann G., Thomae H. (eds.): Persönlichkeit und Entwicklung. Weinheim: Beltz. 191–228.

Giddens A. (1984): The Constitution of Society. Outline of the Theory of Structuration. Cambridge: Polity Press.

Große S. (2008): Lebensbrüche als Chance? Lern- und Bildungsprozesse im Umgang mit kritischen Lebensereignissen – Eine biographieanalytische Studie. Münster: Waxmann.

Harms S., Truffer B. (2005): Vom Auto zum Car Sharing: Wie Kontextänderungen zu radikalen Verhaltensänderungen beitragen. In: Umweltpsychologie 9 (1): 4–27.

Heine H., Mautz R., Rosenbaum W. (2001): Mobilität im Alltag: Warum wir nicht vom Auto lassen. Frankfurt a. M., New York: Campus.

Herde A. (2007): Nachhaltige Ernährung im Übergang zur Elternschaft. Berlin: Mensch & Buch.

Herkommer S., Bischoff J., Maldaner K. (1984): Alltag, Bewusstsein, Klassen. Hamburg: VSA.

Hitzler R., Honer A. (1991): Qualitative Verfahren zur Lebensweltanalyse. In: Flick U., von Kardorff E., Keupp H., von Rosenstiel L., Wolff S. (eds.): Handbuch Qualitative Sozialforschung. 382–385.

Hoerning E. (1987): Lebensereignisse. In: Voges W. (ed.): Methoden der Biographie- und Lebenslaufforschung. Opladen: Leske und Budrich. 231–259.

Jaeger-Erben M. (2010): Zwischen Routine, Reflektion und Transformation. Die Veränderung von alltäglichem Konsum durch Lebensereignisse und die Rolle von Nachhaltigkeit – eine empirische Untersuchung unter Berücksichtigung praxistheoretischer Konzepte. Dissertationsschrift. Berlin: Technische Universität Berlin. [http://opus.kobv.de/tuberlin/volltexte/2010/2816/pdf/jaegererben_melanie.pdf, 22.06.2012]

Jurczyk K., Rerrich M. S. (eds.) (1993): Die Arbeit des Alltags. Beiträge zu einer Soziologie der alltäglichen Lebensführung. Freiburg: Lambertus.

Kaufmann-Hayoz R., Bamberg S., Defila R., Dehmel C., Di Giulio A., Jaeger-Erben M., Matthies E., Sunderer G., Zundel S. (in this volume): Theoretical perspectives on consumer behaviour – attempt at establishing an order to the theories.

Klöckner C. (2005): Können wichtige Lebensereignisse die gewohnheitsmäßige Nutzung von Verkehrsmitteln verändern? – Eine retrospektive Analyse. Umweltpsychologie 9 (1): 28–45.

Lefèbvre H. (1975): Kritik des Alltagslebens. München: Hanser.

Noble G., Stead M., Jones S., Mc Dermott L., McVie D. (2007): The paradoxical food buying behaviour of parents: insights from the UK and Australia. In: British Food Journal 109 (5): 387–398.

OECD (2002): Towards Sustainable Household Consumption? Trends and Policies in OECD Countries. [http://www.oecd.org/dataoecd/28/49/1938984.pdf, 22.06.2012]

Prillwitz J., Harms S., Lanzendorf M. (2006): Impact of life course events on car ownership. In: Transportation Research Record 1985: 71–77.

Schäfer M., Herde A., Kropp C. (2010): Life events as turning points for sustainable nutrition. In: Tischner U., Stø E., Kjærnes U., Tukker U. (eds.): System Innovation for Sustainability 3: Case Studies in Sustainable Consumption and Production – Food and Agriculture. Sheffield: Greenleaf Publishing. 210–226.

Schütz A., Luckmann T. (1979): Strukturen der Lebenswelt (Band 1). Frankfurt a. M.: Suhrkamp.

Strauss A., Corbin J. (1996): Grounded Theory – Grundlagen qualitativer Sozialforschung. Weinheim: Beltz.

Thøgersen J. (2009): Seize the opportunity: The importance of timing for breaking commuters' car driving habits. In: Klein V., Thoresen V.W. (eds.): Making a Difference: Putting Consumer Citizenship into Action. Hedmark: Høgskolen i Hedmark. 87–93.

Voß G. G. (2000): Alltag. Annäherungen an eine diffuse Kategorie. In: Voß G., Holly W., Boehnke K. (eds.): Neue Medien im Alltag. Begriffsbestimmungen eines interdisziplinären Forschungsfeldes. Opladen: Leske + Budrich. 31–77.

Voß G. G. (1991): Lebensführung als Arbeit. Über die Autonomie der Person im Alltag der Gesellschaft. Stuttgart: Enke.

Warde A. (2005): Consumption and Theories of Practice. Journal of Consumer Culture 5 (2): 131–153.

Witzel A. (2000): Das problemzentrierte Interview. Forum Qualitative Sozialforschung/Forum: In: Qualitative Social Research 1(1): Art. 22. [http://www.qualitative-research.net/index.php/fqs/article/viewArticle/1132/2519; 30.07.2010]

Ellen Matthies, Dirk Thomas

4 Sustainability-related routines in the workplace – prerequisites for successful change

If consumption is understood in the general sense of making use of goods (cf. Kaufmann-Hayoz/Bamberg et al. in this volume), it follows that the workplace too is an environment where consumption in general, and sustainable consumption in particular, takes place. The use of energy plays a particular role in the workplace, because in industrialised nations almost every workplace is equipped with energy consuming appliances. This article focuses on energy-consuming, recurring behaviour patterns in office environments, such as regulating the heating, airing the rooms, and the use of computers and peripherals (printers, monitors). It is estimated that a shift in consumer behaviour holds an energy-reduction potential of between 5–15 percent of overall consumption – and this with no loss of comfort (e.g. NRW Energy Agency, 2007; Energy Office 2007).

If the potentials for reducing energy in offices are so high, why are people so reluctant to act? One possible reason, which was investigated as part of the "Change" project, is the fact that it is difficult to assess the potential in the real world, where a variety of factors come into play (particularly in the context of heating energy; cf. Klesse et al. 2010). But even if the potential for reduction is almost exclusively dependent on user behaviour, e.g. on the use of computers in offices, there still remains the question of which steering mechanism can best achieve changes in staff behaviour.

A primary aim of the "Change" project group (cf. profiles of all project groups in the appendix of this volume) was to examine in more detail how changes in user behaviour in the office environment can potentially conserve energy. As a next step, on the basis of psychological findings in relation to habit formation, a range of possible measures were compared and evaluated. Given the availability of reliable estimates for potential savings (Kattenstein et al. 2000) and evaluations of measures designed to change user behaviour (Staats et al. 2000) in universities, it was decided to target these institutions for the intervention study. The following article presents the major results of the intervention study, focusing in particular on habit formation in the context of energy-related behaviour. As a further step, the set of strategies that was developed for promoting energy efficiency in universities was examined (from a perspective of organisational sociology), in terms of the extent to which they could be transferred to other organisations with similar consumption patterns.

4.1 Energy consumption behaviour in office environments and their habitualisation

On the basis of workshops with experts and literature reviews, the following areas of high potentials for energy savings in university office environments were identified: switching off computers and peripherals; correct ventilation (i.e. short bursts of airing instead of tilting windows permanently); setting the heating at a lower temperature (Klesse et al. 2010; Schahn 2007). In terms of the structure of the phenomena of consumer behaviour described in this volume (Kaufmann-Hayoz/Bamberg et al.), these measures can be termed as non-essential consumer acts (non-essential in this context being understood as interchangeable, i.e. energy consumption in the workplace does not have any significance beyond its functional use). At the same time, these behaviours are characterised by a high degree of freedom: when leaving the office at the end of the day, staff members can switch off – or leave on – the various devices. Changes in activities with large degrees of freedom can be adequately described in terms of norm activation theory. Starting from the premise that people are generally capable of altruistic behaviour, this theory can be deployed as a way of explaining sustainable consumption (Schwartz 1977, Schwartz/Howard 1981, Stern 2000). An account of this kind of theoretical approach can also be found in Kaufmann-Hayoz/Bamberg et al. in this volume. According to norm activation theory, an increased awareness of the problem, together with knowledge about alternative courses of action, should trigger a person's sense of commitment to more sustainable behaviour. A prerequisite for behavioural change is, however, that sustainability, or related issues such as climate change, matter to the individual in question. It follows that those interventions intended to raise awareness of the problem and provide knowledge about practical action would be suitable and potentially effective. In accordance with the terminology developed in this volume, these interventions can be termed *motivational communicative instruments* (Kaufmann-Hayoz/Brohmann et al. in this volume). Based on more recent reflections on the process character of voluntary behaviour change (ibid.), a distinction can be made between those communicative tools that are primarily employed in the motivation phase (*motivational communicative instruments*), and those employed in the planning and execution phase of the behaviour (*supportive communicative instruments*). The behaviours considered in the "Change" project are mostly routines that are repeated day after day. They would fall under the heading of potentially habitualised, non-reflected consumer actions. It is the routinisation of behaviours that, from a psychological perspective, represents a strong barrier to intentional behaviour change. In an empirical survey, Ouellette and Wood (1998) demonstrated that, for frequently repeated routines, there is a close link between past behaviour (habits) and future behaviour, and intentions do not seem to have a great impact. By contrast, those actions that are rarely performed are more likely to be influenced by intentions. According to Ouellette and Wood (1998), habits form when an action is performed regularly and in the same context. This applies to energy user behaviour in office environments (ventilation, computer use), given that these behaviours are usually performed several times a day.

To date, the concept of habitualisation has been used mainly to explain people's resistance to switching their means of transport. Verplanken and Aarts (1999) view the habitual use of a means of transport as scripts that are triggered automatically by certain action targets (e.g. "shopping" triggers "use of car"). Aarts and Dijksterhuis (2000) demonstrated in a series of experiments that subjects with highly developed habits found it difficult to give them up once they were established. In an early study on sustainable purchasing behaviour Dahlstrand and Biel (1997) speak of "frozen behaviour", which is particularly resistant to change because new behavioural intentions compete with old habits.

With this in mind, intervention studies designed to change strongly habitualised behaviour (Fujii/Gärling 2003, 2007; Fujii et al. 2001; Matthies et al. 2006) employ techniques that support new behavioural intentions. Examples include temporary, radical changes in particular contexts (e.g. temporary free tickets), encouraging commitment (provision of feedback cards) and strengthening new intentions by picturing the action in its situational context (Bamberg 2002).

4.2 Psychological knowledge about interventions designed to save energy in organisations

Despite Paul Stern's suggestions as early as 1992 that energy consumption behaviour and organisational decision-making should be made the object of psychological research, only a handful of empirical studies (cf. Wortmann 2004) have so far been conducted. Tönjes (2009) reviewed 15 published studies on energy-related interventions in organisations; twelve of these had been conducted within university settings. Reported consumption reductions amounted to 6 percent or more (Griesel, 2004; McClelland/Cook 1980; Schahn 2007; Staats et al. 2000). Studies conducted in the office context almost exclusively made use of communicative instruments, and in particular those designed to enhance motivation, such as problem-focused information campaigns (Siero/Bakker et al. 1996). Some intervention studies aimed to support behavioural changes. Staff members were, for instance, encouraged to commit to certain behavioural changes (Zinn 2002; Griesel 2004), while several studies incorporated feedback (Siero/Bakker et al. 1996; Zinn 2002; Griesel 2004). Feedback can be seen as a motivational instrument, but when it is provided while action is ongoing, it can also serve as supportive instrument because the user's attention is directed towards current behaviour as well as new intentions. Long-term effects (six months or longer) were predominantly noted in the case of combined supportive techniques, such as the provision of information on current behaviour plus feedback (Staats et al. 2000), or provision of information on current behaviour, plus feedback, plus an indication of target behaviours (Siero/Bakker et al. 1996; Siero/Boon et al. 1989). These findings indicate that energy consumption in the workplace is habitualised; changing such behaviour requires strategies that support the user in his or her behavioural changes. The office setting provides ample opportunity for implementing such strategies. A change of context can, for instance, be created by attaching labels to appliances, windows, etc. (prompts/reminders, which the users themselves can apply). An additional resource available to organisations is the exploitation of social interaction and communication. Not only can information days, competitions, group goals and commitments activate positive social norms and individual intentions (i.e. act as motivational instruments), but, by changing the context for behaviour, they can also support people as they change their behaviour.

4.3 Evaluation of the intervention programme "Change" – aimed at changing habitual behaviour within universities

In a separate study, the potentials of two intervention packages – tailored to office settings within universities – were examined. Both focused on identical behaviours (airing in bursts; switching computers off during breaks and disconnecting them from the mains at the end of the day; lowering the heating) and were implemented in comparable settings in terms of the structure of buildings and possibilities for changing behaviour. The first intervention package comprised *motivational* techniques only; the second contained additional strategies *designed to support sustainable behaviour*. The design of the motivational communicative instruments was underpinned by the norm activation model: brochures, posters and flyers were used to address people's concern for climate change and suggest ways of contributing to the university's financial savings. (For a detailed account of the development of intervention materials on the basis of the norm activation model, see Matthies/Hansmeier 2010.) The intervention package contained the following items: posters and flyers illustrating the recommended behaviour and explaining the consequences of energy-saving actions (potential for financial and CO_2-savings for the university); a website detailing the energy consumption of each university, addressing the CO_2 implications and recommending a range of behaviours; e-mailings outlining the problems and linking to the website; a personalised information package with a covering letter from the head of the university; a booklet containing energy saving tips. These largely motivational instruments were targeted at two intervention groups. One group was targeted with additional *communications designed to focus people's attention and to support them in their actions* (i.e. change their habits). These supportive instruments included: prompts (stickers as reminders for airing in bursts and switching off electronic equipment); feedback cards encouraging voluntary commitment; thermometers (for instant feedback on room temperature); vouchers for switchable multiple socket power strips; an information day organised to encourage staff to talk to each other about the interventions.

The overall evaluation included three groups in a total of 15 buildings: two intervention groups, (concurrently) receiving either the standard communication materials (standard motivational only: SIG), or the extended variant (additionally focusing on habits: HIG); a third group receiving no intervention (control group). Three forms of pre-intervention and post-intervention values were recorded: (1) energy consumption for heating/electricity (per building), (2) observed ventilation behaviour (per building), (3) self reported behavioural change – plus a number of further psychological features (on level of the individual staff members). An integrated analysis of all three forms of values can be found in Matthies et al. (2011). The present article focuses on changes in habitualisation. The data for these were gathered on individual level. In order to measure the extent of habitualisation, a version of the *self report index of habit strength* (SRHI, cf. Verplanken/Orbell, 2003) was adapted to energy user behaviour and related to four routine behaviours in the office environment (see Table 1). The results of a pre-intervention/post-intervention comparison of the two intervention groups confirmed the initial hypothesis that the two interventions (standard vs. extended habit focused) would result in different outcomes. Self-reported changes in behaviour towards the desired outcome were considerably higher in the habit intervention group (HIG) than in the standard intervention group (SIG) (see Table 1) – with the sole exception of switching the computers off at the end of the day or at weekends, where no significant changes were recorded for the habit intervention group. It is worth noting that this behaviour exhibited very high initial values, which suggests possible ceiling effects.

Table 1: Overview of changes in habitualisation of selected behaviours in both intervention groups

	Group	n	M$_{Pre}$	M$_{Post}$	Δ (M$_{Post}$ – M$_{Pre}$)	F
Switching off PC	HIG	33	6.03	6.33	0.30	1.47
at the weekend	SIG	26	6.57	6.11	−0.46	3.75
Switching off PC during	HIG	33	2.81	3.64	0.83	5.12*
extended absence	SIG	26	3.24	3.49	0.25	0.57
Frequency of airing with	HIG	34	3.61	2.62	−0.98	12.02**
permanently tilted windows	SIG	27	3.52	3.01	−0.51	1.16
Frequency of airing	HIG	32	3.80	4.82	1.02	7.12*
in bursts	SIG	26	4.58	4.62	0.04	0.01

HIG = habit intervention group; SIG = standard intervention group

Pre = pre-test (October 2008); Post = post-test (January 2009)

Δ reflects the changes in self-reported habitualisation. A seven-point Likert scale was used. For pre/post test differences, a positive value indicates a change as intended with the intervention, with the exception of airing with permanently tilted windows, where a negative value (i.e. less airing with permanently tilted windows) represents the desired effect.

* $p < .05$ ** $p < .01$

The assumption that supportive communicative instruments could break habits was thus confirmed. Furthermore, the results at both observational and consumption level (cf. Matthies et al. 2011) showed that combining motivational and supportive instruments can lead to considerable reductions in energy consumption. Compared to the overall heating and electricity consumption in each building, the electricity savings and heat energy reductions respectively amounted to between 0.9 and 6.6 percent and between 0.7 and 5.4 percent (the values vary according to reference measure). These results indicate that changes in consumption behaviour may have high energy saving potentials.

4.4 How can "Change" contribute to altering routine behaviours in organisations?

In order to assess the feasibility of transferring the intervention strategies used by the "Change" campaign to other organisations, the psychology-based intervention study had to be complemented with a study based on principles of organisational sociology. Many of the conditions present in universities and used in the "Change" campaign may not exist in other organisations. Thus, an organisational study was conducted in parallel to the interventions. The intention was to investigate the initial situation in different types of organisations in order to assess the transferability of "Change". The focus was twofold: (1) the contexts within the institutions (organisational analysis) plus energy-related user behaviour; (2) readiness for action plus extent of motivation on the part of both decision makers and ordinary staff (actor analysis).

4.4.1 Organisational structures according to Mintzberg

Organisations have been the object of sociological research ever since Max Weber (1922/1972) analysed them as permanent apparatuses of domination. Even so, no generally accepted definition of what exactly constitutes an organisation has yet been formulated in the literature (Abraham/Büschges 2009; Endruweit 2004). Allmendinger and Hinz's (2002, p. 10) definition seems particularly appropriate: "An organization is a collective or corporate social system that is intended to primarily solve coordination and cooperation problems [...]. Organisations are secondary actors who bring together resources provided by the primary actors in order to pursue specific purposes." (translated by C. Holzherr).

In order to identify the organisation type that might be suitable for the interventions of the "Change" project (government agencies, financial institutions, non-university research institutions), Mintzberg's (1992) organisational typology was drawn on. It is based on what Mintzberg termed organisational configurations (ibid., p. 35): public authorities and financial services providers belong to what he terms the machine bureaucracy (consisting of public machine bureaucracy and what could be considered employee bureaucracy); universities belong to the professional bureaucracy and other research institutions belong to the adhocracy.

Machine bureaucracies are old, large and hierarchically structured. Decision processes take place according to strictly hierarchical principles (top down), participation is given low priority in decision-making, and informal channels of communication are, whenever possible, discouraged. A strong controlling mentality and a high degree of behavioural formalisation are characteristic of machine bureaucracies. Whilst innovations are generally disapproved of, the strength of the machine bureaucracy is its standardisation of work processes. Since the employees in public machine bureaucracies and employee bureaucracies remain largely anonymous and working conditions are standardised, employees tend to be passive. Personal identification with the organisation is average.

Professional bureaucracies vary in age and size. Formalisation of behaviour is minimal. A prominent feature of the professional bureaucracy is the autonomy of employees at the core of the organisation. Compared to other types of organisation, the management's power to act is very limited. In professional bureaucracies, strategic initiatives are usually started by the professional employees. Decision processes at the core of the organisation tend to run bottom up, which means that participation plays a central role. The professional bureaucracy is highly decentralized, both in a vertical and a horizontal direction. Informal channels of communication are especially important in carrying out administrative functions. Innovation in this organisation type requires a high degree of cooperation. According to Mintzberg, it is rare for employees to feel very loyal towards the organisation. Direct supervision and mutual adjustment are seen as an interference in personal autonomy and are therefore rejected.

Adhocracies are, typically, recently established small organisations. The structure in an adhocracy is such that problem solving and innovation are highly valued; formalisation of behaviour is minimal. Communication is informal and decision-making takes place at all levels. In an attempt to avoid bureaucracy, information flow and decision processes are handled flexibly and informally. Adhocracies value staff participation, and the power of the management is limited. Adhocracies are thus highly democratic organisations.

4.4.2 Necessary preconditions in organisations for the implementation of "Change"

Studies have shown that an initial willingness to implement energy saving measures can be fostered by a corporate culture that embraces innovation (Prose et al. 1999). Following Mintzberg's typology, it can be assumed that, given their conservative attitude towards innovation, machine bureaucracies and employee bureaucracies would be unwilling to implement energy-saving interventions. According to Mintzberg, the types of organisations also differ in a number of other characteristics which are particularly relevant in the context of energy-saving interventions, namely decision-making and participation procedures. In universities, for instance, it is important to ensure well in advance of implementing any strategies that various decentralised sub-systems are in a position to participate and lend their full support to the impending changes. By contrast, in the public machine bureaucracy and the employee bureaucracy decision-making is hierarchically determined.

Planning, implementation and monitoring of energy-saving interventions additionally requires both financial and human resources. Particularly in public machine bureaucracies, where the vertically and horizontally specialized work tasks are narrowly defined, a lack of human resources could prove an obstacle to the implementation of energy-saving measures. When proposing an intervention programme to an organisation, it is crucial that the actors in question are given the opportunity to assess the effectiveness and suitability of the proposed tools. Given the differences in the preconditions across organisations, it should be anticipated that these assessments may differ from one organisation to another.

In order to underpin the theoretical analysis, expert interviews were conducted with a number of actors across twelve organisations. The intention was to explore the various aspects of the "Change" intervention programme and to assess the extent to which the interventions could be adopted in organisations other than universities.

4.5 Empirical analysis of organisations in relation to the transferability of "Change"

The following organisations were selected: German federal and regional authorities, urban public buildings (public machine bureaucracies), financial services providers (employee bureaucracies) and non-university research institutions (adhocracies). To gather expert opinion, three groups of actors (or their chosen spokespersons) within each organisation were questioned by means of oral, semi-standardised interviews: (1) managers (e.g. office managers, branch managers, heads of specialised services, company managers, directors of institutions; (2) heads of department (e.g. section managers, area managers, team leaders, group leaders); (3) technical services managers. A tailored questionnaire was drawn up for each group.

The intention was to determine the sociographic characteristics of the organisation (obtained from the managers), the technological, structural and personnel-related prerequisites for a successful implementation of "Change" (obtained from the various managers and the technical services managers), and the context-specific conditions within each organisation (obtained from the different managers, technical services managers and heads of department). Additionally, interviewees were asked to give their opinion as to the effectiveness and suitability of the intervention tools implemented by "Change". Finally, all three groups were asked to state the level of interest of their organisation in participating in a campaign of a tried-and-tested kind, designed to change user behaviour.

4.5.1 Specific contextual circumstances, technical requirements and organisational framework of "Change"

The expert interview revealed that all recommended behaviours, with the exception of installing a power management system for the computer, could be practically implemented by the majority of staff members across organisations. A tried-and-tested energy-saving campaign would meet with great interest from the public authorities and the research institutes. It was noted that the public authorities and the larger financial services had had previous experience with interventions designed to change staff behaviour. The costs involved in implementing an intervention programme represent

Table 2: Contextual circumstances, technological prerequisites and organisational preconditions for successful implementation of "Change" interventions (from the viewpoint of interviewees)

Organisation types according to Mintzberg	Public machine bureaucracy	Employee bureaucracy		Operating adhocracies
Forms of organisations	Federal/regional authorities and urban public buildings (N=6)	Small financial services providers (N=1)	Large financial services providers (N=2)	Non-university research institutions (N=3)
Size (staff members)	50 to 250	Fewer than 50	250 to 1,000	100 to 400
Extent of interest in a tried-and-tested campaign	Fairly high* 6/6	Fairly low* 0/1	Approx. medium* 1/2	Fairly high* 2/3
Experiences relating to measures for changing behaviour	Tended to have such experiences* 5/6	Tended not to have such experiences* 0/1	Tended to have such experiences* 2/2	Tended not to have such experiences* 1/3
Budget	Barrier: central cost absorption 6/6	Barrier: central cost absorption 1/1	Potential: global budget 2/2	Potential: global budget 3/3
Human resources for monitoring	Tended to be restricted* 3/6	Tended not to be available* 0/1	Tended to be available* 2/2	Tended to be available* 3/3
Monitoring of monthly energy and heat consumption data	Tended not to be feasible* 2/6	Tended not to be feasible* 0/1	Tended to be feasible* 2/2 (electricity) 1/2 (heating)	Tended to be feasible* 2/3

Note: *Estimates are based on statements by interviewees.

an obstacle for the public authorities and the small financial institutions, particularly in view of the fact that the savings do not remain in-house, whilst financial outlay is no barrier for the larger financial services providers and research institutions. These organisations also tend to have at their disposal greater human resources that can help implement such a campaign, making it more feasible for them to monitor, for instance, the monthly energy consumption data (see Table 2).

4.5.2 Assessment of motivational and supportive communicative instruments

Both the managers and the heads of department were asked to comment on how suitable and potentially successful the interventions selected by "Change" (both motivational and supportive communicative instruments) were in terms of promoting energy-efficient behaviour in their respective working environments (see Tables 3 and 4).

Table 3: Assessment of efficacy of supportive communicative instruments

Campaign elements	Federal and regional authorities	Urban public buildings	Financial services providers	Research institutions
Prompts with energy saving tips	Not suitable	Suitable	Suitable	Not suitable
Commitment to adhere to certain behaviours	Fairly weak	Fairly weak	Fairly strong	Average
Price draw/power strip vouchers in exchange for filling in and returning commitment sheet	Average	Fairly weak	Average	Fairly weak
Promotional stand at start of campaign: information on energy situation of organisation, plus advice	Neither suitable nor unsuitable	Neither suitable nor unsuitable*	Suitable	Suitable

Note: * Not all interviewees commented on the point in question.

Index (0–12): 0–4 = fairly weak/unsuitable, 5–7 = average/neither suitable nor unsuitable, 8–12 = fairly strong/suitable. The indices are derived from the three organisations that were interviewed for each organisation type.

Regarding the supportive instruments (see Table 3), prompts (reminders) with energy-saving tips were considered useful only by managers of urban public buildings and financial services providers; they were largely dismissed by research institutions, and to an even greater extent by regional and federal authorities. The chances of success of voluntary staff commitment to adhere to certain procedures (e.g. airing in bursts) were considered fairly high by financial services providers, average by research institutions, and fairly low by all other organisations. The incentive of winning an attractive prize in a lottery draw, or receiving a voucher for a multiple socket power strip in exchange for filling in and sending back a commitment sheet, was viewed as moderately successful by federal and regional authorities and financial services providers, and as fairly low by the other organisations. The provision of a promotional stand at the beginning of the campaign was considered by research institutions and financial services providers as a suitable means of drawing attention to energy-efficient user behaviour in the workplace.

Regarding motivational instruments (see Table 4), posters in corridors were not considered suitable by any of the organisations. By contrast, flyers were considered suitable by the financial services providers and the managers of urban public buildings. E-mailings containing energy efficiency tips were considered useful only by the financial services providers. Creation of internal media (e.g. a specific homepage) with the purpose of publicising the campaign and providing interim reports was viewed

Table 4: Assessment of suitability of motivational communicative instruments

Campaign elements	Federal and regional authorities	Urban public buildings	Financial services providers	Research institutions
Posters with energy-saving tips in corridors	Neither suitable nor unsuitable	Neither suitable nor unsuitable	Neither suitable nor unsuitable	Not suitable
Flyer with energy-saving tips	Neither suitable nor unsuitable	Suitable	Suitable	Neither suitable nor unsuitable
E-mails with energy-saving tips	Neither suitable nor unsuitable	Neither suitable nor unsuitable	Suitable	Neither suitable nor unsuitable
Public awareness campaign/use of internal media (e.g. home page)	Suitable	Suitable	Suitable	Suitable
Provision of information material on energy saving and recommended behaviour for users in the workplace	Fairly high	Average	Fairly low	Average
Information package, personally addressed to each staff member (information brochure, thermometer)	Neither suitable nor unsuitable*	Suitable	Suitable	Suitable

Note: * Not all interviewees commented on the point in question.
Index (0–12): 0–4 = fairly weak/unsuitable, 5–7 = average/neither suitable nor unsuitable, 8–12 = fairly strong/suitable. The indices are derived from the three organisations that were interviewed for each organisation type.

as positive by all organisations. The provision of information on energy saving and recommended behaviour in the workplace was considered fairly useful by the federal and regional authorities only. A personally addressed information package, containing a brochure with energy saving tips and a thermometer distributed to all staff members, was considered useful mainly by managers of urban public buildings, the financial services providers and the research institutions.

4.6 Conclusions

The intervention package "Change" was developed for the promotion of sustainable behavioural routines in office environments. When implemented in universities, interventions concerned with electricity usage achieved savings of up to 6.6 percent of energy expenditure, while interventions concerned with heating and ventilation practices achieved savings of up to 5.9 percent – results that were considered satisfactory. Beyond that, an intervention study confirmed the assumption that *motivational strategies alone were not sufficient* for changing behavioural routines. Changes in self-reported habitualisation were particularly noticeable as a consequence of implementing the complete intervention package, which, in addition to motivational materials, also contained *interventions that supported users in modifying their everyday behaviour* (supportive communicative instruments: prompts, power strips, self-commitment).

However, the empirical analysis conducted in line with organisational sociology principles revealed that those supportive instruments (prompts, self-commitment) were regarded sceptically by respondents in the actor analysis. This could have a number of explanations. For instance, in certain organisations, some of the communicative instruments seem incompatible with the hierarchical communication strategies still persisting in those organisation types. Another explanation is the probable lack of knowledge as to the effectiveness of the interventions. In these instances, the managers of the organisations were informed of the proven effectiveness of the supportive instruments. Given that the effectiveness and suitability of the communicative instruments is paramount to a successful implementation of the "Change" programme, it seems particularly important to test them out in a range of organisation types, where potential false suppositions and incompatibilities can be overcome.

The empirical analysis further revealed that – alongside the practical issues involved in following the recommended behaviours – technological, human and financial resources were seen as paramount to a successful implementation of the "Change" strategies, and that these resources can vary considerably from one organisation to another. The expert interviews revealed that in all organisation types the behavioural changes (with the exception of installing a power management system for the computer) could realistically be adopted by the majority of staff and would therefore be good starting points for changing user behaviour in the workplace. By contrast, the technological requirements for evaluating the interventions and the availability of human and capital resources were regarded as more of a challenge, particularly in the case of the public authorities. Yet, the lack of technological, human and capital resources did not impact negatively on the reported interest of the organisations. Although – and possibly because – the financial means of public authorities and urban public buildings tend to be limited, it is these entities that seem to be particularly interested in a standardised, tried-and-tested energy-saving campaign such as "Change".

References

Aarts H., Dijksterhuis A. (2000): The Automatic Activation of Goal-Directed Behaviour: The Case of Travel Habit. In: Journal of Environmental Psychology 20: 75–82.

Abraham M., Büschges G. (2009): Einführung in die Organisationssoziologie (4. Auflage). Wiesbaden: VS Verlag für Sozialwissenschaften.

Allmendinger J., Hinz T. (2002): Perspektiven der Organisationssoziologie. In: Allmendinger J., Hinz T. (eds.): Organisationssoziologie. Sonderheft 42 der Kölner Zeitschrift für Soziologie und Sozialpsychologie: 9–28.

Bamberg S. (2002): Effects of implementation intentions on the actual performance of new environmentally friendly behaviors – results of two field experiments. In: Journal of Environmental Psychology 22 (4): 399–411.

Dahlstrand U., Biel A. (1997): Pro-environmental habits: Propensity levels in behavioral change. In: Journal of Applied Social Psychology 27 (7): 588–601.

Endruweit G. (2004): Organisationssoziologie (2. Auflage). Stuttgart: Lucius und Lucius.

Energieagentur NRW (2007): Mündliche Mitteilung Frau Hollweg vom 10.05.2007.

Energyoffice (2011): Katalog nicht-investiver Maßnahmen – Energiesparen (fast) ohne Geld. [http://www.energyoffice.org/deutsch/massnahmen/main.html; 22.06.2012]

Fujii S., Gärling T. (2003): Development of script-based travel mode choice after forced change. In: Transportation Research F: Traffic Psychology and Behaviour 6: 117–124.

Fujii S., Gärling T. (2007): Role and Acquisition of Care-Use Habit. In: Gärling T., Steg L. (eds.): Threats from Car Traffic to the Quality of Urban Life. Problems, Causes, and Solutions. Amsterdam: Elsevier. 235–250.

Fujii S., Gärling T., Kitamura R. (2001): Changes in Drivers' Perceptions and Use of Public Transport During a Freeway Closure: Effects of Temporary Structural Change on Cooperation in a Real-Life Social Dilemma. In: Environment and Behavior 33 (6): 796–808.

Griesel C. (2004): Nachhaltigkeit im Bürokontext – eine partizipative Intervention zur optimierten Stromnutzung. In: Umweltpsychologie 8 (1): 30–48.

Kattenstein T., Unger H., Wagner H.-J. (2002): Handlungskonzepte zur wirtschaftlichen Optimierung des Energiebedarfs und der Energieversorgung der Ruhr-Universität Bochum. In: Abschlussbericht zum Vorhaben: IV A4-20600298 Optimierung der bestehenden Energieversorgung der Ruhr-Universität Bochum. Bochum: Ruhr-Universität Bochum, Lehrstuhl für Energiesysteme und Energiewirtschaft.

Kaufmann-Hayoz R., Bamberg S., Defila R., Dehmel C., Di Giulio A., Jaeger-Erben M., Matthies E., Sunderer G., Zundel S. (in this volume): Theoretical perspectives on consumer behaviour – attempt at establishing an order to the theories.

Kaufmann-Hayoz R., Brohmann B., Defila R., Di Giulio A., Dunkelberg E., Erdmann L., Fuchs D., Gölz S., Homburg A., Matthies E., Nachreiner M., Tews K., Weiß J. (in this volume): Societal steering of consumption towards sustainability.

Klesse A., Hansmeier N., Zielinski J., Wagner H.-J., Matthies E. (2010): Energiesparen ohne Investitionen – ein Feldtest in öffentlichen Liegenschaften. In: Energiewirtschaftliche Tagesfragen 60 (4): 8–12.

Matthies E., Hansmeier N. (2010): Optimierung des Energienutzungsverhaltens in Organisationen – Das Beispiel der Ruhr-Universität Bochum. In: Umweltpsychologie 14 (2): 76–97.

Matthies E., Kastner I., Klesse A., Wagner H.-J. (2011): High Reduction Potentials for Energy User Behavior in Public Buildings – How Much can Psychology Based Interventions Achieve? In: Environmental Studies and Sciences 1 (3): 241–255.

Matthies E., Klöckner C. A., Preißner C. L. (2006): Applying a Modified Moral Decision Making Model to Change Habitual Car Use – How can Commitment be Effective? In: Applied Psychology 55: 91–106.

McClelland L., Cook S. W. (1980): Energy conservation in university buildings. Encouraging and evaluating reductions in occupants' electricity use. In: Evaluation Review 4 (1): 119–133.

Mintzberg H. (1992): Die Mintzberg-Struktur: Organisationen effektiver gestalten. Landsberg, Lech: Verlag Moderne Industrie.

Ouellette J. A., Wood W. (1998): Habit and Intention in Everyday Life: The Multiple Processes by Which Past Behavior Predicts Future Behavior. In: Psychological Bulletin 124: 54–74.

Prose F., Clases C., Schulz-Hardt S. (1999): Umweltbewußtes und ressourcenschonendes Verhalten in Organisationen. In: Graf Hoyos C., Frey D. (eds.): Arbeits- und Organisationspsychologie. Ein Lehrbuch. Weinheim: Psychologie Verlags Union. 147–159.

Schahn J. (2007): Projekt Energiemanagement am Psychologischen Institut der Universität Heidelberg: Ein erfolgreicher Fehlschlag. In: Umweltpsychologie 11 (2): 138–163.

Schwartz S. H. (1977): Normative influences on altruism. In: Berkowitz L. (ed): Advances in experimental social psychology. New York: Academic. 221–279.

Schwartz S. H., Howard J. A. (1981): A normative decision-making model of altruism. In: Rushton J. P., Sorrentino R. M. (eds): Altruism and helping behavior. Hillsdale: Erlbaum. 189–211.

Siero F.-W., Bakker A.-B., Dekker G.-B., van den Burg M.-T.-C. (1996): Changing organizational energy consumption behaviour through comparative feedback. In: Journal of Environmental Psychology 16 (3): 235–246.

Siero S., Boon M., Kok G., Siero F. (1989): Modification of driving behavior in a large transport organization: a field experiment. In: Journal of Applied Psychology 74: 417–423.

Staats H., van Leeuwen E., Wit A. (2000): A longitudinal study of informational interventions to save energy in an office building. In: Journal of Applied Behavior Analysis 33 (1): 101–104.

Stern P. C. (1992): What psychology knows about energy conservation. In: American Psychologist 47: 1224–1232.

Tönjes M. (2009): Energienutzungsverhalten in Organisationen: Evaluation psychologischer Interventionsstrategien. Fern-Universität Hagen. (unpublished master's thesis).

Verplanken B., Aarts H. (1999): Habit, Attitude, and Planned Behaviour: Is Habit an Empty Construct or an Interesting Case of Goal-Directed Automaticity? In: European Review of Social Psychology 10: 101–134.

Verplanken B., Orbell S. (2003): Reflections on Past Behavior: A Self-Report Index of Habit Strength. In: Journal of Applied Social Psychology 33 (6): 1313–1330.

Weber M. (1922/1972): Wirtschaft und Gesellschaft. Tübingen: Mohr.

Wortmann K. (2004): Energie als Thema der Umweltpsychologie – Einführung in das Schwerpunktthema. In: Umweltpsychologie 8 (1): 2–11.

Zinn F. (2002): Evaluation und Modifikation umweltgerechten Verhaltens in Organisationen. Hamburg: Dissertationsschrift.

Section C

Social embedding of consumer behaviour

Matthias Barth, Daniel Fischer, Gerd Michelsen, Horst Rode

5 Schools and their 'culture of consumption': a context for consumer learning

Education has always been recognised as a key component in the endeavour to change consumption patterns towards greater sustainability. The task of "reorienting education towards sustainable development" called for in chapter 36 of Agenda 21 is considered of "crucial importance to the success of sustainable development", according to the action plan adopted at the Rio+10 summit in Johannesburg in 2002 (UNDSD 1992, clause 116). In order to change unsustainable consumption patterns, a task force was set up as part of the Marrakech Process[1], with the remit of devising educational programmes in sustainable consumption. The demand for education in this area thus originates in the explicitly stated goal – contained in Agenda 21 – of changing non-sustainable consumption patterns through education.

5.1 Consumer learning in educational institutions

It is imperative that educational institutions contribute to sustainable development not only as facilitators of skills and knowledge, but that – as social institutions – they live by sustainable principles (cf. Bänninger et al. 2007). Educational institutions can influence young people's consumer behaviour in two ways. Firstly, they can contribute with educational programmes related to consumption that make us reflect and render our own consumption patterns more conscious. Secondly, they are places where goods are consumed (e.g. in their cafeterias and canteens). While the true quantity of informal learning is rather contested, it is estimated that over 50 percent of learning takes place in informal settings, i.e. outside formal teaching and learning contexts (cf. Livingstone 2000). A consideration of sustainable consumption as an "informal learning topic" (Tully/Krok 2009, p.181) raises the question of how informal learning environments

---

1 The purpose of the Marrakech Process is to implement the 10-year programme on sustainable consumption and production adopted in Johannesburg. Since its launch in Marrakech in June 2003, the process has been led by UNEP and UN DESA.

in education contexts could be so structured as to maximise both formal learning and accidental learning.

A systematic linking of formal and informal consumer learning has been attempted as part of the research and development project "BINK" (an acronym for Bildungsinstitutionen und nachhaltiger Konsum – educational institutions and sustainable consumption). The aim of the project was to identify contributions by educational institutions to the promotion of sustainable consumption among teenagers and young adults. For the purposes of the project, sustainable consumption was regarded as a metaphorical 'corridor'. Outer boundaries of this corridor are the Earth's carrying capacity, for instance a certain concentration of CO_2 in the atmosphere or a degree of biodiversity (cf. Rockström et al. 2009); in order to maintain the ecological balance, these levels must not be exceeded. The ecologically determined boundaries of the corridor are contrasted with the ethical scenario inside the corridor, i.e. the demand that present and future generations can satisfy their needs and lead a good life (among other criteria also by consuming goods). In a series of workshops, organised for the benefit of research and field partners, a range of sustainable consumption issues were discussed and a common understanding of 'sustainable consumption' as an educational topic was developed (cf. Fischer 2011a). Complementary to these workshops, those "key competencies for sustainable consumption" have been elaborated that were considered prerequisites for individuals (of differing ages and developmental stages) to consume more sustainably (Barth/Fischer 2012). The strategy of the "BINK" project was twofold: by focusing on intentional behaviour change (Fischer et al. in this volume), the aim was a) to develop teenagers' and young adults' competencies and thus enable them to change their behaviour towards greater sustainability, and b) to put in place an organisational framework capable of facilitating their sustainable behaviours. Three areas of consumption – food, transport and energy – were identified as particularly relevant in terms of environmental impact and were thus selected for the intervention programme (cf. Spangenberg/Lorek 2001).

In order to identify starting points for formal and informal consumer learning, a conceptual framework of consumer culture within educational institutions was set up and analysed in terms of its functionality within a learning context (cf. section 5.2). The rationale behind the subsequent development of specific interventions was the endeavour to establish a consumer culture that would foster young consumers' competencies in and facilitate behavioural change towards sustainability (cf. section 5.3). As a last step, the effects of the changes were empirically tested. The findings are presented in section 5.4.

5.2 Educational organisations' 'cultures of consumption': conceptual framework

The "BINK" project proceeds on the assumption that educational institutions embody a specific culture – a culture that is governed by rules promoting some consumption practices and discouraging others. Young people's consumer behaviour is thus not only influenced by overt instruction, but by the "hidden curriculum" of the institutional culture (Gerstenmaier 2008, p. 142). In order to conceptualise and empirically capture the specific consumer cultures within educational institutions, an analytical framework was established. For that purpose, studies of organisational culture (Schein 2004) and school culture (Helsper et al. 2001) were consulted. The framework was presented to and negotiated with field partners, before it was employed for the planning of the intervention measures (for a more detailed account cf. Fischer 2011b). Six domains at three levels can be distinguished within the framework (cf. Figure 1).

The first level – artefacts – includes the domains *resource management, disciplines and themes,* as well as *participation structures*. Artefacts in these domains are basically visible (e.g. solar panels on the roof, teaching units on consumption or a suggestions box in the cafeteria), but cannot be meaningfully interpreted without knowledge of the educational institution and its context. For instance, a suggestions box in one institution may genuinely be used for improving the quality of cafeteria food, whereas in another institution it might be a mere relic of a contractual duty and be ineffective in bringing about changes to the cafeteria food. Artefacts are thus to be understood as the visible tips of an (organisational) iceberg.

The first level of artefacts is underpinned by the second level of values and norms. In terms of consumption, they belong to the domains *performance orientations* and *educational goals and objectives*. As regards values, for instance, educational institutions may differ in the extent to which a) sustainable consumption is part of the institution's educational mission statement, b) traditional and transformative goals are pursued and c) sustainable consumption is part of their examination syllabus. Espoused norms are essentially to be understood as mediating between artefacts and basic assumptions. Basic assumptions are part of the third and lowest level. They are condensed, unconscious and unquestioned cognitions and values that are the result of interaction with the educational environment. In terms of consumption in an educational context, they refer to teachers' and lecturers' *pedagogical assumptions* about young people's values around and attitudes to consumption, as well as their assumptions about educational institutions' power to exert an influence on young people's attitudes to consumption. Since pedagogical assumptions go unquestioned and have the power to influence behaviour, they are capable of influencing educational interaction processes (cf. Page 1987).

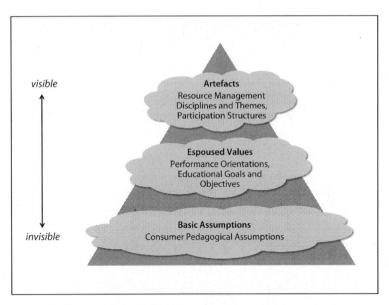

Figure 1: Conceptual framework of educational organisations' 'cultures of consumption' (Fischer 2011b, p. 601)

In these interaction processes the acceptance and rejection of young people's conceptions of identity – reflected in their consumer practices – are being negotiated (cf. Helsper et al. 2001).

5.3 From a conceptual framework to a systematic intervention plan: the "BINK" project

The aim of the "BINK" project was to develop intervention strategies on the basis of the conceptual framework and implement them in collaboration with research and field partners. The interventions were thus jointly developed, locally implemented and evaluated by the project partners. Two secondary schools, two vocational schools and two universities agreed to act as field partners; sociologists, educational researchers, environmental scientists and environmental psychologists contributed as research partners.

The interventions were designed to transform the institutions and the locations where consumption took place in such a way as to allow young people – through formal and informal learning opportunities – to assess the impact of their consumer behaviour, and to encourage them to make *sustainable* consumption decisions. More than three dozen intervention measures, concerning different consumer areas, were employed. The implementation of these intervention measures started in autumn 2009 and ended in spring 2011.

Based on its theoretical framework, the "BINK" project hypothesised that both formal and informal learning settings can support young people in acquiring attitudes, skills and knowledge about sustainable consumption and sustainable development. This hypothesis was tested by way of an empirical study of the consumer cultures within the schools, which was carried out in cooperation with the Institute for Environmental Communication (INFU) of the Leuphana University of Lüneburg and the German Youth Institute (DJI) in Munich. The data from the study is presented in the following sections.

5.4 'Cultures of consumption' as a learning context: empirical findings

5.4.1 Study design

The accompanying empirical study focused on the connection between the perception of educational organisations' cultures of consumption and young consumers' learning and behaviour. Questionnaires were distributed to the students in the four participating schools. The objective was to elicit students' values and attitudes towards consumption. One question also asked students how they thought teachers and staff perceived the educational institution as a learning environment.

The following individual features of students acted as dependent variables: self-assessment of students' consumption-related learning (AK1, AK2); self-assessment of the extent of students' influence on consumption-related matters (E1, E2); relevance of sustainable development and sustainable consumption to them personally – both in terms of personal relevance and self-reported sustainability-oriented consumer behaviour (R1, R2). Regarding organisational culture of consumption within an educational setting, three areas were identified: formal, partially informal[2] and informal learning settings (LS1, LS2, LS3, LS4), educational goals (EG1, EG2), and overall commitment of the institutions to sustainable consumption, both as expressed in their mission statement and in implemented changes (OC1, OC2) (cf. Table 1).

The readability and scope of the questionnaire was initially tested as a two-stage pre-test, by way of a retrospective think aloud protocol: after answering a question,

---

2 "Partially informal" learning settings tend to be situated outside the actual formal instruction context, but are often connected to formal learning in some way or other (e.g. competitions, exhibitions).

Table 1:
Overview of the independent and dependent variables

Abbr.	Content	Example	Scale type
	*Independent variables: organisational features*		
LS1	Learning setting: informal activities around sustainable consumption (sc) / sustainable development (sd)	"Please name any sc/sd activities you have noticed around your school."	Number of coded responses to 3 open questions (value range min = 0, max = 12, M = 1.13, SD = 1.73)
LS2	Learning settings: positively rated, partially informal modes of learning and activities	"Competitions, e.g. environmentally friendly school"	Number of response level – "I liked it" – for 5 items (value range min = 0, max = 5, M = 1.01, SD = 1.27)
LS3	Learning settings: teaching/discussing "consumption" in different subjects	"In which subjects did you come across the topic 'consumption'?"	Number of response level – "once" and "several times" – across all specified subjects (value range min = 0, max = 7, M = 2.20, SD = 1.76)
LS4	Learning settings: teaching/discussing sd in various subjects	"In which subjects did you come across topics of sd"?	Number of response level – "once" and "several times" – across all specified subjects (value range min = 0, max = 7, M = 1.76, SD = 1.87)
EG1	Educational goals: awareness of sustainability issues	"I am meant to learn about the environmental consequences of my shopping habits."	Scale of 5 items (value range min = 0, max = 25, α = .84, M = 20.87, SD = 5.88)
EG2	Educational goals: traditional	"I am meant to learn about my rights as a consumer, so I get what I'm entitled to get."	Scale of 3 items (value range min = 0, max = 15, α = .66, M = 12.99, SD = 3.41)
OC1	Organisational commitment: sc as an objective of the school	"Thinking about your school, how high does sc rank overall in your opinion?"	Single item (value range min = 0, max = 3, M = 2.46, SD = 0.86)
OC2	Organisational commitment towards changes in the school	"In my school, there have been some real shifts towards greater sustainability."	Single item (value range min = 0, max = 5, M = 2.82, SD = 1.63)
	*Dependent variables: individual features*		
AK1	Acquired knowledge: self-assessed learning of sustainability issues	"I have learnt what is generally meant by sd."	Scale of 7 items (value range min = 0, max = 35, α = .90, M = 22.43, SD = 8.72)
AK2	Acquired knowledge: relevance of what has been learnt for own consumption decisions	"Acquired knowledge is important when it comes to buying technical devices."	Scale of 3 items (value range min = 0, max = 15, α = .86, M = 8.51, SD = 4.60)

E1	Effectiveness: influence as a consumer on sustainability of products and production processes (consumer effectiveness)	"Consumers can influence the working conditions under which the products they buy are produced."	Scale of 5 items (value range min = 0, max = 25, α = .81, M = 15.82, SD = 5.71)
E2	Effectiveness: influence as a consumer on the range of foods on offer in the school canteen	"You can influence the school's foods on offer by praising or criticising their offers or coming up with your own suggestions."	Single item (value range min = 0, max = 5, M = 3.65, SD = 1.63)
R1	Relevance: personal relevance of sd	"sc is important for me personally in my own life."	Scale of 3 items (value range min = 0, max = 15, α = .86, M = 10.32, SD = 4.20)
R2	Relevance: self-reported sustainability oriented consumer behaviour	"I specifically buy products whose production and use have low impacts on the environment."	Scale of 5 items (value range min = 0, max = 25, α = .77, M = 13.47, SD = 5.50)

selected respondents were asked to put into words their thought processes as they were answering the question (cf. Someren et al. 1994). The modified questionnaire was then tested in two classes. After a rephrasing of individual items and the exclusion of ambiguous items, the final, machine-readable version of the questionnaire was drawn up.

The survey took place between June and August 2010: students were asked to fill in the questionnaire during one of their lessons – in the classroom (t = 45 min.). A total of 780 students – 201 grade 7 students (n = 201, response rate = 75.2 percent), 267 grade 11 students (n = 267, response rate = 63.9 percent) and 312 students on vocational courses (n = 312, response rate = 69.4 percent[3]) filled in the two-part questionnaire.

After data import and a randomised quality check of the raw data in SPSS scale items were first checked by way of a factor analysis to identify underlying discrete constructs which have been confirmed in Mixed Rasch models where appropriate. Reliabilities of these scales are, with one exception (EG2), in a good to very good range (α = .66–.90). The analysis was performed in two steps: firstly – using multiple linear regression – the relationships between organisational and individual features were investigated. In a second step, the relevance of the individual predictors (independent variables) was assessed with a stepwise regression (cf. Aiken et al. 2003): variables were ranked according to their explanatory power (i.e. each according to its capacity to predict values of other variables). This method allows for simple ('economical'), but

---

[3] In the case of one school the response rate could not be established. This was due to a lack of data on overall class size. The value stated therefore only refers to one of the two field partners.

highly fitting models, in which the observed effects can be explained by a minimal number of variables. The following section will focus on the extent to which the organisational features (independent variables) influenced each individual feature (dependent variable).

5.4.2 The explanatory power of organisational 'cultures of consumption'

In the first part of the analysis, the initial task was to empirically test the influence of organisational features on the various individual features. On the basis of stepwise regression analyses, explanatory models were constructed; these were capable of showing the impact of the independent variables (e.g. educational goals) on the dependent variables (e.g. sustainable consumer behaviour), i.e. they would show which dependent variables were influenced and the extent to which they were affected. In these models, the 'explanatory power' of organisational features is represented as degree of variance in percentages. The value indicates what percentage of variance in each (dependent) variable can be attributed to the organisational features. It is evident from Table 2 that the values of the variables differ – some to a significant extent.

The strongest relationships between organisational and individual features were noted for the self-assessed learning of sustainable issues (AK1). 26 percent of the dif-

Table 2:
Overall models for each dependent variable based on all independent variables

	R^2	Corrected R^{2*}	F	Significance
Self-assessed learning of sustainability issues (AK1)	.263	.254	29.572	.000
Relevance of what has been learnt for own consumption decisions (AK2)	.142	.132	13.682	.000
Influence as a consumer on sustainability of products and production processes – consumer effectiveness (E1)	.091	.080	8.266	.000
Influence as a consumer on the range of foods on offer in the school canteen (E2)	.106	.095	9.802	.000
Personal relevance of sustainable development (R1)	.351	.343	44.942	.000
Sustainability-oriented consumer behaviour (R2)	.167	.157	16.776	.000

All estimated models are significant.

* Corrected R^2 also takes account of negative values of some variables. The almost negligible differences between R^2 and corrected R^2 reflect the quality of the chosen regression models.

ferences between students can be explained by the influence of organisational features. Organisational features have an even greater impact on personal attitudes to sustainable development (R1, 35 percent). With just under 17 percent, self-reported sustainable consumer behaviour (R2) is somewhat less influenced by organisational features. Less powerful, but nonetheless traceable relationships exist for relevance for own consumption decisions of what has been learnt (AK2, just over 14 percent), influence as a consumer on the range of foods on offer in the school canteen (E2, just under 11 percent), and influence as a consumer on sustainability of products and production processes (consumer effectiveness) (E1, 9 percent).

The results may also contain a temporal dimension as discussed in research on transfer of training (Steindorf 2000), namely that student achievement and personal relevance are observable after a short period, while *changes* in consumer behaviour and influence as a consumer take longer to develop and often require further individual commitment. These influences, which act as moderating variables, have, however, not been statistically tested in this study.

5.4.3 The explanatory power of specific aspects of organisational 'culture of consumption'

Whereas the first part of the analysis was concerned with the influence of the *sum* of organisational features on the variation between the appearance of individual features, the second part investigated the influence of *specific* organisational variables. The aim was to identify the impact of the variables and to explain their (additional) impact on the individual variables. Here too, 'economical' models were developed, where the lowest possible number of independent variables could explain the largest possible amount of variance in a given dependent variable.

Of the eight independent organisational variables, two stand out for their strong influence on five dependent variables (cf. Table 3): educational goals to raise awareness of sustainability issues (EG1), and implementation of changes in the school (OC2). The former (EG1) primarily influence the personal relevance of sustainable development (R1) and self-reported sustainability-oriented consumer behaviour (R2). The latter (OC2) is the key influential factor for self-assessment of what has been learnt about sustainability (AK1).

The influence of the remaining organisational features is relatively weak: organisational commitment, i.e. sustainable consumption as an objective of the school (OC1) and learning settings, i.e. positively rated, partially informal modes of learning and activities (LS2) score relatively low; an even weaker impact is noted for addressing consumption and sustainable development issues in various subjects (LS3, LS4), and the implementation of measures towards sustainable consumption and sustainable

Table 3:
Economical models for the explanation of dependent variables (explained variance in percentages)

Dependent variables (individual)	Independent variables (organisational)	Learning settings				Educational goals		Organisational commitment	
		Informal (LS1)	Partially informal (LS2)	Formal consumption (LS3)	Formal SD (LS4)	Transformative goals (EG1)	Traditional goals (EG2)	sc as an objective of the school (OC1)	Implementation of changes (OC2)
Learning outcome	Self-assessment of what has been learnt (AK1)		2.0%		3.1%	5.8%			**14.4%**
	Relevance of what has been learnt (AK2)	1.6%		1.4%		0.8%	2.2%	0.7%	7.0%
Effectiveness	Consumer effectiveness (E1)					7.5%			1.0%
	Influence on range of foods on offer in school (E2)				2.5%				5.8%
Relevance	Personal relevance of SD (R1)		1.2%			**23.7%**		3.1%	6.8%
	Sustainability-oriented consumer behaviour (R2)		1.0%			**12.8%**		1.5%	

development. Furthermore, it is very likely that the effects of different influences overlap. As the influence of these variables cannot be clearly isolated, they must be interpreted with great caution (cf. section 5.4.4).

It has been shown above that the independent variables concerned with learning settings, educational goals and organisational commitment strongly influence the consumption-related student features. It appears that the combination of formal and informal learning settings promotes sustainable attitudes. Both approaches should therefore be employed simultaneously in order to ensure the success of educational interventions.

5.4.4 Interrelationships between individual aspects of a consumer culture

The individual organisational features were additionally examined as to their interrelationships and potential overlapping effects. This was done in order to ensure that the effects measured were not skewed by relationships between the variables. For instance, a significant correlation was found between teaching/discussing consumption (LS3) and teaching/discussing sustainable development (LS4) in different subjects ($r = .65$, $p = .000$, $n = 780$). Given the goal of "BINK" to combine these two approaches, this is unsurprising. A further strong link was noted between traditional educational goals (EG2) and sustainability-oriented transformative educational goals (EG1) ($r = .46$, $p = .000$, $n = 762$). This finding is also convincing: sustainability-oriented consumer education need not run counter to, but can be regarded as an extension of traditional consumer education (cf. McGregor 2005).

Similarly, it is plausible that a school puts in place tangible changes towards greater sustainability (OC2), if the topic sustainability is part of the syllabus[4] (LS4) ($r = .38$, $p = .000$, $n = 713$), and that consumption-related interventions are more likely to be implemented in an informal learning setting (LS1), if the topics consumption and sustainability form part of the subject matter across the school ($r = .31$, $p = .000$, $n = 780$ for L3; $r = .39$, $p = .000$, $n = 780$ for LS4). Furthermore, addressing consumption in class (LS3) and including consumption in partially informal modes of learning (LS2) appear to be distinctly connected ($r = .34$, $p = .000$, $n = 780$). The same holds for sustainability (LS4) ($r = .34$, $p = .000$, $n = 780$). Finally, a reasonably close relationship is discernable between sustainable consumption as an overall objective of the school (OC1) and the implementation of actual changes (OC2) ($r = .37$, $p = .000$, $n = 682$).

The overlapping effects outlined above are plausible and within an acceptable range.

5.5 Conclusions

From a perspective of educational research, the growing wealth of materials and initiatives in the field of sustainable consumer education is entirely welcome. However, this is contrasted by a lack of empirical evidence regarding the extent and quality of consumer education in schools and how these materials and initiatives feed through into individual attitudes and values. The results presented in this article can be seen as a contribution to a more solid empirical basis for developing appropriate educational materials that will foster a generation of more sustainably-minded citizens.

---

4 Students were asked to comment on the inclusion of sustainability issues in a total of eleven subjects groups, some of which were overlapping.

The approach and results outlined in this article lead to the following conclusions and recommendations for practice:

Consumer policy implications: towards a broader education
> *The dominant notion of education as an instrument of consumer information is too narrow in scope. Accordingly, educational institutions should be reframed as more holistic learning settings.*

The results from the empirical study indicate that the holistic approach taken in the "BINK" project (the combination of formal and informal learning settings) strongly shapes both the attitudes of individuals towards sustainability and their consumer behaviour. It is argued here that policy interventions should steer away from focusing purely on formal consumer and sustainability education.[5] Instead, they should take a wider view of educational institutions and support them in providing not only formal classroom instruction, but more informal learning environments located across the institution.

An example: based on a comprehensive study by Tim Jackson (2005) on the prerequisites for the promotion of sustainable consumption, Great Britain has adopted the so-called 4 E model as a policy recommendation – a model that has also become the source of consumer policy strategies for the federal states in Germany (cf. MELR BW 2009). The 4 E model calls for a broad mix of instruments for the promotion of sustainable consumption, with the core strategies of *enabling* and *engaging* individuals and *encouraging* and *exemplifying* behavioural change (cf. DEFRA 2005). Education is seen as an overall 'enabler'. In terms of a change of perspective as suggested in this article, the task of schools is to enable young people to consume sustainably not only through formal teaching, but also by putting in place informal incentivising settings where young people can experience sustainable consumption at first hand.

Consequences for educational policy:
counteracting the marginalisation of consumption
> *If educational policy is to contribute meaningfully to young people's understanding of sustainable consumption issues, it is imperative to overcome the current fragmented approach to teaching this topic (i.e. the lack of systematic curricular inclusion of sustainable consumption, and the fact that canteens and cafeterias are*

---

[5] This notion of consumer education forms the basis of the current European Consumer Policy Strategy (European Commission 2007) and is also reflected by a recent OECD survey (2009) on the practice of consumer education (cf. Fischer 2011b).

managed not by the schools themselves but by the educational authorities and their contractors). To this end, it is necessary to put in place the basic conditions that permit a systematic approach to the subject both inside and outside the classroom.

The current trends towards all-day schools – necessitating students' increased presence on the school premises – offer opportunities for setting up informal learning settings around the schools, where young people can experience sustainable consumption. Yet, the findings of the empirical study indicate that the provision of informal learning activities alone is not sufficient for behavioural change. Sustainable consumption issues need to be integrated in the core mission of educational institutions, namely their formal teaching. It is the task of policy makers to provide an overarching framework within which all educational institutions can integrate sustainable consumption issues into their school life. Faced with the reality that consumption issues such as nutrition have increasingly been marginalised by educational institutions (cf. Heseker/Beer 2004), it would be advisable to support those who call for an educational reform to overcome the "(near) absence of consumer education in secondary schools" (Schlegel-Matthies 2004, p. 7; translated by C. Holzherr). Whether this means creating a separate subject or integrating sustainability issues into existing subjects remains up for debate and requires further research. Both scenarios necessitate the provision of appropriate teacher training. If a separate subject were to be created, universities would have to set up courses accordingly. If sustainability issues were to be integrated into existing subjects, teachers would have to be equipped with the necessary competencies (Haan 2006). Overall – in order to create the necessary conditions for teaching sustainable consumption in schools – teachers need to be trained a) in their initial teacher training courses, b) in a specific preparation phase, and c) on an ongoing basis of professional development.

The findings of this study further suggest that it is not only a matter of raising awareness of sustainable consumption in various learning settings, but that the objectives of such a campaign should be clearly visible. Educational goals regarding sustainable consumption differ from traditional consumer education goals insofar as they should not only enable young people to interact competently with the market in their consumer role, but also encourage them to look critically at the conditions of production and view consumption in the context of a good life for present and future generations (McGregor 2005). It is the authors' view that, at a time when more financial and economic subjects are being added to the school curriculum, the challenge for educators is to empower students not only to maximise their individual benefit, but to enable them to act in the interest of others today and in the future (McGregor 2011).

Coda: the transformation of educational institutions into settings where sustainable consumption is a way of life

The annual report 2010 of the Worldwatch Institute concludes that schools have so far failed to exploit their potential for discussing the background and impact of consumerism and to work towards a more sustainable society (cf. Assadourian 2010). By not actively engaging with and committing to sustainable consumption, schools thus contribute to the perpetuation of current unsustainable consumption patterns.

The empirical findings of the "BINK" project presented in this article underscore the need for schools to engage more intensively with sustainability. If we want young people to learn and care about sustainable consumption, it is crucial that schools initiate changes and make sustainable consumption part of their mission, their operations and their tuition offers. In this respect, the "BINK" project represents a tried-and-tested, participatory approach to facilitating changes in organisational cultures of consumption within schools and universities, and can thereby provide guidance for future policy reform.

References

Aiken L. S., West S. G., Reno R. R. (2003): Multiple regression. Testing and interpreting interactions. Newbury Park, Calif.: Sage.

Assadourian E. (2010): The Rise and Fall of Consumer Cultures. In: Starke L., Mastny L. (eds.): State of the World 2010 – Transforming Cultures: From Consumerism to Sustainability. Washington, D.C.: Worldwatch Institute. 3–20.

Bänninger C., Di Giulio A., Künzli David C. (2007): Schule und nachhaltige Entwicklung. In: GAIA 16 (4): 267–271.

Barth M., Fischer D., Michelsen G. (2010): Bildung für nachhaltigen Konsum. In: GAIA 19 (1): 71.

Barth M., Fischer D. (2012): Key competencies for sustainable consumption. Paper presented at 2nd International PERL-Conference "Beyond Consumption", March 19th–20th 2012, Berlin.

DEFRA – UK Department for Environment, Food and Rural Affairs (2005): Securing the future. The UK Government Sustainable Development Strategy. London.

European Commission (2007): EU Consumer Policy strategy 2007–2013. Empowering consumers, enhancing their welfare, effectively protecting them. Luxembourg: Office for Official Publications of the European Communities.

Fischer D., Michelsen G., Blättel-Mink B., Di Giulio A. (in this volume): Sustainable consumption: how to evaluate sustainability in consumption acts.

Fischer D. (2011a): Der Kompass "Nachhaltiger Konsum". Eine Orientierungshilfe. Bad Homburg: VAS Verlag.

Fischer D. (2011b): Educational Organisations as "Cultures of Consumption": Cultural Contexts of Consumer Learning in Schools. In: European Educational Research Journal 10 (4): 595–610.

Gerstenmaier J. (2008): Der heimliche Lehrplan in Organisationen. In: Genkova P., Abele A. E. (eds.): Lernen und Entwicklung im globalen Kontext. "Heimliche Lehrpläne" und Basiskompetenzen. Perspektiven politischer Psychologie, Bd. 3. Lengerich: Pabst. 142–155.

Haan G. de (2006): The BLK '21' programme in Germany: a 'Gestaltungskompetenz'-based model for Education for Sustainable Development. In: Environmental Education Research 12 (1): 19–32.

Helsper W., Böhme J., Kramer R.-T., Lingkost A. (2001): Schulkultur und Schulmythos. Gymnasien zwischen elitärer Bildung und höherer Volksschule im Transformationsprozeß. Studien zur Schul- und Bildungsforschung, Bd. 13. Opladen: Leske + Budrich.

Heseker H., Beer S. (2004): Ernährung und ernährungsbezogener Unterricht in der Schule. In: Bundesgesundheitsblatt – Gesundheitsforschung – Gesundheitsschutz 3: 240–245.

Jackson T. (2005): Motivating Sustainable Consumption. A review of evidence on consumer behaviour and behavioural change. University of Surrey: Center for Environmental Strategy.

Livingstone D. W. (2000): Exploring the icebergs of adult learning: findings of the first Canadian Survey of Informal Learning Practices. Toronto: Centre for the Study of Education and Work (CSEW).

McGregor S. L. T. (2005): Sustainable consumer empowerment through critical consumer education. A typology of consumer education approaches. In: International Journal of Consumer Studies 29 (5): 437–447.

McGregor S. L. T. (2011): Consumer Education Philosophies: The Relationship between Education and Consumption. In: ZEP – Zeitschrift für internationale Bildungsforschung und Entwicklungspädagogik 34 (4): 4–8.

MELR BW – Ministerium für Ernährung und Ländlichen Raum Baden-Württemberg (2009): Verbraucherpolitische Strategie Baden-Württemberg. Diskussionspapier (Stand: 8. Mai 2009). Stuttgart.

OECD – Organisation for Economic Co-Operation and Development (2009): Promoting consumer education. Trends, policies and good practices. Paris: OECD.

Page R. (1987): Teachers' Perceptions of Students. A Link between Classrooms, School Cultures, and the Social Order. In: Anthropology & Education Quarterly 18 (2): 77–99.

Rockström J., Steffen W., Noone K., Persson Å., Chapin F. S., Lambin E. F., Lenton T. M., Scheffer M., Folke C., Schellnhuber H. J., Nykvist B., Wit C. A. de, Hughes T., van der Leeuw S., Rodhe H., Sörlin S., Snyder P. K., Costanza R., Svedin U., Falkenmark M., Karlberg L., Corell R. W., Fabry V. J., Hansen J., Walker B., Liverman D., Richardson K., Crutzen P., Foley J. A. (2009): A safe operating space for humanity. In: Nature 461 (7263): 472–475.

Schein E. H. (2004): Organizational culture and leadership (3. ed.). San Francisco, Calif.: Jossey-Bass.

Schlegel-Matthies K. (2004): Verbraucherbildung im Forschungsprojekt REVIS – Grundlagen. Paderborner Schriften zur Ernährungs- und Verbraucherbildung, Bd. 2. Paderborn: Universität Paderborn.

Someren M. W. v., Barnard Y. F., Sandberg J. A. C. (1994): The think aloud method: A practical guide to modelling cognitive processes. London: Academic Press.

Spangenberg J. H., Lorek S. (2001): Environmentally sustainable household consumption: from aggregate environmental pressures to priority fields of action. In: Ecological Economics 43 (2–3): 127–140.

Steindorf G. (2000): Grundbegriffe des Lehrens und Lernens. Bad Heilbrunn: Julius Klickhardt.

Tully C. J., Krok I. (2009): Nachhaltiger Konsum als informeller Lerngegenstand im Jugendalltag. In: Brodowski M., Devers-Kanoglu U., Overwien B., Rohs M., Salinger S., Walser M. (eds.): Informelles Lernen und Bildung für eine nachhaltige Entwicklung. Beiträge aus Theorie und Praxis. Leverkusen: Budrich Barbara. 181–189.

UNDSD – United Nations Division for Sustainable Development (1992): Agenda 21 – Rio Declaration on Environment and Development. New York.

Konrad Götz, Wolfgang Glatzer, Sebastian Gölz

6 Household production and electricity consumption – possibilities for energy savings in private households

This article is a background study to the "Intelliekon" project, which is an investigation into electricity usage feedback (by means of smart meters etc.) to consumers. Given that the project group is concerned with electricity consumption in the context of household production, the difference between household production and commercial production will briefly be outlined first. This is followed by an analysis of social trends within households – whereby continued mechanisation is leading to ever-increasing energy consumption. The focus will be on potentials for energy savings, strategies which encourage these and possible reasons for failure. The article ends with a discussion of future perspectives.

6.1 The importance of "household production"

Consumption in modern society is concerned to a large extent with goods and services that are produced in private households. In our context, households are understood as residential dwellings, within which economic production and consumption by one or more individuals take place. On the basis of the shared living arrangements and housekeeping activities of the individual members, households are characterised by specific roles and gender arrangements, interpersonal relationships, shared activities and more or less shared norms, values and expectations of their members (Glatzer 1990).

Alongside companies, state institutions and intermediary organisations, households play an important part in the production of goods and services, i.e. in providing individuals with goods and services. The private household takes on a particularly important role in society when all the socio-psychological tasks routinely performed by household members are considered: they cook meals and do the washing; they raise children and care for the elderly; they contribute to a homely atmosphere; they provide entertainment and deal with conflicts. Concepts such as relationship-building, emotional work, and balancing different demands (cf. Negt/Kluge 1993) reflect these activities. Productive leisure activities can also be considered as household production (Götz 2007).

Whether for the purpose of housework or leisure activities, the fact remains that the more technical the devices and appliances that are used, the greater the demand for electricity (cf. VDEW 1993 and Gruber/Schlomann 2008). Electricity is not only one of the most important means of production in private households, but it is also ubiquitous in everyday life in industrialised countries. Electricity is available for all kinds of purposes, its supply depends on large technical systems, and its special feature is its invisibility. Thus, if electricity saving measures are to be considered, household production has a central place in that.

Compared with other areas of production, household production follows its own distinct rules. In contrast to private, public and collective goods, goods produced in households are always closely related to individual people. While commercial production supplies serial goods to an abstract consumer market, household production is concerned with providing goods for and satisfying the needs of specific individuals within the household or its immediate social network. Access to these products is not a matter of purchasing power (as with private goods) or legal rights (as with public goods), but by virtue of a particular social relation. For that reason, the private household is uniquely effective in its own context: it is exceptionally flexible and creates highly diverse products, destined for specific individuals (Glatzer 1994). In contrast to abstract commercial production, such tangible tailor-made production therefore requires different efficiency criteria and an altogether different rationale. In commerce, the aim is to produce goods with minimum input. An increase in output in relation to input raises profits and thus fulfils the commercial efficiency criterion. Private households, on the other hand, adapt their production (meals, cleanliness, light, warmth, comfort etc.) to the needs of particular individuals, who by their very nature will change throughout their lives. Goods and services produced by households are unique, diverse and tailored to the life stage and lifestyle of the recipient. While households do work to budgets, profit maximisation is not their overall aim. A case in point is the care of children. By a combination of routine and personal attention, the members of a household are able to satisfy individual needs and requests flexibly and selectively.

Official data show the continuing importance of household production for Germany's overall economic balance sheet: both in 1992 and 2001, unpaid working hours (102 and 96 billion) far exceeded paid working hours (59 and 56 billion). Unpaid work amounts to approximately 63 percent of overall annual working hours (Schäfer 2004, pp. 965 and 974). The share of unpaid work in Germany's gross domestic product amounted to a remarkable 40 percent in both years.[1] Nonetheless, these figures indicate a distinctly lower productivity rate for private households. However, without the production of goods and services in private households the overall standard of living would undoubtedly drop. Furthermore, the production of Germany's voluntary sector would also decrease significantly.

When it comes to economising on electricity, household production is crucial – a fact that has to be taken into account in the development of 'electricity saving aids' such as electronic feedback systems.

Household technology and appliances are only productive once they are used by individuals who have learnt how to acquire, use, take care of and possibly repair them. A productive balance is reached once the household member can use the technology with ease. It makes sense to call such an interaction between technology and the individual a socio-technical system[2] (cf. Ropohl 2009 and Geels 2004). Another characteristic of home technology is its integration in "larger technical systems" (Mayntz/Hughes 1988), e.g. connections to power stations and water works, cables and servers. Thus, in order to meet productivity criteria, electrical household technology needs to be integrated at two levels: with the larger technical systems and within the individual household context. The next sections will focus on the changing role of technology in modern households.

6.2 Development trends in household technology and lifestyles

The changes in households and families over time are often related to terms such as mechanisation of everyday life, singularisation and pluralisation of households (cf. Peukert 2008; Hampel et al. 1991).

The principal trends are increasing mechanisation, ever smaller households and a fundamental reorientation in marriage and family arrangements, including the emergence of new types of households. These factors strongly affect the requirements and

---

1 Gross value added of household production was 690 bn. euros in 1992 (gross domestic product 1,613 bn.), 2001 820 bn. (GDP 2,074 bn.) (Schäfer 2004, pp. 965 and 974).

2 Depending on the precise technology in question, different aspects of household mechanisation come to the fore. In terms of action theory, the concept of technology is related to a conscious and methodical use of certain means to reach certain goals. Of central importance in this process is the rationalisation of work processes. The concept of technology need not be linked to the use of household technology. It can be applied to other fields such as educational techniques, and it is to be generally understood as a goal-oriented, methodical action. In systems theory, technology is defined as the interdependence of technical systems and action and is thus termed a socio-technical system (cf. Geels 2004).

performance of households and thereby have an impact on household electricity requirements. The effects on electricity consumption are, however, not uniform across households.

6.2.1 Mechanisation of households

The expansion of electricity put its stamp on the 20th century and decisively influenced its society. At the end of the 19th century, electricity consumer numbers in Germany and other countries were still relatively low. After the turn of the century – in the wake of the development of the small electric motor and the gradual adoption of gas and electricity – the mechanisation of households increased dramatically – but not without encountering obstacles, such as technical shortcomings in the products, lack of electrical connections, expensive tariffs and high purchase costs. Hence, in the 1920s, a mere 59 percent of households were connected to the electricity grid; most had no electrical equipment apart from lighting (cf. Glatzer/Dörr et al. 1991).

The period after the Second World War finally saw a turning point in the take-up of mechanised household goods. The household sector – with housewives as potential customers – became a profitable market segment for the industry, which took advantage of increasing technical opportunities. However, in Germany, the widespread diffusion of household appliances took place at a relatively late stage: only between 1960 and 1980 did household appliances truly conquer the market. These days, typical household technologies such as refrigerator, washing machine and television are considered basic[3] household equipment and are part of virtually all German households.[4] Microwaves, dishwashers and freezers are considered standard equipment and are owned by over half of households, although the proportion of freezers has been declining. Tumble dryers are only slightly on the increase; awareness of their high energy consumption seems to have prevailed over the desire for more convenience.

In terms of information and communication technology (ICT), the landline telephone became widely established, but is now being superseded or complemented by the far more rapid expansion of the mobile phone. Both are considered basic equipment nowadays, while a computer and an internet connection have become standard technologies. ICT has clearly experienced a considerably faster expansion than has been

---

3 In terms of the presence of particular items of equipment within households, the following terms are defined in relation to the proportion of people who own the appliance: basic equipment = over 80 percent; standard equipment = 50–80 percent; extended equipment = 20–50 percent; rare equipment = less than 20 percent. Expansion follows an S-curve, starting with slow take-off, subsequent acceleration phase and final decrease in diffusion when approaching saturation point.

4 The following details on electrical equipment in Germany are based on figures from the German Federal Statistical Office (1993, p. 163; 2003, p. 13; 2008a, p. 17; 2011).

the case for more traditional appliances. Mobile phones are currently evolving into smartphones with built-in internet connection, so that the latter is becoming mobile too. Furthermore, as smartphones are increasingly bundled with applications – so-called apps – they are developing into multifunctional devices capable of providing us with internet information, satnav and entertainment.

In terms of entertainment technologies, televisions, radios and CD players nowadays fall under the heading of basic equipment, and DVD players under standard equipment. Within an increasingly diverse product range, innovations constantly supersede older technologies. Two cases in point are flat screen TVs and smartphones. As well as the devices mentioned above, a huge range of other electronic equipment is increasingly becoming part of our lives: DIY equipment, gadgets for health and personal care, recreational and leisure equipment, lighting, heating and air conditioning systems, security systems capable of detecting movement, surveillance cameras, electronic roller blinds and garage doors.

Technology in private households has become so ubiquitous that the term "machinery park" is not amiss. The following trends can be noted:

Household technology is *continually gaining ground* – even if saturation is reached in certain areas and some appliances or devices are no longer adequate. In contrast to the post-war period, there is, for instance, no more need for stocking large amounts of food in huge freezers. In the face of rapid technological change, replacement purchases are ever more popular. New products at the start of the diffusion curve come onto the market – this is a relatively rare, but nonetheless regular occurrence. A further trend in many modern households is multiple versions of the same product type (it is particularly common to have two, three or even four televisions and/or computers). In the case of televisions, for instance, the equipment inventory of the average household stands at 158 percent.

Another noticeable modern trend is the *qualitative improvement* of technology: products that are considered and marketed as inferior lose out against more advanced products. An example is the development from black and white televisions to colour television and most recently to flat screens. Criteria of what is considered superior and progressive differ widely, ranging from giant screen TVs for one person to energy saving potential for someone else. For some products, increasing specialisation can be observed: a conventional electric drill develops into an impact drill and eventually into rotary hammer. Due to their progammability, electronic devices – especially smartphones with their diverse applications – are particularly versatile and multi-functional.

Two parallel trends are discernable: *professionalisation*, i.e. the introduction of products requiring skilled users, and the *simplification* of professional equipment. Depending on the circumstances, they may assist or overwhelm users.

A trend that first emerged in the 1990s, but whose future development is still unclear, is intelligent metering: the integration of different devices into a single system, such as 'intelligent home' or 'smart home' (Glatzer et al. 1998).[5] Its characteristic feature is the electronically controlled linking of the appliances and devices in question. Its aim is to allocate resources efficiently, increase technical reliability, convenience and user-friendliness, and integrate households more effectively in their social environment. The interactive smart metering systems developed by the "Intelliekon" project are examples of this trend. Micro-electronic technology provides users with devices that monitor their energy use[6] and provide them with feedback, on the basis of which they can adjust and economise their consumption.

The mechanisation of households thus leads to two opposing tendencies: on the one hand, the increase in the number of electrical appliances inevitably results in a higher demand for electricity, on the other hand, the complementary developments described above (such as the use of laptops instead of desktop computers or smartphones instead of laptops) may lead to a more economical use of electricity.

6.2.2 Shrinking households

The size of households is exhibiting a long-term shrinking tendency, i.e. the average number of household members diminishes. Around the turn of the 19th/20th century, three-fifths of the population lived in households with four or more people – in 2008, one- or two-person households accounted for three-fifths of German households, and this trend is still continuing (German Federal Statistical Office 2000, p. 38; German Federal Statistical Office 2010a). The once clear dominance of larger multi-person households has thus given way to a relative prevalence of single-person households. In terms of electricity users, a small number of large households have been replaced by a large number of small households. In 2009, Germany recorded a still growing number of 40 million households requiring electricity. The electricity demand in larger households does not grow proportionally to the number of members, but – owing to the economies of scale – tends to flatten out. In 2005, a person in a household of four consumed only 60 percent of the electricity compared with a person in a single-person household (Forsa 2005). Therefore, as households shrink, they contribute to higher overall electricity consumption.

---

5 Cf. offers available from: http://www.cuculus.net/drupal/smarthome [16.06.2012]

6 Cf. http://www.digitalstrom.org [16.06.2012] and http://www.ecowizz.net [16.06.2012]

6.2.3 Pluralisation of household structures

Households may be made up of family members or non-family members. Alongside the diverse types of family households (e.g. married couples with child/children, single parents, multi-generation families) there exists a range of non-family living arrangements: unmarried and same-sex partnerships, shared accommodation arrangements and one-person households (including singles). Multi-generation households are relatively uncommon nowadays – in contrast to the rising number of multi-location households.

Significant changes in household and family structures have led to an increasing variability of 'old' and 'new' forms of households or a "plurality of individual living patterns" (Meyer 2008, p. 340; original: Pluralität von Privatheitsmustern), which is regarded as perfectly normal these days. The nuclear family, although still in the majority among household types, has lost considerable ground in recent times: slightly less than two-fifths of households consist of married couples with children (below and over 18 years). Electricity requirements match these diverse living arrangements and lifestyles, and new pluralistic patterns of both electricity requirements and consumption are emerging.

In terms of the allocation of household tasks, a clear gender division is identifiable: in households with both sexes present, certain chores (washing, cooking, cleaning, child care) are chiefly done by women, while repairs and technical tasks are performed by the male household members (German Federal Statistical Office 2008b). With regard to the use of information and communication technology (ICT), the overall trend is towards gender equality, although the older people are, the greater the gender differences. However, young women are overtaking men in the use of ICT. While electricity is consumed in equal amounts by men and women, different appliances and devices are used. Children's use of ICT is increasing rapidly (German Federal Statistical Office 2011).

The change in household structures has brought with it a considerable improvement in the standards of comfort, which in turn has led to a sharp increase in the demand for electricity. Almost half of households now live in their own homes, especially in single family houses. Living space is constantly expanding, whereby the standard to which dwellings are fitted out varies considerably. The average living space of 20 square metres per person in 1960 rose to 43 square metres in 2006 – when the living space per person for three or more person households amounted to 28.5 square metres, for two-person households 43.4 square metres, and for single households as much as 62.5 square metres (German Federal Statistical Office 2010a). There is hardly a more graphic indicator of Germany's ever increasing prosperity on the one hand, and the extent of individual differences on the other.

Despite decreasing family size, the widespread use of electrical appliances and devices in households has not resulted in a detectable reduction of time spent on housework. Although improved household appliances have undoubtedly resulted in time savings for individual processes, most studies have come to the conclusion that time spent on housework has not diminished – at least not in the last century. The fact that the expectation of time savings through household mechanization has not been fulfilled is commonly referred to as the 'household paradox' (cf. Schwartz Cowan 1985).

6.3 Saving electricity in household production

It is a feature of industrial societies that the electricity consumption of private households rises in the long term. In Germany, the household share in overall electricity consumption rose from 19.4 percent in 1964 (Energiewirtschaft 1971) to 27.2 percent in 2009 (German Federal Ministry of Economics and Technology 2010). While overall electricity consumption between 1990 and 2005 rose by 14 percent, private household electricity consumption over the same period rose disproportionately by almost 22 percent (IFEU 2007).

Since the beginning of the environmental and anti-nuclear movement in the 1970s and 1980s – and increasingly since climate change has been more widely acknowledged – the trend described above has gradually become a focus of public debate. Although it is widely recognized by now that current consumption levels are incompatible with sustainable development, the public debate on how best to curb electricity consumption in private households is ongoing.

Two basic options are under discussion. The first proposes to maintain current household production (i.e. the production of goods and services in private homes) at existing levels, while minimising the use of electricity – which in effect amounts to increasing energy efficiency. The other proposal, which has enjoyed a recent revival – particularly in the wake of the Japanese nuclear accident in Fukushima 2011 – is to reduce our reliance on electricity-consuming appliances. This option would entail a cutback in household technology and thus remains somewhat taboo in political circles, as it is unlikely to find ready acceptance in the population at large.

6.3.1 Energy efficiency in household production

In a consumer study conducted as part of the "Transpose" project (Bürger 2009; cf. Brohmann/Bürger et al. in this volume) electricity-consuming appliances are divided into eight categories (see Figure 1). In the context of the present article they can be referred to as areas of domestic production. One part of the circle in Figure 1 contains the traditional areas: kitchen, heating, hot water, lighting and washing (75 percent); the other part contains the new areas: information and communication technology (ICT) and entertainment technologies. They make up 18 percent of the total electricity consumption. Thus, the majority of electricity is consumed by traditional household appliances, whereas the more recent electronic devices are relatively modest in their consumption. Nonetheless, it is that new technology sector that is currently under rapid expansion (German Federal Statistical Office 2007; 2010b).

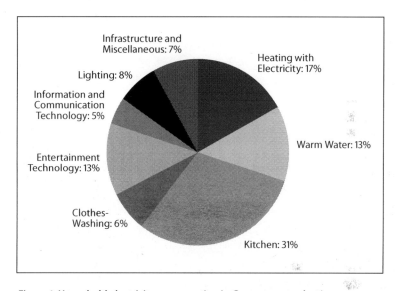

Figure 1: Household electricity consumption in Germany – production processes and categories of devices (updated 2004)
Source: Bürger 2009, p. 18

6.3.2 Product-related vs. behaviour-related electricity savings potentials

In terms of electricity savings in the context of household production, several practical options present themselves. Apart from cutting back on household technology (outlined above), a second option relates to the socio-technical context of energy consumption and relies on technological progress: efficiency is built into – and thus determined by – the product. Not least as a consequence of policy requirements, recent years have seen major technological progress – especially in the PC sector – which has led to the development of appliances that are more energy-efficient. In terms of the relationship between electricity consumption and household production, ultra-modern household equipment is far more efficient than older equipment, i.e. for the same result, the consumption of primary energy is considerably lower.

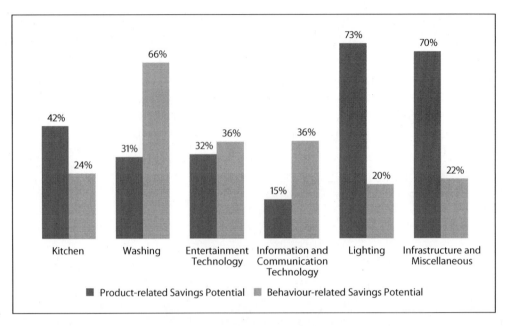

Figure 2: Product-related vs. behaviour-related savings potential for Germany (in percent)
Source: Bürger 2009, pp. 50–83 (author's diagram). The savings potentials are based on 2005 data.

A third option for saving electricity is again situated in the socio-technical context, but relates to user behaviour, i.e. household members contribute to a more efficient use of energy through their considered use of appliances.

The last two options, i.e. product-related vs. behaviour-related electricity savings potentials, are compared in Figure 2; the categories of energy consumption are taken from Figure 1 (Bürger 2009). The diagram illustrates the vast potentials for efficiency savings. In view of these results, it would be advisable to examine all processes in household production as to their product-related and behaviour-related savings potentials. In the case of ICT and clothes-washing, behavioural changes seem to be more effective than technical product improvements. Well known examples of behavioural changes are a) optimal loading of washing machines and dishwashers and b) switching off the standby functions of appliances. In some production areas – especially with technical product improvements – savings of over 70 percent are theoretically possible. Comparable behaviour-related savings can be achieved only with clothes-washing, namely by washing at low temperatures (i.e. not above 40 °C), and by loading the machine to full capacity. In the areas of heating, warm water, preparation of food, and washing, the study has found that the majority of savings involve a change of energy supplier. For instance, cooking can be switched from electricity to gas, although unlocking some of these potentials depends on the local infrastructure, i.e. on the externally available supply systems. Furthermore, product-related and behaviour-related savings potentials are not additive.

6.3.3 Social and socio-cultural barriers to saving energy

It was outlined in the introduction of this article how household and commercial production are each governed by distinct rules. The social functions and objectives of household production can be generally defined as the fulfilment of household members' needs in the most flexible manner possible – an endeavour that has to take account of a range of considerations: individuals' objective needs and subjective desires come into play, as do socially mediated ideas about the fulfilment of these needs and desires (cf. Di Giulio et al. in this volume). Together, these needs, desires and expectations feed into both individual and socially shared beliefs of what constitutes a good life – beliefs which in turn manifest themselves in a specific lifestyle (cf. Götz et al. 2011). The way in which household production and sustainability initiatives operate must be compatible with this lifestyle. If incentive programmes designed to promote energy efficiency ignore these considerations, and focus instead on the hypothesis that individuals are primarily motivated by financial savings, they are doomed to failure. Numerous studies point in this direction: in his review of evaluation studies of incentive programmes, Stern (1986) concludes that the frequent failure of incentive strategies is due to the fact that, where energy consumption decisions are concerned, individuals also take into account social dimensions, such as concerns over status, ethical norms or the avoidance of CO_2 emissions. Lutzenhiser et al. (2002) report that behavioural change as a consequence of financial incentives is highly variable and that economic decisions are dependent on social group membership. In a study of individuals living in two high-rise buildings, Hackett/Lutzenhiser (1991) were able to demonstrate that user behaviour cannot be reliably inferred from energy costs.

The above findings indicate that household production follows different rules from commercial production: while the latter is governed by market forces, household production is geared towards fulfilling individual needs. The findings of Wilhite et al. (1999) point in the same direction. They reconceptualise the technical terms of base and peak load as 'social base and peak load'. Drawing on the example of a typical Norwegian living room, in which a certain level of lighting and heating is considered socially appropriate, they illustrate what is meant by a social peak load. What is considered a wasteful use of energy in technical terms may be appropriate from a social perspective. In Norway, cosy warmth and bright light is considered part of a hospitable home. Guests should not be put in a position where they have to ask the homeowner to turn up the heating or switch on more lamps. Another example of Wilhite et al. (1996) demonstrates how in Japan, the housewife's switching on the air conditioner just before her husband's return from work also represents a social and cultural need. The authors draw an analogy between the concept of 'loading' in electrical engineering, and the culturally and socially necessary adjustment of the temperature as illustrated by the above examples; this phenomenon is termed 'social loading'.[7] Based on Wilhite et al. (1999), the following important social functions determining energy use can be identified. (This research tallies with findings in lifestyle research.)

- Social status and symbolic value,
- Social integration and convention,
- Safety and convenience,
- Embeddedness in structures and systems,
- Gender-specific negotiated norms of social care and child rearing (paraphrased from Wilhite et al. 1999).

These social functions can be seen as imperatives, strictly to be adhered to by household production. The socially 'correct' temperature, food, lighting, clothing, education, ICT equipment – all these prerequisites for social appropriateness, well-being and the realisation of people's lifestyles have to be put in place by household members. These imperatives take precedence over any efficiency considerations.

In relation to saving energy in household production the following can be concluded: Even though certain saving strategies make perfect sense technically and rationally, one should not assume that public acceptance is the natural consequence. For example, if the recommended lowering of the washing temperature does not conform to the household-specific hygiene standards or if switching off electronic devices (rather than leaving them on standby) is considered uncool or bothersome among friends, one should not expect that energy savings will be made in those areas.

6.4 Conclusions and prospects

The supply of private households with electricity represents a means of production that supports comprehensive production processes in private households. This production serves a range of purposes: individual consumption, needs fulfilment and socially

---

[7] This is not the only example that illustrates the gender specificity of the various areas of household production. We cannot elaborate on the issues here, but further information can be found under Alcántara et al. and Offenberger/Nentwich in this volume.

crucial tasks such as child rearing and care of the elderly. There might be isolated cases where the objective and subjective, social and societal benefits are questionable (critique of consumerism), but overall, the benefit to society is undisputed. Household production has become an appliance theme park that has steadily grown over the last hundred years and has inevitably led to an increase in electricity consumption. An unchecked continuation of this growth is incompatible with sustainable development. Not only the irreversible effects of CO_2 emissions, but also technological risks (e.g. Fukushima) require that energy production and consumption – including that of private households – is kept in check.

The approach to household production taken in this article may thus be summarised as follows: many strategies for behavioural change towards a more economical use of energy (e.g. pricing incentives, information campaigns) are ultimately based on the premise that household production follows similar guidelines of utility maximisation and motivations as does commercial production. According to this rationale, household members should primarily be concerned with rationalising production. In reality, however, households place far more value on the quality of their goods and services. Therefore, from the perspective of the household members, performing the daily (technically assisted) household activities under strict economic criteria would not be compatible with their understanding of housekeeping – and might even bring into question the core character of a household as a productive and cohesive social system. It is therefore hardly realistic to expect households to make extensive use of rational savings strategies. Instead, one should opt for strategies and measures that support household members in saving energy without losing out on production. (These strategies could be of technical or other nature.) This does not mean that households behave irrationally – they simply play by different rules.

It can thus be concluded that energy-saving measures should be compatible with household routines, and that, ideally, they themselves should become routines. Above all else, their implementation should not require additional effort on the part of household members.

The aim of the "Intelliekon" project (of which this present study is a part) is to install digital metering systems in households across Europe. Households will thus be provided with detailed and current data on their energy usage. In addition to simplified reading and billing procedures for the utility companies, smart metering offers households precise feedback on their consumption behaviour. The visualised feedback allows them to draw conclusions as to where they can save energy and become more efficient in their usage of energy (see also Gölz/Biehler 2008). However, feedback systems represent an additional home technology, which requires additional attention. The trends described in this article are thus continued: a new technology fails to reduce household members' workload and generate time savings; even if the new technology is of overall benefit to the household, it leads to new tasks that previously did not form part of the household responsibilities.

Fortunately, some suppliers of metering equipment have already become aware of this problem and have developed services for installing and monitoring the technology for private households. These services even come with a promise to carry out the necessary installation free of charge, and then to recoup their costs subsequently from the savings (cf. Starzacher 2010). As yet, it is unclear if these enticing offers will convince consumers to allow external companies to take control of these important household functions. There is no doubt that, to encourage a widespread diffusion of smart metering systems, strong incentives – or new regulations mandating certain levels of energy-efficiency for households – are essential.

References

Alcántara S., Wassermann S., Schulz M. (in this volume): Is "eco-stress" associated with sustainable heat consumption?

Brohmann B., Bürger V., Dehmel C., Fuchs D., Hamenstädt U., Krömker D., Schneider V., Tews K. (in this volume): Sustainable electricity consumption in German households – framework conditions for political interventions.

Bundesministerium für Umwelt, Naturschutz und Reaktorsicherheit (2010): Umweltbericht 2010. Umweltpolitik ist Zukunftspolitik. Berlin. [http://www.bmu.de/files/pdfs/allgemein/application/pdf/umweltbericht_2010.pdf; 14.06.2012]

Bundesministerium für Wirtschaft und Technologie (2010): Energie in Deutschland. Trends und Hintergründe zur Energieversorgung. Broschüre des Bundesministeriums für Wirtschaft und Technologie. Berlin.

Bürger V. (2009): Identifikation, Quantifizierung und Systematisierung technischer und verhaltensbedingter Stromeinsparpotenziale privater Haushalte. TRANSPOSE Working Paper No. 3, Freiburg. [http://www.uni-muenster.de/Transpose/en/publikationen/index.html; 14.06.2012]

Di Giulio A., Brohmann B., Clausen J., Defila R., Fuchs D., Kaufmann-Hayoz R., Koch A. (in this volume): Needs and consumption – a conceptual system and its meaning in the context of sustainability.

Energiewirtschaft (1971): Statistik der Energiewirtschaft 70, Heft 16.

Forsa (2005): Erhebung des Energieverbrauchs der privaten Haushalte für das Jahr 2005. Essen.

Geels F. W. (2004): From sectoral systems of innovation to socio-technical systems. Insights about dynamics and change from sociology and institutional theory. In: Research Policy 33: 897–920.

German Federal Statistical Office (Statistisches Bundesamt) (1993): Statistisches Jahrbuch 1993 für die Bundesrepublik Deutschland. Stuttgart: Metzler Poeschel.

German Federal Statistical Office (Statistisches Bundesamt) (2000): Datenreport 1999. Zahlen und Fakten über die Bundesrepublik Deutschland. Bundeszentrale für politische Bildung, Schriftenreihe, Band 365. Bonn.

German Federal Statistical Office (Statistisches Bundesamt) (2003): Wirtschaftsrechnungen. Einkommens- und Verbrauchsstichprobe – Ausstattung privater Haushalte mit langlebigen Gebrauchsgütern. Fachserie 15, Heft 1. Wiesbaden.

German Federal Statistical Office (Statistisches Bundesamt) (2007): Statistisches Jahrbuch 2007 für die Bundesrepublik Deutschland. Wiesbaden.

German Federal Statistical Office (Statistisches Bundesamt) (2008a): Wirtschaftsrechnungen. Einkommens- und Verbrauchsstichprobe – Ausstattung privater Haushalte mit ausgewählten Gebrauchsgütern. Fachserie 15, Heft 1. Wiesbaden.

German Federal Statistical Office (Statistisches Bundesamt) (2008b): Datenreport 2008 – ein Sozialbericht für die Bundesrepublik Deutschland. Bundeszentrale für politische Bildung. Bonn.

German Federal Statistical Office (Statistisches Bundesamt) (2010a): Statistisches Jahrbuch: 46 [https://www.destatis.de/DE/Publikationen/StatistischesJahrbuch/Bevoelkerung.pdf?__blob=publicationFile; 25.06.2012]

German Federal Statistical Office (Statistisches Bundesamt) (2010b): Wirtschaftsrechnungen. Laufende Wirtschaftsrechnungen – Ausstattung privater Haushalte mit ausgewählten Gebrauchsgütern 2009. Fachserie 15, Reihe 2. Wiesbaden.

German Federal Statistical Office (Statistisches Bundesamt) (2011): Wirtschaftsrechnungen. Laufende Wirtschaftsrechnungen – Ausstattung privater Haushalte mit ausgewählten Gebrauchsgütern. [https://www.destatis.de/DE/Publikationen/Thematisch/EinkommenKonsumLebensbedingungen/LfdWirtschaftsrechnungen/AusstattungprivaterHaushalte2150200117004.html, 25.06.2012]

German Federal Statistical Office (Statistisches Bundesamt) (2011): Wirtschaftsrechnungen. Private Haushalte in der Informationsgesellschaft – Nutzung von Informations- und Kommunikationstechnologien. [https://www.destatis.de/DE/Publikationen/Thematisch/EinkommenKonsumLebensbedingungen/PrivateHaushalte/PrivateHaushalteIKT2150400117004.pdf?__blob=publicationFile; 14.06.2012]

Glatzer W. (1990): Die Rolle der privaten Haushalte im Prozeß der Wohlfahrtsproduktion. In: Heinze R.G., Offe C. (ed.): Formen der Eigenarbeit – Theorie, Empirie, Vorschläge. Opladen: Westdeutscher Verlag. 15–29.

Glatzer W. (1994): Haushaltsproduktion und Haushaltstechnisierung. In: Schlosser H.D. (ed.): Gesellschaft – Macht – Technik. Vorlesungen zur Technikgenese als sozialer Prozess. Frankfurt am Main: Verlag GAFB. 154–179.

Glatzer W., Dörr G., Hübinger W., Prinz K., Bös M., Neumann U. (1991): Haushaltstechnisierung und gesellschaftliche Arbeitsteilung. Frankfurt, New York: Campus.

Glatzer W., Fleischmann G., Heimer T., Hartmann D.M., Rauschenberg R.M., Schemenau S., Stuhler H. (1998): Revolution in der Haushaltstechnologie. Die Entstehung des Intelligent Home. Frankfurt, New York: Campus.

Gölz S., Biehler M. (2008): Von der Energiesparforschung zur Energiepsychologie – Mögliche psychologische Perspektiven zur Gestaltung des künftigen Energiesystems am Beispiel "Smart Metering". Umweltpsychologie 12 (1): 66–79.

Götz K. (2007): Freizeit-Mobilität im Alltag oder Disponible Zeit, Auszeit, Eigenzeit – warum wir in der Freizeit raus müssen. Berlin: Duncker & Humblot.

Götz K., Deffner J., Stieß I. (2011): Lebensstilansätze in der angewandten Sozialforschung – das Beispiel der transdisziplinären Nachhaltigkeitsforschung. In: Otte G., Rössler J. (eds.): Lebensstilforschung. Kölner Zeitschrift für Soziologie und Sozialpsychologie, Sonderheft 51: 86–112.

Gruber E., Schlomann B. (2008): Stromsparen im Haushalt: Potenziale und Probleme. In: Fischer C. (ed.): Stromsparen im Haushalt. München: oekom. 22–41.

Hackett B., Lutzenhiser L. (1991): Social Structures and Economic Conduct: Interpreting Variations in Household Energy Consumption. Sociological Forum 6: 449–470.

Hampel J., Mollenkopf H., Weber U., Zapf W. (1991): Alltagsmaschinen. Die Folgen der Technik in Alltag und Familie. Berlin: edition sigma.

IFEU (2007): Innovative Stromrechnungen als Beitrag zur nachhaltigen Transformation des Elektrizitätssystems. Im Auftrag des DIW. Heidelberg. [http://www.ifeu.org/energie/pdf/Bericht_ Innovative_Stromrechnung_Okt07_221107_fin.pdf; 14.06.2012]

Kaufmann-Hayoz R., Bamberg S., Defila R., Dehmel C., Di Giulio A., Jaeger-Erben M., Matthies E., Sunderer G., Zundel S. (in this volume): Theoretical perspectives on consumer behaviour – attempt at establishing an order to the theories.

Lutzenhiser L., Harris C. K., Olsen M. E. (2002): Energy, society, and environment. In: Dunlap R. E., Michelson W. (eds.): Handbook of environmental sociology. Westport: Greenwood Press. 223–271.

Mayntz R., Hughes Th. P. (1988): The Development of Large Technical Systems. Frankfurt, New York: Campus.

Meyer S., Schulze E. (1993): Technisiertes Familienleben – Blick zurück und nach vorn. Berlin: edition sigma.

Meyer T. (2008): Private Lebensformen im Wandel. In: Geißler R. (ed.): Die Sozialstruktur Deutschlands. Wiesbaden: VS Verlag für Sozialwissenschaften. 331–338.

Negt O., Kluge A. (1993): Geschichte und Eigensinn. Frankfurt am Main: Suhrkamp.

Offenberger U., Nentwich J. (in this volume): Socio-cultural meanings around heat energy consumption in private households.

Peukert R. (2008): Familienformen im sozialen Wandel. 7. Auflage, Wiesbaden: VS-Verlag für Sozialwissenschaften.

Richarz I. (1991): Oikos, Haus und Haushalt. Ursprung und Geschichte der Haushaltsökonomik. Göttingen: Vandenhoeck & Ruprecht.

Ropohl G. (2009): Allgemeine Technologie – Eine Systemtheorie der Technik. 3. Auflage, München, Wien: Hanser.

Schäfer D. (2004): Unbezahlte Arbeit und Bruttoinlandsprodukt 1992 und 2001. In: Wirtschaft und Statistik 9: 960–971.

Schwartz Cowan R. (1985): More Work for Mother: The Ironies of Household Technology from the Open Hearth to the Microwave. New York: Basic Books.

Starzacher N. (2010): Meßstellenbetreiber – neuer Ansprechpartner für Endkunden. In: Smart Metering. Geschäftsmodelle mit Kundennutzen. Dokumentation OTTI-Fachforum Augsburg, 30. Juni 2010.

Stern P. C. (1986): Blind Spots in Policy Analysis: What Economics Doesn't Say about Energy Use. Journal of Policy, Analysis and Management 5: 200–227.

Stieß I., van der Land V., Birzle-Harder B., Deffner J. (2010): Handlungsmotive, -hemmnisse und Zielgruppen für eine energetische Gebäudesanierung. Frankfurt am Main. [http://www.enef-haus.de/fileadmin/ENEFH/redaktion/PDF/isoe_Handlungsmotive_ -hemmnisse_und_Zielgruppen_Bericht_EnefHaus2010_end_kompr.pdf; 25.06.2012]

VDEW (1993): Analyse und Prognose des Stromverbrauchs der privaten Haushalte 1970–1990– 2005–2010. Auswertungsbericht alte Bundesländer. Frankfurt am Main.

Verplanken B., Orbell S. (2003): Reflections on past behavior: A self-report index of habit strength. Journal of Applied Social Psychology 33 (6): 1313–1330.

Wilhite H., Høivik A., Olsen J.-G. (1999): Advances in the use of consumption feedback information in energy billings: The experience of a Norwegian energy utility. Mandelieu: Proceedings of the Summer Study of the European Council for an Energy Efficient Economy ECEEE, panel III/02.

Wilhite H., Nakagami H., Masuda T., Yamaga Y., Haneda H. (1996): A cross-cultural analysis of household energy-use behavior in Japan and Norway. Energy Policy 24: 795–803.

Wood W., Quinn J.M., Kashy D. A. (2002): Habits in everyday life: Thoughts, Emotion, and Action. Journal of Personality and Social Psychology 83 (6): 1281–1297.

Melanie Jaeger-Erben, Ursula Offenberger, Julia Nentwich, Martina Schäfer, Ines Weller[1]

7 Gender in the focal topic "From Knowledge to Action – New Paths towards Sustainable Consumption": findings and perspectives

7.1 Introduction: focus and objectives of our analysis

The relevance of gender to both sustainability (e.g. Weller 1999; Schultz 1999; Martine/Villarael) and sustainable consumption (e.g. Weller 2004) has been a subject of academic study for many years. One notable outcome was the thesis of the "feminisation of environmental responsibility" – aptly illustrated by the example of waste separation (Schultz/Weiland 1991), which was shown to be mainly carried out by women. The relevance of gender is also implicit in the very definition of sustainability. For instance, in a purely economic conception of sustainability – which is largely focused on the market economy – no attention is given to unpaid housework (cf. Schön et al. 2002), an oversight that has negative environmental and social consequences (cf. Spitzner 1999). Despite some acknowledgement that the topic is neglected, the lack of relevant research is discernible (Schultz/Stieß 2009), and "gender gaps" persist in sustainability-oriented policy instruments (Vinz 2009). This has occurred partly because measures aimed at promoting sustainability are rarely considered in terms of their separate consequences for men and for women.

It is therefore a welcome development that, as part of the announcement of the focal topic "From Knowledge to Action – New Paths towards Sustainable Consumption" (Federal Ministry of Education and Research 2006), applicants were asked to incorporate into their proposals a "gender perspective" with "relevant insights that could feed into recommendations for social and governmental policy-making". In this article we provide an outline of how six projects fulfilled this brief, classifying their approaches

---

[1] Melanie Jaeger-Erben and Martina Schäfer are members of "LifeEvents", Ursula Offenberger and Julia Nentwich are members of "Seco@home". The following members of other project groups contributed to this article: Victoria van der Land ("ENEF-Haus"), Saskia-Fee Bender and Birgit Blättel-Mink ("Consumer/Prosumer"), Sabrina Gebauer and Susanne Ihsen ("User Integration") and Katy Jahnke ("Heat Energy").

into different types of social construction of gender. We then discuss the projects' outcomes in terms of their scientific and practical relevance. Their findings will be used for recommendations for future projects.

7.2 Background: a multi-dimensional perspective on the social construction of gender

Our classification and synthesis of the gender-relevant project outcomes is based on the concept of "doing gender" (West/Zimmermann 1987), according to which gender is not a feature of individuals (i.e. something that is "inherent" in individuals), but the result of social construction (i.e. something individuals "do"). As such, gender becomes part of institutions, bodies, artefacts and knowledge bases. Drawing on Harding (1986), we distinguish between three dimensions of social construction:
- *Individual dimension*: What gender differences motivate individuals' attitudes, actions and orientations?
- *Structural dimension*: What is the role of a gendered division of labour, e.g. a division between paid work and unpaid, reproductive work?
- *Symbolic dimension*: What is the relevance of interpretive paradigms, bodies of knowledge, and cultural artefacts that have either feminine or masculine connotations? For instance, what is the significance of technology as a "masculine domain"?

In the actual construction of gender the above three dimensions are closely linked, but they can be treated separately for the purpose of gender analysis. The multi-dimensional approach to gender shows parallels to dimensions adopted in research on sustainable consumption, where consumption can be analysed at both an individual and a structural level (cf. Jackson 2005). Recent research into transformation processes suggests that the interactions between individual and structural changes should receive particular attention (Brand 2010; Spaargaren/Oosterveer 2010; also cf. Kaufmann-Hayoz et al. in this volume).

7.3 Gender-related outcomes of the projects

For the present synthesis the results of six project groups were put at our disposal.² Three projects focused on heat and electricity consumption in domestic households: "Heat Energy" looked at everyday decisions around heating and ventilation, as well as efforts to renovate one's home; the "ENEF-Haus" project focused on general refurbishing decisions by private homeowners, and the "Seco@home" project analysed homeowners' decisions when buying a new heating system. The "User Integration" project was concerned with user involvement in innovation processes; the "Consumer/Prosumer" project looked at motivations, attitudes towards the environment, and green behaviour in the context of online second-hand trading; and the "LifeEvents" project examined changing patterns of consumer behaviour related to nutrition, transport/mobility and energy use after relocation or the birth of a first child.

We now present the findings of these projects based on the three gender dimensions listed above. Whilst the individual projects often relate to several dimensions, we focus on the key results of each project. Table 1 provides an outline of the six projects.

The *individual dimension* is predominant in the following project groups: "Heat Energy", "ENEF-Haus", "Consumer/Prosumer" and "User Integration".³

The **"Heat Energy"** project group conducted two surveys involving 136 homeowners and 185 tenants. In the homeowner survey, householders were asked who (man, woman or both) was responsible for a) the washing, cooking and childcare, b) the initiation of the renovation and c) the organisation of the renovation.⁴ Additional questions related to who took responsibility for regulating the heating and ventilation, and who decided on room temperature in the living room and bedrooms. The tenant survey confirmed the hypothesis of a "traditional" division of labour in areas such as washing and cooking (i.e. women are in charge of these areas). Whilst men and women gave very similar responses to the questions relating to household chores, their responses relating to renovation varied considerably: half the men claimed that they had initiated a project, but according to women a mere 13 percent of men did so. In comparison, 39 percent of men and 48 percent of women said they had initiated it together. The researchers found a similar pattern in the organisation of the refurbishing: 56 percent of men and 29 percent of women claimed they did the organising, whereas 3 percent of

---

2   The results of the other four project groups were not relevant to the context of this article.

3   Some results also concern the structural level of gender relations and division of labour. However, in view of the fact that the standardised surveys of the three projects focused on individuals rather than households or couples, the results do not allow for inferences relating to the structural level.

4   For other results of this project, see Alcántara et al. in this volume.

Table 1: Outline of projects and results in terms of gender

Project name	Key gender dimensions	Focus of investigation	Key result	Recommendations for action
Heat Energy	Individual	Division of labour in household, responsibility for renovation	Joint responsibility for activities in the context of renovation	Both women and men should be addressed in the context of renovation.
ENEF-Haus	Individual and structural	Responsibility for renovation	Renovation as a joint project	Do not approach users in a gender-specific way.
Consumer/ Prosumer	Individual	Use of eBay	Significance of specific life phases, e.g. parenthood in use of eBay	Gender-specific exploitation of sustainability potential is possible.
User Integration	Individual	Development of sustainable products	In product development, internally diverse user groups generate more creative ideas than homogenous groups.	Integration of users into development of sustainable products should be oriented towards diversity categories.
LifeEvents	Structural and symbolic	Changes in everyday routines after birth of first child or relocation	Parenthood often entails a return to traditional gender relations.	Gender-specific division of labour should be taken into account in the planning of intervention measures.
Seco@home	Structural and symbolic	Heat consumption in own home	Heating has different gender connotations as "home making" or "facility management".	Development, design and marketing of technologies should consider home heating as a fundamental part of practices of "home making".

men and 35 percent of women said it had been their partner's responsibility. Regarding the responsibility for heating and ventilation, responses diverged considerably when only one person was in charge, but were much more similar when both partners were involved.

The project group "ENEF-Haus" interviewed homeowners who had carried out an energy-saving project within the three years leading up to the survey. A standardised survey was conducted, comprising 1008 refurbishers, of whom 877 lived in a household of at least two people.[5] Similar to the respondents of the "Heat Energy" project, the sub-sample indicated joint responsibility for most project decisions:[6] 70 percent of men and 78 percent of women indicated shared responsibility for commissioning contractors and tradespeople (bricklayers, plumbers etc.). Somewhat fewer indicated shared responsibility for project management and liaison with tradespeople (59 percent of men and 67 percent of women); more men than women took on those responsibilities (36 vs. 11 percent). The authors of the "ENEF-Haus" study concluded that "renovation tasks are generally shared among couples. This applies particularly to the decisions relating to the type of renovation and to financing" (Stieß et al. 2010, p. 40; translated by C. Holzherr). Joint responsibility appears to be the norm, as women tend to refer to it more frequently than men. A complementary qualitative study addressed issues of how such decisions were jointly reached and negotiated.

The survey conducted by the **"Consumer/Prosumer"** project group involved 2,500 users of an online second-hand trading platform. Amongst other questions, they were asked to state their motives for online trading and to indicate how much they bought and sold. The five consumer types identified (cf. Blättel-Mink et al. 2011) were not gender-specific. Men and women did, however, differ in their product preferences. Women bought and sold more "fashion, clothes, accessories", "children's and baby articles" and "books", whereas men were more interested in "motor vehicles" and "entertainment technology". Both women and men traded in new and second-hand products. The products bought and sold by women were assessed as more sustainable as their use did not involve any further resources (energy, water). Men and women diverged significantly in their motives for trading on eBay. When asked to rate their answers to the question of why they traded on eBay, women agreed more strongly ("I agree" or "I strongly agree") with the following statements: "because it is enjoyable", "because it helps protect the environment", "because I can buy products at reasonable/reduced prices", and "because I do not have to observe conventional trading hours". Moreover, more women (69 percent of women vs. 60 percent of men) were prepared to choose, and pay a bit extra for, climate-neutral environmentally friendly shipping options (49 percent vs. 41 percent). Divergences were not found for any other motives. Finally, men did not agree more strongly than women with any one motive. On the basis of these findings, the "Consumer/Prosumer" project proposed a gender-specific approach to promoting sustainable consumption, addressing women and men in terms

---

[5] Further information relating to the composition of the samples, e.g. in terms of age and income, can be found in Stieß et al. 2010, pp. 26 ff.

[6] 68 percent of men and 76 percent of women indicated joint responsibility for gathering information; over 80 percent of both sexes reported to have reached joint decisions on suitable measures.

of the sustainability potentials that they were most likely to fulfil. Furthermore, the data suggested an association between online trading and particular life phases. For instance, a large proportion of respondents who placed value on the green aspect of online trading and on the freedom to shop at all hours of the day were mothers of young children. In a complementary qualitative survey involving twelve parents, it emerged that parents used online trading for the particular purpose of buying branded products at reasonable prices and selling them again once their children had outgrown them. Parents were thus identified as a target group for second-hand sustainability-oriented online trading.

The sub-project on gender and diversity conducted as part of the **"User Integration"** project group also considered the individual dimension of gender, but had a different focus than the above projects. Its objective was to explore how users were involved in different phases of sustainable innovations in housing, product design and transport. One objective was to find out what impact – in terms of gender, age, education and expertise of participants – the composition of a workshop had on group collaboration and on the generation of ideas. Drawing on their empirical findings on gender-specific consumer behaviour and attitudes (as well as differences between younger and older people), the authors hypothesised that a heterogeneous group would be more inclined to take account of different needs and contexts for implementation and would generate more creative ideas. After conducting twelve innovation workshops with different groupings of participants (e.g. lead-user/non-lead user, heterogeneous/homogeneous groups in terms of age and gender), the researchers interviewed 53 workshop participants and evaluated 157 questionnaires. They also asked experts to rate the creativity of the ideas generated in workshops. The heterogeneous groups came up with highly creative ideas – particularly when the workshop participants were aware of the diversity of skills within the group. The participants in the more heterogeneous groups also tended to be more sensitive to the potential for controversy and the necessity for compromise. This had a beneficial effect on the workshop outcome as long as the workshop atmosphere was positive. In the more homogeneous workshops (e.g. only men or older people) group composition was discussed, with some participants perceiving homogeneity as a disadvantage. The project findings thus confirmed what had been found in previous studies: under favourable conditions, a heterogeneous group can raise the potential for innovation (cf. Kutzner 2010).

The *structural dimension* of doing gender was mainly considered in qualitative sub-projects of the project groups "ENEF-Haus", "LifeEvents" and "Seco@home"; the latter two also looked at the *symbolic dimension*.

Before the **"ENEF-Haus"** researchers conducted their quantitative surveys, they interviewed a number of homeowner couples who had indicated they took joint responsibility for efforts at refurbishing their homes (van der Land 2010). During these interviews the researchers noted an association between the particular gender-specific division of labour and the time within which the refurbishing had to be completed. In cases where both partners worked full-time, the men took on the masculine tasks (e.g. liaison with tradespeople, physical activities), whereas women who had more time at their disposal, because they worked part-time, assumed responsibility for those masculine tasks. Overall, gender appeared to play less of a role in refurbishing projects than did lifestyle, education and income (van der Land 2010). Attitudes towards renewable energies, the environment and climate change, for instance, played a crucial role in these renovation decisions. Such attitudes tended to diverge frequently between unmarried partners, but rarely between married partners. Given that both partners were involved in the refurbishing, a gender-specific approach was considered unsuitable. Overall, people who had similar lifestyles, attitudes and reservations towards energy-efficient refurbishing projects were identified as the ideal target group, and the researchers recommended that advice and information should be designed for both women and men.

The focus of the "LifeEvents" project group was on the changes in social practices and everyday routines in the context of life-course transitions.[7] The researchers conducted 40 interviews with individuals who had recently relocated and with parents who had recently had their first child. They found that with new parents, particularly during the child's first months, the household division of tasks had become more traditional. For instance, more female respondents had decided to stay at home for an extended period after the birth of their first child – even at the potential cost of an uncertain professional future. A similar trend towards conventional gender roles was noted for people who had relocated: Women tended to move into the town or city where their male partner had found work. It was further noted that mothers tended to perform more household tasks (e.g. shopping and food preparation), even if these tasks had previously been shared; they explained that since they were "at home anyway" they could easily carry out those tasks.[8] In this context, consumption and consumer behaviour assume a key role as part of social practices. An analysis of the everyday routines of young mothers revealed that feeding the baby, transporting it safely and providing it with a cosy and warm home were of key importance, and that mothers organised their daily lives around these objectives. The project group was able

---

7   See Schäfer/Jaeger-Erben in this volume for further findings of this project group.

8   In the literature such developments are described as "traps of traditionalisation" (Rüling 2007). They tend to be associated not only with structural inequalities (e.g. lower wages for women), but also with competence-related gender stereotypes.

to identify the significance of socially constructed, consumption-related models for living. Many mothers had been exposed to gender-specific "appropriate" attitudes and behaviours – e.g. the "caring mother" – through media (e.g. books, magazines, guidelines), midwives and doctors, social networks, internet forums, parent groups etc. A review of several parenting publications revealed that information and products tend to be targeted at mothers, thereby pre-emptively allocating responsibility in areas such as baby care and nutrition, transport, the use of certain devices and the use of energy for heating. Among the relocators it was observed that if couples moved in together, their negotiation of household tasks often resulted in very conventional outcomes. For example, men took on responsibilities for installing and maintaining electrical appliances while women were in charge of cooking. Turning to the newly established shared households of people who were neither in a relationship nor had children, the researchers found that they negotiated tasks along less conventional lines: individuals chose those tasks that suited their individual competences and preferences. It appears that couples and families are particularly prone to conventional consumption patterns: women and men take on and ascribe to each other very traditional gender roles. The symbolic dimension of gender was apparent in the role models (of mothers, fathers and couples) propagated in parenting magazines etc. and alluded to in the interviews, where particular areas of responsibility and activities were associated with masculinity or femininity. The respondents used these associations as they negotiated and reoriented their task allocation. For instance, a common reference in interviews was "motherly intuition", which apparently invested women with a particular flair for looking after children. The above results suggest that anyone formulating strategies aimed at promoting sustainability should pay attention to stereotypes of persistent gender roles. Tasks should not be allocated in ways that enhance existing inequities, for example, allocating responsibility to women for healthy eating.

A sub-project of **"Seco@home"** also employed a qualitative research method to look at the interweaving of the symbolic and structural dimensions of doing gender.[9] The analysis showed that couples distinguished between household tasks in terms of the categories "affinity with technology" as a male domain and "aesthetics" as a feminine domain. The objective of the subproject was to investigate domestic, renewable energy heating systems. The analysis was based on participant observation at consumer fairs, expert interviews, interviews with homeowner couples and marketing brochures. The analysis revealed two facets of the gender coding of "affinity with technology" and "aesthetics". In the marketing brochures, heating and warmth carried differing connotations depending on the location of the heating (e.g. a wood-burning stove in the

---

9   For further findings from this project see Offenberger/Nentwich in this volume.

living room or a heating boiler in the basement): the feminine notions of emotionality and cosiness coupled with the idea of "home making" contrasted with the male functional and technology-oriented association of "facility management" (Offenberger/Nentwich 2009). These symbolic gender distinctions were also made relevant in the interviews with the homeowners: technological competence was associated with masculinity whereas low-tech, aesthetically-oriented "home making" was associated with femininity. It should be noted that the stereotypical gender perceptions that appeared during the interviews did not translate consistently into behaviour. Thus, it showed that women were, in fact, competent users of everyday technologies and men did contribute to "home making" (e.g. by starting the fire in the wood stove). Overall, the findings demonstrated that people did not merely consider heating the home in its functional dimensions (e.g.: is it efficient or not?); instead, it was part of the group of practices that ensured a cosy home in which individuals could live and interact with each other. As such, the management of home heating conformed, like other social practices, to the same principles of a gendered division of labour. Based on their findings, the authors recommended that heating technology development should increasingly take into account aspects that are associated with the everyday practices of "home making".

7.4 Summary and discussion of results

The projects presented above investigated a range of gender dimensions and generated recommendations for promoting and implementing sustainable consumption patterns. The results show that in the implementation of sustainable consumption patterns, the individual, structural and symbolic dimensions of gender are relevant. The gender dimension focused on in each project depended on the theoretical understanding of gender in the specific context and on the research questions. In the individual dimension of gender, the focus tends to be on gender-specific aspects of user behaviour and issues around addressing users, but gendered attitudes in product design are also taken into account. This dimension received particular attention by the project groups "Consumer/Prosumer", "ENEF-Haus", and "User Integration".

Problems can arise, however, if measures aimed at changing individual consumer behaviour focus one-sidedly on (possible) problems created by gender differences. Thus, it remains a challenge to accommodate the interests and requirements of consumers without propagating the same gender stereotypes that deny the existence of variety within gender groups. A possible solution is linking gender with other categories of social inequality, along the lines of intersectional analysis (Winker/Degele 2009). That approach focuses on the interactions of categories such as class, race, gender, age, sexuality, nationality and religion, which normally differentiate between groups of individuals.

The findings relating to age and life phases, generated in the projects "Consumer/Prosumer" and "ENEF-Haus", illustrate the interaction of gender with other categories. Both projects demonstrate that gender is always linked to other areas of people's lives and that those promoting sustainable consumption should always keep in mind how gender interacts with other factors such as lifestyle, life phase, age, sexuality, nationality and religion. In fact, focusing on gender provides access to a range of other differences within a group of consumers and makes it possible to formulate more targeted recommendations.

Incorporating the structural dimension of doing gender into the analysis, however, reveals how the differences between the sexes have emerged and evolved and are reinforced. The projects "ENEF-Haus", "LifeEvents", and "Seco@home" investigated such issues, e.g. how women and men divide up their labour and negotiate the allocation of tasks. The findings illustrate the key role that such negotiation processes play in consumer decisions within multiple-person households. When a structural dimension is incorporated, the gendered characteristics or activities emerge as not inherently female or male, but as relative in terms of how they are aligned with or delineated from each other.

Evaluating gender data can be problematic in surveys where multi-person households are involved, but only one person – or in a worst-case scenario the perceived "head of household" – is interviewed. The responses are inevitably given by either a man or a woman and are therefore of limited validity. Turning to the projects involved in our comparison, we therefore suspect that responses relating to who was responsible for what and which decisions were taken jointly depend on the perceptions of the particular household member being interviewed. With this in mind, the data collection methods in future studies should always take account of the dynamics within couples and other relationships in a given household.

As part of everyday consumption practices, gender distinctions at the individual, structural and symbolic levels are always interlinked. One option is to focus exclusively on individual differences between men's and women's attitudes and behaviours. The structural and symbolic dimensions, however, should not be neglected. They contribute to an understanding of the conditions for, and the context in which, particular consumer behaviours arise and provide a platform from which to initiate changes. Moreover, they can indicate where the promotion of sustainable consumption might risk reproducing or reinforcing unequal divisions of labour and structural inequalities. Thus, a research perspective that incorporates the individual, structural and symbolic dimensions is well placed to contribute towards appropriate and realistic transformation processes.

7.5 Possible consequences for future research on gender and sustainable consumption

Apart from a brief reference to the "gender sensitivity" of sustainable consumption and a broad requirement that proposals should incorporate a "gender perspective", the announcement for the research programme within the focal topic contained no further specifications as to how exactly the gender dimension should be assessed. That is, it gave no pointers to existing gender-related theoretical approaches or research findings. The variety of ways in which the various projects chose to approach gender reflects this fact.

In an effort to consolidate the research area of sustainable consumption, or social-ecological research, we propose that future focal topic research agendas that include gender research specify more clearly how gender analyses might usefully be carried out. We propose that such a specification include two dimensions.

First, we need a classification of gender issues in terms of their relevance to various areas within sustainable consumption. This could be the result of a synthesis process which would include theoretical approaches, existing research projects and indications of gaps and blind spots in current research. Project groups could also be supported in generating gender-related information as part of their research schedule. The findings could then be embedded in the overall agenda of the focal topic.

Second, knowledge around gender should ideally flow into the planning and development of a focal topic. If research programmes want to expressly incorporate gender aspects, it is essential to consult gender experts in the planning stage of a programme and to implement their suggestions during the review process. With a view to raising awareness of gender issues and demonstrating to project groups how gender can be incorporated into a project, gender issues should be included from the very beginning of a project (cf. Schäfer et al. 2006). For instance, each group could be required to develop gender-specific research questions with the help of the accompanying research project team; by the end of the project, the leaders and a number of the team members would agree on a common perspective to take forward. The project group would then be more likely to support the chosen perspective and collaborate on integrating the gender-related, and other, findings in the course of the project.

The two measures proposed above can also help to channel learning from previous experiences and improve the quality of research on sustainability.

References

Alcántara S., Wassermann S., Schulz M. (in this volume): Is "eco-stress" associated with sustainable heat consumption?

Blättel-Mink B., Bender S.-F., Dalichau D., Hattenhauer M. (2011): Nachhaltigkeit im online gestützten Gebrauchtwarenhandel: empirische Befunde auf der subjektiven Ebene. In: Behrendt S., Blättel-Mink B., Clausen J. (eds.): Wiederverkaufskultur im Internet. Chancen für nachhaltigen Konsum auf Basis einer Analyse der Handelsplattform eBay. Berlin: Springer. 69–126.

Brand K.-W. (2010): Social Practices and Sustainable Consumption: Benefits and Limitations of a New Theoretical Approach. In: Gross M., Heinrich H. (eds.): Environmental Sociology: European Perspectives and Interdisciplinary Challenges. Springer Business and Media. 217–235.

Bundesministerium für Bildung und Forschung (2006): Bekanntmachung: Richtlinien zur Förderung von Forschungs- und Entwicklungsvorhaben im Rahmen der sozial-ökologischen Forschung zum Themenschwerpunkt "Vom Wissen zum Handeln – Neue Wege zum nachhaltigen Konsum". Berlin. [http://www.sozial-oekologische-forschung.org/_media/Richtlinien_Nachhaltiger_Konsum.pdf.; 14.06.2012]

Harding S. (1986): The science question in feminism. Ithaca: Cornell University Press.

Jackson T. (2005): Motivating Sustainable Consumption – SDRN briefing 1. London: Policy Studies Institute.

Kaufmann-Hayoz R., Bamberg S., Defila R., Dehmel C., Di Giulio A., Gölz S., Jaeger-Erben M., Matthies E., Sunderer G., Zundel S. (in this volume): Theoretical perspectives on consumer behaviour – attempt at establishing an order to the theories.

Kutzner E. (2010): Diversity im Innovationsprozess – erste Ergebnisse einer Untersuchung. In: Jacobsen H. S., Schallock B. (eds.): Innovationsstrategien jenseits traditionellen Managements. Stuttgart: Fraunhofer Verlag. 41–58.

Martine G., Villarael M. (1997): Gender and Sustainability. Re-assessing Linkages and Issues. [www.fao.org/waicent/faoinfo/sustdev/wpdirect/wpan0020.htm; 14.06.2012]

Offenberger U., Nentwich J. C. (2009): Home Heating and the Co-construction of Gender, Technology and Sustainability. In: Women, Gender and Research, Kvinder, Kon & Forskning 3–4: 83–89.

Offenberger U., Nentwich J. (in this volume): Socio-cultural meanings around heat energy consumption in private households.

Rüling A. (2007): Jenseits der Traditionalisierungsfallen. Frankfurt am Main u.a.: Campus.

Schäfer M., Jaeger-Erben M. (in this volume): Life events as windows of opportunity for changing towards sustainable consumption patterns? The change in everyday routines in life-course transitions.

Schäfer M., Schultz I., Wendorf G. (eds.) (2006): Gender-Perspektiven in der Sozial-ökologischen Forschung: Herausforderungen und Erfahrungen aus inter- und transdisziplinären Projekten. München: oekom.

Schön S., Keppler D., Geißel B. (2002): Gender und Nachhaltigkeit. Sondierungsprojekt zur theoretischen und methodischen Weiterentwicklung der Forschungsansätze zum Themenfeld Gender und Nachhaltigkeit. Technische Universität Berlin, Zentrum Technik und Gesellschaft, Discussion Paper 01/02, September 2002. [http://www.tu-berlin.de/uploads/media/Gender_01.pdf; 14.06.2012]

Schultz I. (1999): Eine feministische Kritik an der Studie Zukunftsfähiges Deutschland. Statt einer ausschließlich zielorientierten Konzeptualisierung erfordert nachhaltige Entwicklung eine prozeßorientierte Konzeptualisierung. In: Weller I., Hoffmann E., Hofmeister S. (eds.): Nachhaltigkeit und Feminismus. Neue Perspektiven – Alte Blockaden. Bielefeld: Kleine. 99–109.

Schultz I., Stieß I. (2009): Gender aspects of sustainable consumption strategies and instruments. Frankfurt am Main: ISOE.

Schultz I., Weiland M. (1991): Frauen und Müll. Frauen als Handelnde in der kommunalen Abfallwirtschaft. Studie im Auftrag des Frauenreferats des Magistrats der Stadt Frankfurt am Main. Sozial-ökologisches Arbeitspapier 40. Frankfurt am Main.

Spaargaren G., Oosterveer P. (2010): Citizen-Consumers as Agents of Change in Globalizing Modernity: The Case of Sustainable Consumption. In: Sustainability 2 (7): 1887–1908.

Spitzner M. (1999): Krise der Reproduktionsarbeit – Kerndimension der Herausforderungen eines öko-sozialen Strukturwandels. Ein feministisch ökologischer Theorieansatz aus dem Handlungsfeld Mobilität. In: Weller I., Hoffmann E., Hofmeister S. (eds.): Nachhaltigkeit und Feminismus. Neue Perspektiven – Alte Blockaden. Bielefeld: Kleine. 151–165.

Stieß I., van der Land V., Birzle-Harder B., Deffner J. (2010): Handlungsmotive, -hemmnisse und Zielgruppen für eine energetische Gebäudesanierung – Ergebnisse einer standardisierten Befragung von Eigenheimsanierern. Frankfurt am Main. [http://www.enef-haus.de/fileadmin/ENEFH/redaktion/PDF/Befragung_EnefHaus.pdf; 18.06.2012]

van der Land V. (2010): Die Rolle von Geschlecht im Sanierungsprozess. Ergebnisse einer qualitativen Befragung von Eigenheimsaniererinnen und Eigenheimsanierern. Frankfurt am Main. [http://www.enef-haus.de/fileadmin/ENEFH/redaktion/PDF/Die_Rolle_von_Geschlecht_im_Sanierungsprozess.pdf; 14.06.2012]

Vinz D. (2009): Gender and Sustainable Consumption. A German Environmental Perspective. In: European Journal of Women's Studies 16: 159–179.

Weller I. (1999): Einführung in die feministische Auseinandersetzung mit dem Konzept Nachhaltigkeit. Neue Perspektiven – Alte Blockaden. In: Weller I., Hoffman E., Hofmeister S. (eds.): Nachhaltigkeit und Feminismus. Neue Perspektiven – Alte Blockaden. Bielefeld: Kleine. 9–32.

Weller I. (2004): Nachhaltigkeit und Gender. Neue Perspektiven für die Gestaltung und Nutzung von Produkten. München: oekom.

Weller I., Hoffmann E., Hofmeister S. (1999): Nachhaltigkeit und Feminismus: Neue Perspektiven – Alte Blockaden. Bielefeld: Kleine.

West C., Zimmerman D. H. (1987): Doing Gender. In: Gender & Society 1 (2): 125–151.

Winker G., Degele N. (2009): Intersektionalität. Zur Analyse sozialer Ungleichheiten. Bielefeld: Transcript.

Sophia Alcántara, Sandra Wassermann, Marlen Schulz

8 Is "eco-stress" associated with sustainable heat consumption?

8.1 Introduction

In Germany, domestic consumption of heating and hot water is responsible for over 80 percent of final energy consumption (Tzscheutschler et al., 2009). Against this background, the "Heat energy" project group – which ran from March 2008 to June 2011 – set out to identify barriers to and incentives for "sustainable heat consumption".[1] Given that gender differences were not the primary objective of this project, they were investigated only in a second phase. The thesis of the "feminisation of environmental responsibility" (Schultz & Weiland 1990), which served as a point of departure for the gender analysis, posits that uptake of sustainable consumption practices results in additional work for women (Weller 2005, p. 174; Dörr 1993), which, in turn, tends to cause what Schwartau-Schuldt (1990) terms "eco-stress". As a consequence of symbolically loaded gender attributions (cf. "doing gender"[2]), women are generally regarded as more environmentally aware than men, and it is indeed women who tend to take on the extra work involved in green shopping or the separation of household waste. As a consequence of persisting attitudes favouring a gender-based division of labour (the structural dimension of "doing gender"), women generally take more everyday consumption decisions than men. With this in mind, the objective of our analyses was to a) identify gender-specific attitudes and b) examine whether the feminisation of environmental responsibility, and thus "eco-stress", extended to sustainable heat consumption.

---

[1] Whilst the main objective of the project was an analysis of behavioural patterns, aspects of the buildings and the sustainability of different supply options were also investigated.

[2] Cf. Jaeger-Erben/Offenberger et al. in this volume for the theoretical background to the concept of "doing gender" (West/Zimmermann 1987).

8.2 Analysis of focus groups

In altogether six "focus groups"[3], consisting of homeowners and tenants from Leipzig and Stuttgart, attitudes towards and behaviours around domestic heating were discussed. The composition of each focus group was established on the basis of a quantitative questionnaire distributed to tenants and homeowners in Sachsen and Baden-Württemberg. Factor and cluster analyses were employed to identify lifestyle types in the context of heat consumption, and the resulting socio-demographic variables were used to determine the composition of the focus-groups (see Table 1).[4]

Table 1: Focus groups

No.	consumers	groups	location
A	female tenants	younger women (up to 35) with children[5]	Stuttgart
B	female and male tenants	middle-aged fathers and mothers with at least one child per household	Stuttgart
C	male tenants	middle-aged men (35–60) without children in the household	Leipzig
D	female and male homeowners	older people whose children had moved out	Stuttgart
E	female and male homeowners	fathers and mothers with at least one child per household	Stuttgart
F	male homeowners	fathers with children in the household	Leipzig

Since the composition of the focus groups was not originally conceived with gender in mind, it was not possible to evaluate the data with the help of pre-formulated gender hypotheses. Therefore, instead of proceeding according to strict categorisation, a more open-minded approach was adopted. The advantage of such an approach is that the responses have greater validity: the fact that people refer to gender during discussions and resort to gender stereotypes without having been specifically prompted for gender

---

3 A moderated small group discussion of a topic with the help of information input (Dürrenberger/Behringer 1999).

4 For further details on the written survey see Schulz et al. 2010, Jahnke 2010 and Gallego Carrera et al. 2012.

5 This focus group was originally conceived as a pretest group. However, since the trial run had been successful, the results were included in the study.

validates the existence of gender stereotypes in society. Focus groups tend to be characterised by particular group dynamics: a topic brought up by one group member is generally picked up and elaborated on by another. A drawback of focus groups is that the outcomes do not lend themselves to generalisation and that findings cannot easily be reproduced and are therefore not representative (Dürrenberger/Behringer 1999).

The evaluation method used in this study draws on Mayring's (2002) qualitative content analysis, and in particular on his procedure for summarising content. A selection was made of those passages in the transcription material that related to role models or gender. Drawing on the discussions by the focus groups[6], four categories for analysis were established: ecological awareness, information management, parenting and the need for warmth.

8.3 Gender-specific analysis

Ecological awareness

Here the objective was to find out if and to what extent men and women differed in their attitudes towards sustainable heat consumption. Both genders justify their opinions and actions on the grounds of the need to protect the environment. Some men mentioned environmental motivations as one factor in their investment decisions. More generally, collective responsibility towards future generations was counterbalanced with economic considerations:[7]

> Man: "[...] in terms of ecology, I just had a thought. I would say, I haven't really got the money to be totally ecological, but if I do something green in one area, even if I fail to do something else in another, then, at least, it's good for my conscience, and I've done something for my children – that's how I justify the expense." (F:112)[i]

---

6 A number of topics were identified and suitable stimuli were used to give structure to the group discussions. The following topics were discussed: provision of energy advice (for tenants: a promotional flyer advertising an "energy detective", for homeowners and landlords: a short film about energy advice services); behaviour around heating (paper thermometer to check room temperature), use of water (water saving device) and information (two brochures).

7 The reference at the end of each citation refers to the section of the focus group transcript in the MAX.QDA database, which was set up in order to simplify data evaluation. "[...]" refers to a shortened passage. The Roman numerals refer to the original German citations listed at the end of this article.

Apart from financial motivations for pro-environmental behaviour, men came up with social and political arguments, e.g. climate change or pros and cons of renewable energies.

By contrast, women referred to the protection of the environment as a universal value ("our environment, our air", B: 330); additionally, their focus was on everyday behaviour:

Woman: "[...] well, I tell my son to turn of the tap when he brushes his teeth, because it's just not right, [...] it's not a matter of money, it's just a waste of water [...]." (A: 226)[ii]

Woman: "[...] I don't know. Nowadays, if you just leave the tap on, you immediately think 'I got to turn it off' – it's internalised. We've been living with the promotion of pro-environmental behaviour for so long now. Even on holiday, we won't throw our empty batteries in the bin; but there, people do throw them away, and that's when we notice how internalised our behaviour is." (E: 153)[iii]

Both male and female participants in the focus groups were ecologically aware. Women associated pro-environmental behaviour more with their everyday routines and appeared to have "internalised" green values to a greater extent. Previous studies on environmental awareness (Zelezny et al. 2000; Empacher et al. 2001) confirm these findings. In order to assess if these differences between men and women actually lead to eco-stress, it is necessary to take a closer look at the circumstances and tasks that present themselves in people's daily lives.

Information management

In order to consume heat sustainably, people need to know how much they consume and how they could reduce their consumption. With this in mind, the study looked at the strategies used to gather and make use of available information. Given that in real life, sustainable heat consumption is largely dependent on structural contexts and financial considerations, tenants and homeowners were assigned to separate groups.

The discussion topic for homeowners was energy-efficient refurbishment. Homeowners appeared well informed and even made a distinction between information materials aimed at themselves and those aimed at tenants:

Man: "As I said before, I think this brochure is more suited to tenants or other people. For us this is of no interest, there is nothing here that we don't know already. [...]" (D: 402)[iv]

In the context of investment decisions, the transcription material does not reveal gender-specific differences, but it is noteworthy that householders often used "we" when referring to investment decisions. In order to unpack this "we" and investigate who decides on what, one would have to conduct "couple interviews" (see Offenberger/Nentwich in this volume).

One aspect discussed in the homeowner groups was clearly gender-divided: a number of men appeared very keen and pro-active in gathering and dealing with information materials, whereas no women expressed such commitment:

Man: "I've been advised by three energy consultants: one looked at my house. He loved my roof [...], and then I asked someone else who I know personally. He said that it wouldn't necessarily make a difference, but that it was a question of ventilation. The first one didn't bother to ask about ventilation though. He said, it's all great, but it wasn't great [...]. Then I asked a third one who gave me different advice altogether." (D: 49)[v]
Woman: "At least you didn't give up altogether. Well done." (D: 50)[vi]

Despite being given different – and some decidedly unconvincing – expert opinions, these men did not appear to be stressed out by or put off from gathering information on energy-efficient refurbishments.

The statements in the tenant group clearly pointed towards gendered differences. Here, two types were identified. There was one group of disinterested tenants who were quite open about their lack of curiosity:

Man: "[...] I would never go inside such a silly eco-bus just to be waffled at. I just wouldn't be interested." (B: 428)[vii]

Other group members stated a keen interest in green refurbishment and appeared to have actively sought out information material in the past – especially when they were confronted with a specific problem:

Man: "Eleven years ago, I moved from an old building into a brand new flat. Everything was totally different: bathroom, district heating, everything was new. So I had to get my head round all this, and that's exactly what I did." (C: 231)[viii]

The theory that gathering information on sustainable consumption is perceived as a similar "green overhead" to separating and recycling household waste or ecological cleaning methods and therefore causes eco-stress (Gestring 2000; Röhr 2002) was not confirmed by the male respondents of the tenant group. The disinterested tenants simply failed to actively seek information, but did not see this as a problem. Judging by the reactions from the homeowners and the motivated tenants, information gathering was seen not so much as a burden, but as intrinsically necessary, and pursued in the same way as a hobby would be.

The contributions by the female tenants focused much more on everyday routines than on particular events. Here and there, the discussion flagged up stereotypes. In the context of minor energy efficiency measures, one woman talked about being overwhelmed when faced with a seemingly endless range of products in a large DIY store. Her statement implied that other women had the same problem:

> "I mean, with thousands of shelves, especially being a woman [...], I'm just totally overwhelmed [...]." (A: 304)[ix]

The last utterance reveals the woman's unwillingness to take on this "excessive demand". Generally, the female members of the tenant focus group were of the opinion that people should ideally be able to integrate information management into their everyday activities (e.g. water saving devices available in local chemist). It was also felt that people's specific needs should be taken into account. One suggestion was the provision of an "eco-bus" that would travel into local areas, offer advice to adults, some form of educational entertainment for children, and demonstrate the use of green heating-related products:

> Woman: "This is not a bad idea, because it raises awareness. People go shopping in their local area and if something's going on, [...], if there is an "eco-bus" for instance, where you can actually buy those gadgets, I think it would really focus people's minds, [...], but come to think of it, there would have to be some sort of attraction for the children too; they could give balloons to children, couldn't they [...]." B: 427, 231)[x]

It was not possible to establish whether some women's need for greater convenience meant that they found the gathering of information more stressful than men. Whatever the case may have been, the women who came up with innovative proposals for linking everyday routines with information management were clearly not willing to take on the overheads of current information gathering. This finding differs from other accounts in the literature (see Carlsson-Kanyama/Lindén 2007, p. 2170).

Parenting

In the context of parenting, male and female attitudes towards sustainable heat consumption did not appear to diverge. Pointing to the need for parents to act as role models, both men and women saw the instilling of sustainable heat consumption practices as part of their educational role. Pointing to the central role of schools, teaching children to consume heat sustainably was regarded by men and women as a collaborative effort that should be shared by parents and schools.

Apart from educational aspects, the transcripts revealed that households with young children found it particularly hard to consume heat sustainably. Other studies confirm these findings (Clancy et al. 2004, p. 12; Carlsson-Kanyama/Lindén 2007). Both male and female focus group members mentioned that – in the context of heating – their children's needs came first:

> *Man: "[…] We have underfloor heating. We have four children, one's flown the nest. When they were very young, the underfloor heating in our flat was invaluable, so we decided to go for underfloor heating in our house too." (F: 110)[xi]*

Both male and female participants emphasised the conflicting requirements of ample heating for the benefit of young children on the one hand and sustainable heat consumption on the other. It was not established whether men and women dealt with this dichotomy in different ways. Even though both genders appear to be aware of their responsibility in this context, empirical data point to a gendered division of labour, i.e. that women are still mostly responsible for looking after children (BmFSFJ 2008), even when both partners work full-time (Henninger et al. 2007; Schulz 2011). This structural dimension of "doing gender" was alluded to by some members of the focus groups. Pointing to everyday situations, women tend to be less willing to embrace sustainable heat consumption:

> *Woman: "And especially with a child. I mean, my daughter is so small, she can only move at floor-level, she hasn't learnt to crawl yet. I can't heat the room to just 20 °C, because then I would have to wrap her up to such an extent she couldn't move anymore. That won't do. […]." (A: 108)[xii]*

The need[8] for warmth
When it came to people's requirements for warmth, attitudes appeared to diverge – both amongst women and men. As for hot water consumption, the majority of participants agreed that they would not want to go without a long hot shower. By contrast, opinions diverged as to adequate room temperatures (individual needs vs. saving energy). The thermometer (equipped with a function for indicating savings potentials) distributed to all group members was not positively rated by all: some saw it as a potential incentive, whilst most – especially women – put their personal need for warmth first.

> Woman: "[…] Even if my husband would want me to, I won't start dressing up in ten cardigans, and a blanket wrapped round in order to keep me from freezing. Obviously, it wouldn't be right to go round the flat in just a T-shirt, but I do draw the line somewhere. […] I want to be comfortable and will make concessions to sustainability." (A: 111, 114)[xiii]

Whilst it was mentioned that, in principle, the need for warmth depended on a number of factors – individual perception of warmth, activity, time of day or night etc. – both genders agreed that women were more prone to feeling cold:

> Man: "I find it odd that men should refer to this, because it is usually women who have cold hands and feet. […] It depends on the situation and on the people involved. I don't think we can generalise here." (C: 104)[xiv]
> Woman: "You know men. They're always too hot." (E: 115)[xv]

The reproduction of this stereotype was also evident in the so-called exceptions to the above rule. When a woman mentioned that her husband was more likely to feel the cold than herself, the group agreed that such a phenomenon was "rare" and that "normally it was women who felt the cold" (D: 128–133). This sparked a discussion on a possible association between nationalities and sensitivity to cold. It was noted that particularly in the context of the need for warmth, both men and women tended to refer to collective attitudes towards gender and their needs.

Heat consumption is one aspect of everyday life that has been under permanent negotiation in households (Schultz/Stieß 2009, p. 22). Numerous passages in the transcription material reflected how such negotiation can lead to conflict. A good propor-

---

8   When referring to 'need' in this context, we do not adhere to the distinctions made by Di Giulio et al. (in this volume) between objective needs, subjective desires and ideas about degree and range to which these should be satisfied.

tion of men – who generally showed more interest in this topic – said it was them who decided on (or would like to decide on) room temperature, and that they had tried to get their partner to save heat energy. This largely male need for controlling the room temperature was brought up by both genders:

> Man: *"If I feel it's right, I simply tell everybody in the household that now the heating will be set to 18 degrees."* (E: 208)[xvi]

Whilst room temperature is generally negotiated in households, it remains a subject of considerable contention. This is reflected in the response of a participant to a contribution by a group member about the potential for saving energy by using a thermometer:

> Man 1: *"I simply use the thermometer …"* (E: 214)[xvii]
> Man 2: *"But if your wife divorces you because she's too cold you need a good lawyer."* (E: 215)[xviii]

The reactions by the other focus group members (not included above) to the above statement suggest that – in the context of room temperature – such intense feelings are not unusual. One participant reported that his wife "lambasted" him after he had switched the heating completely off for a certain period (D: 421).

8.4 Conclusion and future perspectives

In the context of heat energy, the thesis of the "feminisation of environmental responsibility" cannot conclusively be confirmed. Possible "green overheads" were noted in relation to information management and parenting. However, not only did both genders take responsibility for those tasks, but – with regard to instilling sustainable practices in children for instance – both men and women called for greater involvement of schools. Some men took an interest in information management, regarding it almost as a hobby, while others were entirely disinterested. Women generally found information management more of a burden and suggested ways in which such overheads could be reduced and integrated into their everyday lives.

The hypothesis that sustainable heat consumption leads to eco-stress was not confirmed by the focus groups. The transcription material clearly showed that women – especially young mothers – were more focused on everyday routines. Several women described situations in their lives that directly conflicted with sustainable heat consumption. There is an inherent conflict between an internalised conviction to save heat energy on the one hand and ample heating for the benefit of children on the other; however, neither men nor women mentioned this as a cause of eco-stress. The fact that this dichotomy was even mentioned indicates that people are aware of the incongruity of the situation. Especially mothers attempted to justify their "wasteful" practices (e.g. "my daughter needs to be able to move" versus "saving on heating"). They clearly sacrificed "green" behaviours for very specific individual reasons.

Eco-stress was overtly addressed in relation to the negotiation process that couples undergo when deciding on room temperature. This kind of eco-stress is not directly attributable to the feminisation of environmental responsibility, but it is clearly a reality. Interestingly, the majority of female group members contested an association of their personal need for warmth and the feminisation of environmental responsibility. Despite internalised pro-environmental convictions, those women who had reported being more prone to feeling cold than men, appeared to make sure that their requirement for more warmth was met. Whether negotiations of room temperatures lead to eco-stress and whether men or women suffer more under this stress depends on the specific circumstances. It is proposed in this study that stereotyping "freezing women" will only lead to unchallenged apportioning of blame.

Finally it is concluded that both individual needs and the presence of children in a household can represent barriers to sustainable heat consumption. Strategies and campaigns designed to promote sustainable heat consumption should therefore be designed with care and sensitivity. Without resorting to stereotypes such as "freezing women", future strategies should aim to give due consideration to individuals' diverse needs.

Endnotes

i   Mann: "[…] der Gedanke kam mir eben bei diesem ökologischen Hintergrund. Dass man sagen kann, also um rundum ökologisch aktiv zu werden, dazu fehlt mir das Geld, wenn ich aber Plan A mitnehme, das heißt C und B streiche ich einfach etwas weg und setzt nur A um, dann habe ich nicht nur was für mein Gewissen sondern auch für die Kinder getan, was dann auch in den Geldbeutel passt." (F: 112)

ii Frau: "[…] also ich sage meinem Sohn er soll das Wasser ausmachen beim Zähneputzen, weil ich das einfach falsch finde, […] auch wenn es nicht ums Geld geht, weil es einfach eine Wasserverschwendung ist […]." (A: 226)

iii Frau: "Also ich weiß nicht. Heutzutage wenn bloß das Wasser laufen lässt 'Huch schnell aus', nebenher – man hat das schon so drin. Wir sind einfach schon so mit dem Umweltschutz die ganzen Jahre. Selbst im Urlaub wenn man Batterien in der Hand hat 'Nein die kann ich nicht da reinwerfen, ich kann es einfach nicht'. Aber es ist dort so üblich und da merken wir, wie wir das eigentlich schon verinnerlicht haben." (E: 153)

iv   Mann: "Also wie gesagt ich würde diese Broschüre eher Mietern oder sonstigen Leuten geben, weil das ist so und so uninteressant. Weil die sagt uns nichts mehr Neues. […]" (D: 402)

v   Mann: "Ich habe auch mit drei Energieberatern zu tun gehabt, nur ein Beispiel ich hatte Jemand da der mein Haus untersucht hat. Und vom Dach war richtig begeistert, […] und dann bin ich zu jemand anders gekommen, den ich persönlich sehr gut kenne, dann sagt der, dass hat gar keine Aussagekraft. Es ist die Frage, wie ist das Dach belüftet. Das hat der mich aber nicht gefragt. Der sagt, das ist alles wunderbar, das ist gar nicht wunderbar. […]. Dann hatte ich noch mit einem zu tun, der hat wieder etwas anderes behauptet." (D: 49)

vi   Frau: "Immerhin haben sie nicht gleich aufgegeben. Sehr entschlossen." (D: 50)

vii   Mann: "[…] ich würde niemals in so ein blödes Ökomobil einsteigen und mir was erzählen lassen wollen, weil es würde mich nicht interessieren." (B: 428)

viii   Mann: "Also ich muss sagen vor elf Jahren habe ich meine Wohnung bezogen aus einem Altbau. Und für mich war an für sich das alles komplett neu, Bad, Fernheizung alles war komplett neu. Und da musste ich mich erst mal mit dieser Materie beschäftigen und da habe ich es dann richtig gemacht." (C: 231)

ix   Frau: "Ich meine, da gibt es 100,000 Regale und gerade als Frau ist es ja wirklich so […], also ich finde das erschlägt mich […]." (A: 304)

x   Frau: "Das ist glaube ich auch nicht so schlecht, weil dadurch kommt noch mehr Bewusstsein dazu, weil man geht ja sehr in sein Viertel und dann ist da einfach was los. […] wenn ich mir dann noch vorstelle, da ist dann so ein Wagen, wo man dann noch die Sachen eben einkaufen kann. Das ist dann so ein konzentrierter Zeitpunkt, wo das dann nochmal, nochmal in das Bewusstsein kommt. […] aber jetzt wenn ich überlege meine Kinder würden dann noch irgendwas tolles sehen oder einen Luftballon kriegen […]." (B: 427, 231)

xi   Mann: "[…] Wir haben eine Fußbodenheizung, wir haben vier Kinder, eins ist schon ausgezogen, weil es eben einfach im Säuglingsalter total super war, da haben wir in der Wohnung gute Erfahrung gehabt, da haben wir gesagt das machen wir im Haus auch." (F: 110)

xii   Frau: "Und vor allem mit Kind. Ich meine, meine Tochter die ist so klein, die ist nur am Boden unterwegs, kann noch nicht mal krabbeln, da kann ich den Raum nicht auf 20 °C lassen. Dann muss ich die so warm einpacken, dass sie sich kaum mehr bewegen kann. Das ist ja irgendwie auch nicht so super. Also das geht halt nicht. […]" (A: 108)

xiii   Frau: "[…] Weil ich werde nicht anfangen in meiner Wohnung – auch wenn mein Mann das manchmal gerne so hätte – mir zehn Jacken überzuziehen, weil ich friere und dann auch noch die Decke… das mache ich nicht! Das ist mir einfach zu blöd. Ich sehe es ein, dass ich nicht im T-Shirt rumlaufen muss, das ist ok, aber dann hört es irgendwann auf. […]. Ich will mich wohlfühlen, dann mache ich auch auf fünf." (A: 111, 114)

xiv   Mann: "Es ist mir ungewöhnlich, dass jetzt gerade Männer hier das sagen, meistens frieren ja doch die Frauen an den Händen, das ist ganz klar – oder an den Füßen. […] Aber ich denke es kommt auf die Situation darauf an und auf die einzelnen Personen. Ich denke man kann es nicht verallgemeinern." (C: 104)

xv   Frau: "Sie kennen die Männer, denen ist es immer zu warm. Immer." (E: 115)

xvi   Mann: "Das heißt wenn ich das möchte, dann gebe ich die Parole zu Hause aus, jetzt wird auf 18 Grad eingestellt." (E: 208)

xvii   Mann: "Ich brauche da nur das Thermometer …" (E: 214)

xviii   Mann: "Doch Sie brauchen einen guten Anwalt, wenn sich Ihre Frau von Ihnen scheiden lässt, weil es ihr zu kalt ist." (E: 215)

References

Bundesministerium für Familie, Senioren, Frauen und Jugend (2008): Familienmonitor 2008. Repräsentative Befragung zum Familienleben und zur Familienpolitik. [http://www.bmfsfj.de/ RedaktionBMFSFJ/Abteilung2/Pdf-Anlagen/monitor-familienleben-allensbach-2008,property= pdf,bereich=bmfsfj,sprache=de,rwb=true.pdf; 14.06.2012]

Carlsson-Kanyama A., Lindén A. (2007): Energy efficiency in residences – Challenges for women and men in the North. In: Energy Policy 35: 2163–2172.

Clancy J., Oparaocha S., Röhr U. (2004): Gender Equity and Renewable Energies. Thematic Background Paper, Bonn: Internationale Konferenz für Erneuerbare Energien. [http://www.renewables2004.de/pdf/tbp/TBP12-gender.pdf; 14.06.2012]

Di Giulio A., Brohmann B., Clausen J., Defila R., Fuchs D., Kaufmann-Hayoz R., Koch A. (in this volume): Needs and consumption – a conceptual system and its meaning in the context of sustainability.

Dörr G. (1993): Die Ökologisierung des Oikos. In: Schultz I. (ed.): GlobalHaushalt. Globalisierung von Stoffströmen – Feminisierung der Verantwortung. Frankfurt: Verlag für interkulturelle Kommunikation. 65–81.

Dürrenberger G., Behringer J. (1999): Die Fokusgruppe in Theorie und Anwendung. Stuttgart: Akademie für Technikfolgenabschätzung in Baden-Württemberg.

Empacher C., Hayn D., Schubert S., Schultz I. (2001): Analyse der Folgen des Geschlechtsrollenwandels für Umweltbewußtsein und Umweltverhalten. [http://nachhaltiges-wirtschaften.net/ftp/ubagenderend.pdf; 14.06.2012]

Gallego Carrera D., Wassermann S., Weimer-Jehle W., Renn O. (eds.) (2012): Nachhaltige Nutzung von Wärmeenergie. Eine soziale, ökonomische und technische Herausforderung. Wiesbaden: Vieweg+Teubner.

Gestring N. (2000): Soziale Dimensionen nachhaltiger Entwicklung. Das Beispiel des ökologischen Wohnens. In: Informationen zur Raumentwicklung 1: 41–49.

Henninger A., Wimbauer C., Spura A. (2007): Zeit ist mehr als Geld – Vereinbarkeit von Kind und Karriere bei Doppelkarriere-Paaren. In: Zeitschrift für Frauenforschung und Geschlechterstudien 25: 69–84.

Jaeger-Erben M., Offenberger U., Nentwich J., Schäfer M., Weller I. (in this volume): Gender in the focal topic "From Knowledge to Action – New Paths towards Sustainable Consumption": findings and perspectives.

Jahnke K. (2010): AP3 Endbericht Wärmeenergiekonsum. Ergebnisse einer Befragung von Hauseigentümern. Bremen: Bremer Energie Institut. [http://www.uni-stuttgart.de/nachhaltigerkonsum/de/Downloads.html, 14.06.2012]

Mayring P. (2002): Einführung in die qualitative Sozialforschung – Eine Anleitung zu qualitativem Denken. Weinheim: Beltz.

Offenberger U., Nentwich J. (in this volume): Socio-cultural meanings around heat energy consumption in private households.

Röhr U. (2002): Gender and Energy in the North. Background Paper for the Expert Workshop "Gender Perspectives for Earth Summit 2002: Energy, Transport, Information for Decision-Making". [http://www.earthsummit2002.org/workshop/Gender%20&%20Energy%20N%20UR.pdf; 14.06.2012]

Schultz I., Stieß I. (2009): Gender aspects of sustainable consumption strategies and instruments. eupopp Working Paper 1. Institute for Social-Ecological Research (ISOE). Frankfurt/Main. [http://www.eupopp.net/docs/isoe-gender_wp1_20090426-endlv.pdf; 14.06.2012]

Schultz I., Weiland M. (1990): Frauen und Müll. Frauen als Handelnde in der kommunalen Abfallwirtschaft. Studie im Auftrag des Frauenreferats des Magistrats der Stadt Frankfurt am Main. Sozial-ökologisches Arbeitspapier (40). Frankfurt/Main.

Schulz M. (2011): Marginalisierung von Intimität? Eine explorative Studie über WissenschaftlerInnen in festen, kinderlosen Doppelkarrierebeziehungen. Opladen & Farmington Hills MI: Budrich UniPress.

Schulz M., Gallego Carrera D., Alcántara S., Hilpert J. (2010): Arbeitspaket 3: Konsumanalyse – Nutzung der Wärmeenergie. Stuttgart: ZIRN. [http://www.uni-stuttgart.de/nachhaltigerkonsum/de/Downloads.html; 14.06.2012]

Schwartau-Schuldt S. (1990): Ökostress im Haushalt. In: Arbeitsgemeinschaft Hauswirtschaft e.V. (ed.): Haushaltsträume. Ein Jahrhundert Technisierung und Rationalisierung im Haushalt. Königstein: Verlag Langewiesche Nachfolger. 119–126.

Tzscheutschler P., Nickel M., Wernicke I. Buttermann H. (2009): Energieverbrauch in Deutschland. In: BWK 61 (6): 6–14.

Weller I. (2005): Inter- und Transdisziplinarität in der Umweltforschung: Gender als Integrationsperspektive? In: Kahlert H., Thiessen B., Weller I. (eds.): Quer denken – Strukturen verändern. Gender Studies zwischen Disziplinen. Wiesbaden: VS Verlag. 163–181.

West C., Zimmermann D. H. (1987): Doing Gender. In: Gender & Society 1 (2): 125–151.

Zelezny L. C., Chua P., Aldrich C. (2000): Elaborating on Gender Differences in Environmentalism. In: Journal of Social Issues 56 (3): 443–457.

Ursula Offenberger, Julia Nentwich

9 Socio-cultural meanings around heat energy consumption in private households

9.1 Introduction

The way that heat is supplied in German households – still largely by burning gas or oil – has increasingly become the subject of measures that aim to increase the use of renewable energy sources. A recent incentive programme launched by the German government, for instance, contained a proposal to "increase the share of renewable energies in heat generation to 14 percent by 2020" (Federal Ministry for the Environment, Nature Conservation and Nuclear Safety (BMU), 2011). Currently, renewable sources supply approximately 7 percent of the country's home energy needs, with biomass (e.g. wood pellets, biogas) representing the largest segment within that fraction (Schmidt/Jinchang 2010, p. 77).

The aim of the qualitative study we conducted as a part of the "Seco@home" project was to investigate the decision-making processes of homeowners who had recently bought a heating system that was (partly or fully) fuelled by renewable energy.[1] Semi-structured interviews with heterosexual couples focused on a) reconstructing the decision-making process that led up to the couple's purchase of a new heating system, and b) the mundane practices of the household members who actually use the system.[2] Purchasing decisions and the use of new technologies were regarded as part of more general social practices that are important for the construction of individual and private spaces: making a 'home'.[3] It is only through such practices that a house which has been bought, built, or renovated becomes 'home' to the new owners.

---

1 For another study on domestic heat consumption conducted as part of the focal topic, see Alcántara et al. in this volume. Also Götz et al., in this volume, analysed electricity consumption in households.

2 Given that the survey was geared towards individuals who were in a position to buy their own heating system, we included only homeowners, and no tenants. Given that homeownership correlates with net income (Federal Statistical Office 2009), this means that certain sections of the population, i.e. economically less privileged individuals, were not included. For a discussion on energy and poverty cf. Walker 2008.

3 For a theoretical elaboration of social practices cf. Kaufmann-Hayoz/Bamberg et al. in this volume.

In multi-person households (in our study, families), living together means creating a community, a sense of togetherness. When members of heterosexual households divide up tasks (including those relating to energy consumption), gender divisions inevitably arise.

In fact, gender becomes a relevant issue in energy consumption. Another issue that emerged in the course of the interviews was the desire for independence from energy utilities, as an important motive for using renewable energies. Interviewees' "ideal home" was constantly associated with notions of autonomy, freedom and self-determination.

In this case study we scrutinise the complex ways that consumption practices are interwoven with the logics of family life. We show how domestic energy consumption belongs to the wider social practices around home making: a home is perceived as a space where a family can live together with a certain amount of autonomy. In studies of science and technology that have explored the relationships between technology and society, the stage when a technology comes into contact with its context of use and its users is sometimes referred to as "domestication" (Lie/Sörensen 1996; Silverstone/Hirsch 1992). The term is used to describe the appropriation processes involved when users adopt a new technology, ascribe meaning to it and make it part of their everyday lives. Several researchers have concluded that the use of a technology is not predetermined, for instance, by the external form or the function that designers or engineers attributed to the technology. Instead, it is subject to socio-cultural negotiation processes involving a range of actors (see e.g. the contributions in Oudshoorn/Pinch 2003; for an example of how the concept of domestication is applied to electricity consumption, see Aune 2007).

Given that our study focused on couples and families, we were interested in how household members – and particularly (married) couples – *jointly* domesticated their new heating system.[4] Drawing on the thesis of the construction of reality in marriage (Berger/Kellner 1965), we proceeded from the assumption that decision-making processes and consumption practices in these households could only be understood in the context of the dynamics of couple relationships. The thesis states that

> *"in a marriage (…) all the actions of one partner [have to] be conceived in relation to the other partner. One partner's definitions of reality have to be constantly*

---

[4] Couples and families are the second largest group of homeowners in Germany. In early 2008 one- and two-person households were mostly tenants (76 and 52 percent respectively). The majority of three- and four-person households were homeowners: 57 percent of three-person households and over 70 percent of four-person households (cf. Federal Statistical Office 2009, p. 23).

aligned with the other partner's definitions. (...) For the sake of psychological efficiency, each partner perceives the other as his or her ideal 'significant other' – as having the most important and decisive effect on their lives." (Berger/Kellner 1965, p. 226; translated by C. Holzherr)

When the essay by Peter Berger and Hansfried Kellner was first published in 1965, marriage was still the main form of stable relationship for a couple. Given the changed realities of modern living, we have equated stable relationships lasting several years with marriage. Underlying the idea of a construction of personal reality in marriage is the more fundamental notion that reality is always socially structured, and that it is continuously re-constructed and affirmed through social interactions. The perceptions, attitudes and actions of individuals are part of this interactional context. Thus, in this study, we regarded consumption decisions and user practices from this perspective, and our examination of the dynamics of couple relationships focused on the dense form of interaction within households. We paid particular attention to the role of (heterosexual) gender differences, i.e. how "doing gender" (West/Zimmerman 1987) was performed in everyday consumption practices.

9.2 Data collection and analysis

To investigate the couple dynamics involved in the decision-making, in the qualitative portion of the "Seco@home" project, we conducted semi-structured interviews with ten couples who used different kinds of heating technologies fuelled by renewable energy. Some of these households additionally used oil or gas; renewable systems included pellet-fired heating systems, wood burners and wood-fired boilers, geothermal and solar installations and air heat pumps.[5] The interview partners were heterosexual couples aged between 40 and 60 living in both rural areas and various major cities in southern Germany. Nine of the couples were married and seven had at least one child. As one element in joint interviews, both partners were asked to reconstruct the decision making that led up to their purchase of their new heating system and to recall their experiences with the new system.

The couples were encouraged to jointly relate their experiences and come up with a shared perspective. Their challenge was to negotiate contradictions and differing ver-

---

5 This study proceeded from an assumption that had been confirmed by other project findings within the focal topic: decisions relating to major energy efficiency investments were jointly taken by partners (cf. Jaeger-Erben/Offenberger et al. in this volume).

sions of their experience and come to an agreement by way of mutual confirmation. These couples tended to 'iron out' the reconstructive accounts of their experiences by making small adjustments of content and perspective; the process of recounting decisions often provides opportunities to inadvertently legitimise them. Despite this phenomenon – or maybe precisely because of it – we were able to draw inferences about relational structures, the division of labour (e.g. when the heating broke down), potential areas of conflict, and common priorities of the partners around the purchase and use of heat technologies.[6]

Our analysis was based on Grounded Theory, developed by Anselm Strauss and Juliet Corbin (1990), and in particular on the methodology of "constant comparison".[7] By examining individual incidences in great depth and working out similarities and differences between incidences, the 'typicality' of a case emerges:

"An individual case always partakes of a general dimension – provided that the sociologist can interpret the case as such. In principle, an individual case is capable of revealing as much about what is socially 'determinable' as is revealed by a whole series of cases." (Honer 1991; translated by C. Holzherr)

A comparison of cases brings to light typical patterns, which in turn can be illustrated by the individual case. Therefore, our analysis of one case is to be understood as representative of our entire sample. It illustrates how renewable energies are associated with notions of increased independence, which interdependencies arise from using renewables, and how their use is adjusted to the everyday routines of couples and families.

Our case illustrates several topics that were relevant throughout all our interviews: the relevance of domestic energy consumption, the social practices of "home making", and the significance of the home as a place of relative autonomy and self-determination where the couple or family can engage in identity construction.

---

6 For a methodology for interviewing couples and empirical findings on the division of labour in the context cf. for example Behnke/Meuser 2004 and Offenberger 2008.

7 For further details on data analysis cf. Offenberger/Nentwich 2012.

9.3 Case study: Heat energy in the context of fostering a sense of family and identity

At the time of our interview, Mr. and Mrs. S. and their two children had lived for about a year in their house, which had been built in 1920, near a medium-sized industrial city in southern Germany. As part of their renovation project, they had had a new, multi-component heating system installed: a logwood boiler and a solar heating system, both connecting to a buffer storage tank which contains the water for household use, both cold and hot water. Occasionally, they would switch to the oil-fired boiler in the basement (e.g. while on holiday). In the living room was a stove powered by the flue gases from the boiler.

Independence as a key notion in images of the "ideal home"
When they were choosing a new heating system, Mr. and Mrs. S. opted for wood because it provided them with the greatest possible degree of independence from global supply infrastructures. They emphasised their interest in not being dependent on gas or oil, as did most other interviewees in this context. Although they mentioned their concern about depleting fossil resources, they gave far more weight to geopolitical uncertainties and dependence on oil and gas price fluctuations. The following passage expresses this concern:[8]

> *Mrs. S.:* "We won't have to depend on some Arab matador [MAN LAUGHS BRIEFLY] who dictates prices."
> *Mr. S.:* "And then there is the dependence. My goal is to be completely self-sufficient. But that's probably impossible [LAUGHS]. It's too expensive. I wanted to buy an emergency generator, but I haven't bought it yet. Because, if anyone cuts me off, I won't be able to use my wood-fired heating either. […] This dependency makes me quite angry […]. I'd rather work in the woods and prepare my own wood, at least, I know what it'll cost me, i.e. my labour. With oil, gas, or electricity, I don't know. When everything gets turned off, I just don't like it, this opaqueness. People tell you what to do and then you get a bill where nobody has a clue how all the numbers add up." (passage 14)[i]

---

8   The reference at the end of the transcript passage refers to the numbering of the interview in the atlas.ti database, which was set up to facilitate our analysis of the data. The transcription symbols have the following meanings: [TEXT] = comment by the interviewer. […] = Shortened passage. (…) = Unintelligible passage. / = sentence interruption. Text = spoken with emphasis.
The Roman numerals refer to the original German citations listed at the end of this article.

Thus, they expressed a certain unease with being tied into global economic cycles and large utility infrastructures without having any control over them. They also saw a risk in depending on other people's decisions. Thus, energy self-sufficiency as an ongoing lifelong project became an aspirational goal. While other interviewees voiced similar opinions about self-sufficiency – mentioning financial reasons and other constraints – they were not prepared to go to such lengths to reach their ideal.

In the above case, the "self-sufficiency project" appeared to be mainly driven by the husband: his observations were more detailed, and he talked more than his wife – mostly in the first person singular. Even during longer pauses in the conversation Mrs. S. did not add any information. Nor did she interrupt or contradict her husband, which in turn meant that he did not have to defend his views. From these clues it can be assumed that it was largely Mr. S. who had thought in depth about becoming self-sufficient and that Mrs. S. shared her husband's views and supported him in his endeavour.[9] While domestic energy supply was clearly a joint project, it was the husband who, as the driving force, displayed his knowledge about the new technology during the interview. A comparison with other interviews showed the same pattern: the man took the lead on the technical aspects of energy efficiency and heating. This is consistent with the widely held view that technology is a "masculine cultural domain" (cf. for example Wajcman 2002).

The above quotation expresses a sense of unease and threat, due partly to concerns about spending money on something (oil or gas) without knowing how the price is calculated. On the one hand, this concern simply expresses a desire to save money. On the other hand, the interviewees went to considerable length and expense to be self-sufficient, indicating that they were not primarily concerned with saving money:

Interviewer: "You mentioned that you had your reasons for wanting a solar system. […] Your argument was to save money during the summer. But have you actually done the calculations on this?"
Mr. S.: "Not really. I didn't calculate whether it made sense, no I didn't do that […]. I didn't calculate cost-efficiency, neither for the entire system nor for the separate parts. I simply wanted to have a solar system." (passage 218)

---

9 In another interview in our sample, all these issues arose when the partners did not agree about the value of a solar heating system they had purchased. As a result, we were aware of an undercurrent of disagreement between the two partners throughout the interview. In the present case, however, the partners showed no such disagreement.

In another passage:

> Mr. S.: "It is true that for a single-family house all this new technology costs a lot of money. [LAUGHS] But my thinking is that in the long term it'll make us independent, and that was _my_ aim." (passage 29)[ii]

Mr. S. did not compare the costs of different energy carriers – probably because he saw energy self-sufficiency as a worthwhile end in itself. It appears that people are willing to pay a price for not having to depend on oil.[10] With long-term security and protection against risks being of prime concern, the value of a new heating system is therefore not simply dependent on current energy prices. Finally, for this couple and for other respondents in our sample the idea of heating with renewable energies was closely associated with their notion of autonomy, freedom and self-determination as an expression of "the ideal home".

Oil and the fear of the global economy

In addition to dependency on oil, several interviewees expressed their fear of having to rely on foreign countries for their energy supplies. The "Arab matador" exemplifies this enemy stereotype. Respondents linked their desire for energy independence with their resentments about existing economic relationships and their concern for Germany as a global market player. When probed about his attitude towards sustainability, Mr. S. answered,

> "We've got our own energy right here in Germany, in our forests, and that's what we should exploit; that's more sustainable in my opinion than buying energy from Russia. [...] They take our sustainability; employing cheap foreign workers from Eastern Europe for our building and infrastructure projects and leaving our own people without work is not sustainable – just as unsustainable as importing gas from Russia. Sustainability is regional. I want to stay put where I live and make sure that people from the region have work and all that. For me, that is sustainability." (passage 273)[iii]

---

10 The "ENEF-Haus" project, where homeowners were asked about their motives for renovation, reached similar conclusions (also compare Weiß et al. in this volume). Here, the significance of "perceived efficiency" with the major objective of safeguarding against risks was evident. The most keenly felt risks appeared to be future energy price fluctuations and the risk of not being able to rely on oil and gas supply in the future ("ENEF-Haus" project 2010, p. 11).

Above, sustainability was understood primarily in terms of economic policy. Mr. S. mentioned using local renewable resources in the same breath as his criticism of globalisation and his concern for national prosperity and domestic jobs. His description of obtaining wood from the forest almost assumed a political dimension: "I'd rather work in the woods and prepare my own wood; at least, I know what it'll cost me, i.e. my labour […]" (passage 14). In contrast to the idea of the unknown and faraway world, wood and the act of working in the forest appeared to symbolise homeland, familiarity and clarity.

When independence is understood in this sense, regionalism assumes a key role as a driver of wealth creation and economic activity within manageable boundaries, and expresses solidarity with local people. Energy consumption is associated with community building and a specific locality where such a sense of community is a priority. Using wood from local forests is associated with fulfilling these ideas. We noted such symbolic associations in other interviews too, e.g. in families with wood stoves. Respondents who used wood pellets expressed this affinity for their locality by pointing out that the raw materials for the pellets came from local forests.

Wood and lumbering as the epitome of the 'comfort zone'

Using wood to heat the home was a marker of identity for Mr. and Mrs. S. Chopping the wood, purchasing a stove and the daily task of making a fire: all these were elements of conjugal and family bonding. The fact that Mr. S. came from a family that owned a forest meant that he had had an affinity with the forest all his life. Thus, for him, chopping wood as an adult was one way of keeping this connection alive: the process of felling trees and then chopping the wood was an expression of attachment to family traditions. It also served as a context for bonding: fraternisation in the most literal sense of the word:

> *Interviewer:* "Did you know where you would get the wood from?"
> *Mr. S.:* "I did actually. My brothers all have similar heating systems or wood stoves. They all get their own wood and sell some of it too. It gives them an extra income, or they just keep the wood. They might buy entire trunks that have been felled and are stored […] in the forest. I've got involved with that too now. Either I'll have the wood delivered or I'll do it all myself. […] I've got my own chainsaw and, well, the tools to prepare the wood. I've always had the tools, the knowledge and the skills. […] Transport and machinery is all within the family and can be shared. Felling and transporting the wood isn't too much of a problem. Many people just have a small car and a trailer, but our family has proper equipment." (passage 194)[iv]

For Mr. S., burning wood represents participation in traditions. His brothers played a crucial role, and their handling of wood (chopping, sometimes even felling) became a model for his own behaviour. Because he associates wood with family traditions, harvesting and preparing firewood became a question of identity formation for him.

It is interesting how his descriptions of acquiring and preparing firewood contain a plethora of masculinity stereotypes. He referred to an all-male community, to controlling the environment, to physical strength, and to handling tools and machinery.[11] No one, in any of our interviews, referred to responsibility for obtaining and preparing the wood as a woman's task. By contrast, the description of those tasks was used as a context for alluding to masculine identity. This gender component has to be taken into account when interpreting the significance that this couple ascribed to "independence" (as described above): greater independence is attained through social practices that are closely tied to perceptions of masculinity.[12] This example shows how the use of a particular energy resource is linked to gender relations and the formation of gender identity in heterosexual relationships.

The wood stove as instrumental in family bonding

This couple's wood stove appeared to fulfill several functions. At the time they purchased it, selecting the right stove allowed them to put their individual stamp on the new home, thus transforming it into a meaningful "family home". Additionally, simply by being the warmest place in the house and by being "hungry for wood", the stove became the focal point in the family's life and thus fulfilled a key function in bonding.

In the passage below the partners describe the careful planning that went into settling for the "right" stove (passage 48–52):

Mr. S.: "Well, it's easy/ it's like a piece of furniture, it is/ because it'll be there forever/ not something you replace every two years or whatever, it is a fixed piece of furniture, and, well, what should it look like, well"
Mrs. S.: "We wanted to (…)" [LAUGHS]
Mr. S.: "It was a joint decision, wasn't it?"

---

11 The strong male connotation of gathering and preparing firewood is also reflected in purely masculine timber-related professions such as forest ranger and lumberjacks. According to "Berufe im Spiegel der Statistik" in 2009, women accounted for a mere 5.6 percent of forest workers and gamekeepers and a slightly more respectable 13.8 percent of wood rangers and woodland managers (cf. http://bisds.infosys.iab.de/; accessed 02.07.2012). Given that under 20 percent of the employees are women, these professions can be regarded as highly segregated (cf. Achatz 2005, p. 277).

12 For further analyses of both gendered and gendering practices in the context of heat consumption cf. Offenberger/Nentwich 2009 and Offenberger/Nentwich 2012.

Mrs. S.: "We wanted plenty of seating area, because we have two children, and everyone wants to sit on the stove in winter." [SMILES]
Mr. S.: "Since I'm an engineer, I eventually drew a plan and we took it to the stove manufacturer. The manufacturer's initial suggestion was for a standard stove. They show you a few tiles, you choose, they put it all together, with the result that all stoves look the same. We weren't keen on that idea. So we thought carefully about what our stove should look like. I then drew it all and back we went to the manufacturer, and he built the stove to our specifications – just as it is here on paper." (passage 48–52)^v

While Mrs. S. talked about what their family expected and demanded of a stove (e.g. plenty of seating area to accommodate all family members), Mr. S. displayed his competence by drawing a plan. It appeared that all the manufacturer had to do was build the stove according to his plan. The stove thus became an important artefact, symbolising both the technical expertise of the man and the individuality of this household, in that they opted not for a standard model but for their very own design. Both the plan and the stove symbolised the fact that this household had found an unusual and very individual solution to heating their home. The stove thus became a prestige object of representation.

A further reason this couple invested so much time and effort in choosing the right stove was permanence: it would be a permanent fixture. As such, it represented a bridging object between their own taste in fittings and furniture and the house they had bought. They observed that they "hadn't wanted to make too many changes" to the historic "character" of the house (passage 33) and had therefore avoided taking on a major renovation. After they bought the house, they had to undertake a process of matching it clearly with the identity of the family. The installation of a unique stove – designed to their own specifications – presented one crucial opportunity to personalise their new home. The work they put into the design thus symbolises an active engagement with the house, helping the family to identify with the new home. We observed a similar phenomenon in some of the other interviews with homeowners who had renovated old buildings. By carrying out the work themselves (e.g. installing a heating system, insulating the building envelope or refitting a bathroom), they engaged in a similar process of matching the material substance of the house to their individual needs. This "domestication" represents an important step in appropriating a new house:

Interviewer: [ADDRESSES WOMAN] "And what was important to you in terms of looks?"
Mrs. S.: "I did not want any tiles. Because it doesn't…"

> Mr. S.: "Those traditional green tiles you know…"
> Mrs. S.: "Go with this house. If it was an old farmhouse, beautiful green tiles would have been suitable. But in our house, it wouldn't have looked right. So we had it all plastered and used a stone slab to sit on, and that's all. The whole thing is quite simple, with three levels; we like it and it suits the house. Each house reflects the time it was built in, and we felt that this stove best suited the character of this house." (passage 53–56)[vi]

In addition to the planning/designing stage, the stove itself and the work involved in tending and "feeding" the fire all contributed to the creation of a cosy family atmosphere. The use of the word "feeding" illustrates a reality in which the stove almost assumes the status of a pet, indicating the esteem and emotional attachment in which the family holds it. Interviewees who had a stove in their home all stressed the sensual dimension of heat energy; elsewhere in our sample we noted observations like the one cited below.

> Mr. S.: "We actually like feeding our stove. It's a bit like"
> Mrs. S.: "We even like to carry the wood up from the basement. [LAUGHS] It's work but I really like doing it." [SMILES]
> Mr. S.: "We all chip in, and/ well I don't see it as/"
> […]
> Interviewer: "And who is in charge of bringing the wood up?"
> […]
> Mrs. S.: "Whoever can spare the time."
> Mr. S.: "It depends."
> Mrs. S.: "Yes."
> Mr. S.: "Nobody is specifically in charge."
> Mrs. S.: "Everybody still enjoys fetching wood, even though we've had the stove for almost a year now." [SMILES] (passage 242–257)[vii]

The stove embodies the notion of family life: caring attitudes, bonding in a place of (symbolic and physical) warmth, and shared work in (and on) a cosy and warm home. If the functions of wood described above are added to its symbolic significance in terms of being independent from global economic cycles, then the traditional wood stove – with its connotations of privacy and the interior of the home – becomes a powerful alternative to the far-off "world": the chopped wood, the personal design of the stove and the work involved in heating with wood enable people to move a little closer to their dream of an "independent" lifestyle. Activities such as harvesting wood and making the fire satisfy their desire for self-sufficiency, and the dependencies that still exist (on tradespeople, the heating system and the availability of wood) become less significant.

9.4 Summary and conclusion

The consumption of heat energy in private households is one area within housing that is currently under policy review with regard to promoting sustainability. For the qualitative study conducted as part of the "Seco@home" project we interviewed ten homeowner households (families or couples) to reconstruct the decision-making processes involved in purchasing their heating system, and the daily routines involved in using it. Our detailed analysis of one interview showed how the buying and using of a heating system is an integral part of people's social practices.

First, purchasing decisions are the result of negotiations between the two people in a couple, as well as recourse to expert advice. They depend on a range of contextual factors: the characteristics of the household, the type and quality of advice service chosen, the homeowners' knowledge of and previous experiences with heating systems, and people's personal preferences for particular energy carriers. Before people make their decisions, they must consider all these factors.

Second, it appears that the consumption of heat energy in private households fulfils far more functions than merely satisfying the physical need for warmth. If we are to fully understand the significance of heat consumption in private households, it is crucial to consider the associated socio-cultural dimensions. These include a warm and cosy home and a private and intimate space where individuals come together and where bonding occurs within the household (cf. studies by Bartiaux 2003; Gram-Hanssen/Bech-Danielsen 2004; Shove 1999). Our analysis of interviews revealed that homeowners have a particularly strong desire for self-determination, independence and freedom. Renewable energies become an attractive heating option once they are perceived to fulfil those ambitions.

Third, the fact that homeowners were eager to be involved in the process of heating their house (e.g. through renovations, repairs, obtaining firewood) reflected the importance of the domestication process: integrating the new heating system into the homeowners' daily routines. (For the significance of do-it-yourself efforts, see Hitzler/Honer 1988; Honer 1991; note that those analyses were from the late industrial era of the 1980s.)

The objective of our study was to demonstrate that housing, as a key element of everyday life in multi-person households, is always linked to the process of household members becoming even more of a community. In single-family houses this process occurs mostly within the family. Identity building and the performance of gender differences are crucial parts of this process. The gendered responsibilities in the process of deciding on a new heating system thereby create a "structure of opportunity" for male gender performances (cf. Hirschauer 2001, p. 224). One key reason for this is the male connotations connected to established notions of self-sufficiency (i.e. independence from technical infrastructure), technological expertise and wood and forest work. (For another perspective on the interaction between gender and technology, see e.g. Cockburn/Ormrod 1997.) Thus, it is evident that home making and learning how to use a new technology are inextricably linked with gender differentiation.

With regard to the diffusion of new technologies, consumers are an important link in the chain. It is thus important to understand how consumers "domesticate" the new technologies and how their handling of them is embedded in a wider context. In order to better predict the impact of sustainability-related intervention measures, the entire socio-cultural context should be taken into account. One insight from the present study, for example, is that any future interventions should consider the still widely held view of technology as a "male domain" (e.g. Grint/Gill 1995), and the equally pervasive conception of the intimacy of the private sphere as a "female domain", with its own intrinsic logic as well as its specific "currencies" of material and symbolic exchange.

Endnotes

i Frau S.: "Man ist nicht unbedingt abhängig von irgendwelchen arabischen Matadoren" [MANN LACHT KURZ], "die einem den Preis diktieren."
Herr S.: "Und die Abhängigkeit. Mein Ziel ist eigentlich komplett autark zu werden. Aber das ist wahrscheinlich unmöglich" [LACHT]. "Das ist zu teuer. Also ich hatte das Ziel, das ist noch nicht umgesetzt, irgendwann ein Notstromaggregat zu kaufen, falls mir irgendeiner den Strom abdreht, kann ich meine Heizung aufgrund dem Holz auch nicht benutzen. [...] [D]iese Abhängigkeit ist für mich relativ ärgerlich [...]. Vorher arbeite ich im Wald und mache mein Holz selber, dann weiß ich, was es kostet, nämlich meine Arbeitskraft, und beim Öl weiß ichs nicht, und beim Gas weiß ichs nicht, und beim Strom weiß ichs nicht. Und das ist alles zu/ so was mag ich einfach nicht, dieses Undurchsichtige, wo man diktiert bekommt und ne Rechnung kriegt und keiner mehr weiß, wie sich das zusammensetzt." (Abs. 14)

ii Interviewerin: "Sie haben ja jetzt gesagt, Sie haben auch Ihre Gründe gehabt, warum Sie ne Solaranlage wollten. [...] Und Sie sagen als Argument um eigentlich gern zu sparen im Sommer. Und haben Sie das für Ihre eigene Berechnung praktisch gegen die Kosten aufgewogen?"
Herr S.: "Nö. Nö. Ich hab nicht ausgerechnet, ob das sinnvoll ist, oder/ das hab ich nicht gemacht. [...] Also ich hab keine Wirtschaftlichkeitsberechnung getätigt über diese Gesamtanlage, weder die Einzelteile/ Bauteile oder über das Gesamte. Das war einfach etwas, was ich haben wollte." (Abs. 218)
Und an anderer Stelle:
Herr S.: "Das ist also für ein Einfamilienhaus einfach viel Technik wo viel Geld gekostet hat."
[LACHT] "Aber ich denk, dass es irgendwann langfristig einfach uns unabhängiger macht von allem anderen, und das war eigentlich <u>mein</u> Ziel." (Abs. 29)

iii Herr S.: "[...] die Energie ist hier in Deutschland und hier vor Ort, die steckt im Wald und da müssen wir was dafür tun, dann hat man das, und das ist für mich nachhaltiger als wenn ich irgendwas in Russland bestelle. [...] Die nehmen unsere/ Nachhaltig, das ist genauso nachhaltig wie wenn man auf dem Bau Billiglohnländer mit beschäftigt und Kontingente in den Ostblock verschickt und die deutschen Bauarbeiter immer weiter abgebaut werden, das ist genauso un-nachhaltig wie wenn ich das Gas aus Russland importiere. Nachhaltigkeit ist regional und da wo ich wohne, gucke ich, dass ich da bleibe, und dass ich die Leute, die mit mir wohnen, die Arbeit und sonstwas davon was haben. Das ist für mich nachhaltig." (Abs. 273)

iv Interviewerin: "War dann auch schon klar, woher Sie das Holz mal beziehen würden?"
Herr S.: "Mmh, eigentlich schon. Meine Brüder haben alle auch so ähnliche Heizungssysteme oder Holzofen. Und die machen alle selber ihr Holz und verkaufen teilweise auch Holz. Um damit einen Nebenverdienst zu haben oder selber einfach das Holz zu haben, kaufen dann stehendes Holz oder Langholz [...] im Wald. Und da klink ich mich immer ein und entweder lass ich das machen oder kümmer mich dann selber drum. [...] Also ich hab selber eine Motorsäge und selber genügend/ also die Materialien, um das Holz zu machen und so was, das ist eigentlich schon immer da gewesen, und das Wissen und das Können auch. [...] Und Transportmittel und Maschinen oder sonstwas gibt es in der Familie auch, so dass es locker allen reicht, das Holz zu fällen und zu transportieren, das ist überhaupt kein Problem. Wenn du denkst, mancher hat nur ein kleines Auto und einen kleinen Holzanhänger, aber das gibt es bei uns halt/ bei uns sind dann die Dimensionen bisschen anders." (Abs. 194)

v Herr S.: "Also das ist einfach/ das ist für uns wie ein Möbelstück, das ist/ weil das ja immer da drin bleibt, das ist/ erneuert man ja nicht alle zwei Jahre oder sonstwas, sondern das ist ein fest eingebautes Möbelstück, und wie das einfach auszusehen hat, ist äh"
Frau S.: "Wir wollten, (...)" [LACHT]
Herr S.: "Gemeinsame/ gemeinsame Entscheidungsfindung gewesen"
Frau S.: "Und viel Bank, weil wir noch zwei Kinder haben, im Winter ja jeder auf das Bänkle liegen möchte." [LEICHTES LACHEN]
Herr S.: "Und da ich ja Ingenieur bin, habe ich halt irgendwann meinen Plan gezeichnet, wie der auszusehen hat, und mit dem sind wir dann zum Ofenbauer gegangen. Weil die erste Resonanz vom Ofenbauer waren so Standardöfen, die man da/ kriegt man fünf Platten gezeigt und dann kann man sich aussuchen, dann wird das zusammengebaut, dann sehen die alle gleich aus, das wollten wir irgendwie nicht haben, und irgendwann haben wir uns Gedanken gemacht, wie der auszusehen hat, und das hab ich dann zu Papier gebracht, und mit dem sind wir dann wieder zum Ofenbauer gegangen, und dann hat er den Ofen so gebaut, wie ich ihn da aufgemalt hab." (Abs. 48–52)

vi Interviewerin: [ZUR FRAU] "Und was war Ihnen bei der Optik wichtig?"
Frau S.: "Ich wollte keine Kacheln. Weil das in unser Haus"
Herr S.: "Solche typische grüne"
Frau S.: "nicht passt, wenn man jetzt ein altes Bauernhaus hat, dann gehören so schöne Kacheln, find ich, dazu. Aber bei unserem Haus hat das nicht gepasst. Und wir haben eigentlich alles verputzt und haben eine Steinplatte zum Draufsitzen, und mehr haben wir gar nicht. Also das Ganze ist ziemlich schlicht, mit drei so Ebenen, und/ wir finden einfach, das passt zum Haus. Weil jedes Haus hat ja ein wenig seinen Charakter, von der damaligen Zeit, und da hat uns das am besten dazu gefallen." (Abs. 53–56)

vii Herr S.: "Und dann machts auch noch Spaß. Den Holzofen zu füttern und so. Also das ist auch bisschen"
Frau S.: "Wir tragen sogar gern das Holz in den Keller" [LACHT], "das ist ja Arbeit, aber das macht mir eigentlich richtig Spaß." [LEICHTES LACHEN]
Herr S.: "Man macht was zusammen, und/ also ich finde das gar nicht so/"
[...]

Interviewerin: "Und wer ist dann dafür zuständig, das hochzuholen und/ ähm"
[...]
Frau S.: "Der grad Zeit hat."
Herr S.: "Unterschiedlich."
Frau S.: "Ja."
Herr S.: "Gibts keinen."
Frau S.: "Machen immer noch, obwohl wir schon bald jetzt ein Jahr da wohnen, immer noch alle gern." [LEICHTES LACHEN] (Abs. 242–257)

References

Achatz J. (2005): Geschlechtersegregation am Arbeitsmarkt. In: Abraham M., Hinz T. (eds.): Arbeitsmarktsoziologie. Probleme, Theorien, empirische Befunde. Wiesbaden: VS Verlag für Sozialwissenschaften. 263–302.
Alcántara S., Wassermann S., Schulz M. (in this volume): Is "eco-stress" associated with sustainable heat consumption?
Aune M. (2007): Energy Comes Home. In: Energy Policy 35: 5457–5465.
Bartiaux F. (2003): A socio-anthropological approach to energy-related behaviours and innovations at the household level. In: ECEEE 2003 Summer Study – Time to turn down energy demand. Panel 6. Dynamics of consumption.
Behnke C., Meuser M. (2004): 'Immer nur alles am Laufen haben.' Arrangements von Doppelkarrierepaaren zwischen Beruf und Familie. Arbeitsbericht des Projekts 'Doppelkarrierepaare'. Dortmund: Typoskript.
Berger P., Kellner H. (1965): Die Ehe und die Konstruktion der Wirklichkeit. Eine Abhandlung zur Mikrosoziologie des Wissens. In: Soziale Welt 16: 220–235.
Bundesministerium für Umwelt, Naturschutz und Reaktorsicherheit (BMU) (2011): Heizen mit erneuerbaren Energien – Jetzt umsteigen mit Fördergeld vom Staat. Berlin.
Cockburn C., Ormrod S. (1997): Wie Geschlecht und Technologie in der sozialen Praxis "gemacht" werden. In: Dölling I., Krais B. (eds.): Ein alltägliches Spiel. Geschlechterkonstruktionen in der sozialen Praxis. Frankfurt a. M.: Suhrkamp. 17–48.
Götz K., Glatzer W., Gölz S. (in this volume): Household production and electricity consumption – possibilities for energy savings in private households.
Gram-Hanssen K., Bech-Danielsen C. (2004): House, Home and Identity from a Consumption Perspective. In: Housing, Theory and Society 21 (1): 17–26.
Grint K., Gill R. (eds.) (1995): The Gender-Technology Relation. Contemporary Theory and Research. London: Taylor & Francis.
Hirschauer S. (2001): Das Vergessen des Geschlechts. Zur Praxeologie einer Kategorie sozialer Ordnung. In: Heintz B. (ed.): Geschlechtersoziologie. Wiesbaden: Westdeutscher Verlag. 209–235.
Hitzler R., Honer A. (1988): Reparatur und Repräsentation. Zur Inszenierung des Alltags durch Do-It-Yourself. In: Soeffner H.-G. (ed.): Kultur und Alltag. Soziale Welt, Sonderband 6. Göttingen: Otto Schwartz & Co. 267–284.
Honer A. (1991): Die Perspektive des Heimwerkers. Notizen zur Praxis lebensweltlicher Ethnographie. In: Garz D., Kraimer K. (eds.): Qualitativ-empirische Sozialforschung. Konzepte, Methoden, Analysen. Opladen: Westdeutscher Verlag. 319–342.

Jaeger-Erben M., Offenberger U., Nentwich J., Schäfer M., Weller I. (in this volume): Gender in the focal topic "From Knowledge to Action – New Paths towards Sustainable Consumption": findings and perspectives.

Kaufmann-Hayoz R., Bamberg S., Defila R., Dehmel C., Di Giulio A., Jaeger-Erben M., Matthies E., Sunderer G., Zundel S. (in this volume): Theoretical perspectives on consumer behaviour – attempt at establishing an order to the theories.

Lie M., Sörensen K. (eds.) (1996): Making Technology Our Own? Domesticating Technology into Everyday Life. Oslo.

Offenberger U. (2008): Stellenteilende Ehepaare im Pfarrberuf: Kooperation und Arbeitsteilung. Münster: Lit-Verlag.

Offenberger U., Nentwich J. (2009): Home Heating and the Co-construction of Gender, Technology and Sustainability. In: Women, Gender and Research 18 (3–4): 83–91.

Offenberger U., Nentwich J. (2012): Home Heating, Technology and Gender: A Qualitative Analysis. In: Rennings K., Brohmann B., Nentwich J., Schleich J., Traber T., Wüstenhagen R. (eds.): Sustainable Energy Consumption in Residential Buildings: Physica-Verlag. 191–208.

Oudshoorn N., Pinch T. (eds.) (2003): How Users Matter: The Co-Construction of Users and Technologies. Cambridge MA, London: the MIT Press.

Projektverbund ENEF-Haus (ed.) (2010): Zum Sanieren motivieren. Eigenheimbesitzer zielgerichtet für eine energetische Sanierung gewinnen. [http://www.enef-haus.de/fileadmin/ENEFH/redaktion/PDF/Zum_Sanieren_Motivieren.pdf; 02.07.2012]

Schmidt M., Jinchang N. (2010): Potentiale erneuerbarer Energien in der Gebäudetechnik. In: Themenheft Forschung – Erneuerbare Energien. Universität Stuttgart 6: 76–83.

Shove E. (1999): Constructing Home. A Crossroads of Choices. In: Cieraad I. (ed.): At Home. An Anthropology of Domestic Space. Syracuse: Syracuse University Press. 130–143.

Silverstone R., Hirsch E. (eds.) (1992): Consuming Technologies: Media and Information in Domestic Spaces. New York: Routledge.

Statistisches Bundesamt (2009): Zuhause in Deutschland. Ausstattung und Wohnsituation privater Haushalte. Ausgabe 2009. Wiesbaden: Statistisches Bundesamt.

Strauss A., Corbin J. (1990): Basics of Qualitative Research. Grounded Theory Procedures and Techniques. Newbury Park, London, New Delhi: Sage.

Wajcman J. (2002): Gender in der Technologieforschung. In: Pasero U., Gottburgsen A. (eds.): Wie natürlich ist Geschlecht? Gender und die Konstruktion von Natur und Technik. Wiesbaden: Westdeutscher Verlag. 270–289.

Walker G. P. (2008): Decentralised systems and fuel poverty: are there any links or risks? In: Energy Policy 36 (12): 4514–4517.

Weiß J., Stieß I., Zundel S. (in this volume): Motives for and barriers to energy-efficient refurbishment of residential dwellings.

West C., Zimmerman D. H. (1987): Doing Gender. In: Gender & Society 1 (2): 125–151.

Section D

Consumers in new roles

Cordula Kropp, Gerald Beck

10 How open is open innovation? User roles and barriers to implementation

10.1 Can open innovation improve the diffusion of sustainable innovations?

Nowadays, the failure of sustainable consumption and the failure to slow down the depletion of resources are due not to the unavailability of sustainable products or services[1], but to their *inadequate diffusion*, i.e. poor market penetration and take-up of sustainable alternatives. In terms of consumption, there are two dimensions to inadequate diffusion: either there is insufficient demand for sustainable products and services, i.e. they have not penetrated the mass market and remain niche – or they are in demand, but are not perceived as true alternatives to non-sustainable options, i.e. instead of *replacing* traditional products and services, they represent *additional* consumption options, thereby causing the dreaded rebound effect.[2]

The purpose of *sustainable innovation* is to find solutions to technical, organisational and institutional-cultural problems – solutions that are compatible with healthy and environmentally friendly consumption practices in the long term and on a global scale (cf. Clausen et al. 2011). Sustainable innovations will counteract the depletion of resources and hazardous emissions only if their impact – from original conception to market penetration and diffusion – leads to an improvement in the overall environmental balance, and if they have a so-called "exnovation effect" (cf. Paech 2005), i.e. if they are capable of *replacing* unsustainable consumption patterns.[3]

---

1 In accordance with the use of the term in this volume and the Oslo definition, we understand sustainable consumption to be the acquisition, use and disposal of products and services that respond to users' needs – both practically and symbolically. These practices should neither limit people's quality of life in the future or in other locations nor contribute to a reduction of this limitation (cf. Fischer et al. in this volume).

2 Cf. discussion about impacts and assessment of consumption (section E and Defila et al. in this volume).

3 Nico Paech (2005, pp. 251 ff.) calls for reforms that *innovatively* fulfil sustainability criteria and *exnovatively* lead to the elimination of non-sustainable practices.

Open innovation promises to support sustainable innovation in a number of ways: firstly, products and services that have been developed as a result of a participatory innovation process tend to enjoy higher user acceptance. Secondly, the desirable ecological and social features of the eventual end products can be continually evaluated and adapted during the innovation process. Thirdly, early contact with consumers may help to identify potential barriers to diffusion and to develop marketing strategies that will eliminate rebound effects.

The overall aim of the project "Fostering Sustainable Consumption by Integrating Users into Sustainability Innovations" was to improve sustainable consumption. Within this general aim, the specific remit was to investigate how user involvement in innovation processes can contribute to the successful development and diffusion of sustainable products and services. The article by Ulf Schrader and Frank Belz (in this volume) focuses on the overall findings of the same project and on how different user groups might contribute. Our aim is to present the findings of one of the sub-projects, "scenarios for diffusion", and establish if, and to what extent, user integration can boost the diffusion of ecological products.

10.2 "I guess I look at it as a bit of volunteering"

> "I guess I look at it as a bit of volunteering, as my contribution to (…) improving public transport for everyone. (laughs) I give them my advice, my criticisms and they do with it what they want, I guess they can take it into account for improving public transport, so that more people can use it." (quote from an interview held during an innovation workshop in Frankfurt, 22.11.2008)[i]

Mr. L., 65, gave up his weekend to take part in an innovation workshop. Together with seventeen other "lead users" he had been identified in a screening process as someone who was particularly conversant with the services of the public transport provider Rhein-Main-Verkehrsbund (RMV), and critical of the existing level of service. In the workshop, he worked on his own and as part of smaller and larger groups. The brief was to state in detail users' transport needs, to report on user experiences, and to devise and draft potential improvements. Lead users such as Mr. L. have for some years been much in demand by companies who are keen to drive forward open innovation. Lead users are regarded as innovative users whose needs are not met by existing products and services and who are therefore finding their own solutions (cf. von Hippel 1986; 2005). They are not considered as a reference group that sets the specifications of a system, but as a source of inspiration that can foster innovation in its own right.

This capacity for innovation is so highly regarded by companies that a growing sector of innovation management is dedicated to researching the advantages and limitations of user involvement in open innovation. Under the heading of "interactive value creation" (cf. Reichwald/Piller 2006; von Hippel 2005), and with the goal of creating novel products and services, potential users are invited to take part in open innovation processes, by way of internet competitions, prototype toolkits or innovation workshops. It is hoped that their creative ideas, specific needs and everyday experiences will feed into the innovation process and help to accelerate market penetration (cf. Arnold 2011; Fischer et al. in this volume; Zwick et al. 2008 for a critical assessment; Hellmann 2010).

As a pro-active public transport user, Mr. L. was not only a member of the customer advisory council of RMV, but had for some years also developed travel ideas for senior citizens, which he had made publicly available online. Given this background, he was invited to take part in our lead user workshops on transport. The objective was to explore ways of linking different public transport modes and providers and come up with sustainable travel ideas. The workshop was led by two of our researchers, who used a range of creative methods to not only let participants share their explicit innovative knowledge, but to also tease out their implicit "sticky" knowledge (von Hippel 1994) around transport (cf. Steiner/Diehl 2011). In the workshops the participants gradually came up with ideas for new packages and services, thereby reducing insecurities on the part of RMV about customers' needs and market demands.[4] The expectation on the part of the participants was that they would eventually benefit from a smoother-running transport service tailored to their own and the general public's needs. The overall objective of the weekend workshop was the joint development of innovative sustainable transport solutions, whereby the participants were requested to keep in mind in equal measure economic, ecological and social criteria. A good proportion of the participants appeared to be motivated not only by potential personal benefits, but by a genuine desire to counteract climate change and the depletion of natural resources, and to contribute to a cleaner environment. Further concerns included improved safety and fair pricing in the public transport system (cf. Hoffmann 2007).

As representatives of the transport provider (RMV), an inter-regional transport services company, Mr. K. from the marketing department and his young colleague took part in our workshop. Both were keen to stress their interest in jointly developing innovative, user-friendly travel ideas. At the end of the workshop they were,

---

4   Requirements and needs of consumers are to be understood in this context as a comprehensive fulfilment of "techno-functional" and "symbolic-communicative" functions on the one hand, and the resulting overall well-being on the other. They are not to be understood as separate needs as suggested by Di Giulio et al. (in this volume).

however, not convinced by the results. The proposed innovations were not considered "feasible"[5], particularly in comparison to a marketing strategy they had recently commissioned from an advertising agency. While conceding that they had a clearer picture of how "consumers tick", they did not consider the innovations to be viable, neither in terms of value for money, the use of off-peak travel periods nor the use of existing underused routes. To our knowledge, no innovations developed in any of the twelve workshops organised as part of this project have actually been implemented.

Given these (somewhat disappointing) outcomes, it may be prudent to reassess the relevance of open innovation in the context of sustainability. Using Mr. L.'s description of his participation as "a bit of volunteering" as the basic attitude of the average participant, we will sketch out in the following section the roles that users adopt in open innovation, the possibilities and limitations that come with these roles and the impacts that participants' contributions may have. This analysis will be contrasted with three successful sustainability innovations that were driven by user input (Ornetzeder/Rohracher 2006; 2011). Linking back to the issue of user involvement in social-ecological transformation processes, we will finally discuss whether a shift of focus from development to implementation and diffusion might be necessary within open sustainability innovation.

10.3 User roles in open innovation processes

Marlen Arnold (2011, p. 31) explains that, depending on the goals and targets of the project, the roles of users in open innovation are either *informative*, *advisory* or *determinative*. She adds that these three roles are to be regarded as part of a continuum. Depending on users' precise involvement and interaction with others during the innovation process, different kinds of information, user knowledge and feedback on existing products and services will be imparted, particular ideas will be focused on, and new inventions will be jointly or individually developed. A scenario where companies involve outsiders purely for the purpose of acquiring information is seen by Arnold as the least intensive form of user integration. Judging by Mr. L.'s quote and the company's evaluation (Mr. K. and his colleague) of the workshop on transport (section 10.2) it appears that, from their perspective, their roles were merely informative. By contrast, Arnold – who helped design and plan the transport workshop – classifies the same workshop as determinative. She explains this by arguing that the users were given the

---

[5] When prompted, the marketing expert revealed that it was his euphemism for proposals that were not cost-effective.

opportunity to develop "their own concepts and solutions" (ibid., p. 32). She correctly adds that as a prerequisite for user innovations "the management has to be prepared to support the ideas that have been developed and the decisions that have been taken in the open innovation process" (ibid., p. 33; translated by C. Holzherr). And yet, external innovation proposals, propagated as the be-all and endall of open innovation, are often rejected by companies in what Katz and Allen call "notinvented-here syndrome" (1982): company managers do not identify (sufficiently) with user innovations, or they reject them on the basis of those features that deviate from what has been tried and tested within the company. By contrast, "genuine" user integration should promote a long-term, fertile collaboration between company-internal innovators and external users. Unfortunately, no case studies exist for such ventures.

A different typology of user roles has been devised by Klaus Fichter. Drawing on Cornelius Herstatt's (1991) earlier-developed user role classification, Fichter identifies six types of integrated users: *articulators*, who bring particular requirements and needs into the innovation process and articulate them; *ideas people*, who generate ideas; *evaluators*, who evaluate ideas and prototypes; *(co-)developers*, who contribute to the development of prototypes, new products and services; *testers*, who test the innovations prior to launch; *marketing supporters*, who act as reference customers, opinion leaders and pioneer customers (Fichter 2005, p. 29). The involvement of articulators, ideas people, evaluators and testers is relatively "easy".[6] They provide the developers with more or less innovative contributions, which are then taken up or not taken up, as the case may be. Overall, the company obtains valuable information on what potential users might want from a product or service, and they may use this information in the further development of the product and in their marketing strategies.

The involvement of these user types does, however, not go far beyond traditional (but nevertheless interactive) market research methodology. The roles of (co-)developers and marketing supporters are – by comparison – much harder to fill. For collaborative development to succeed it is essential that *all* developers and decision makers are able to work in partnership. One of the few areas where "genuine" user integration has been practised is communication and information technology. Unsurprisingly, this is often cited as a successful example of open innovation with external user involvement. Doris Blutner (2010) analyses the roles of innovative users in order to illustrate how consumers turn into prosumers who help create innovations that match consumer needs. As essential prerequisites for collaborative production she cites the "defining of

---

6   We would not want to give the impression that it is generally easy to integrate users in innovation. The more intensive forms of user integration are especially time intensive and costly; preparation and implementation is complex. That is why their actual net benefits are frequently called into question.

a space for problem solving" and "provision of resources" (ibid., p. 88). In her conclusion she makes a sharp distinction between the methods employed in open innovation and open source innovation (which tends to be IT-related). It is only in open source innovation that users take control of the organisation of resources. Genuine "open" innovation is possible only when users are given the freedom to define their own areas of concern, negotiate and establish their own rules during the development and decision-making process and deploy the necessary resources.

While users may be involved in testing and evaluating innovations, and may even develop their own products in innovation workshops, they still tend to be perceived and employed by companies merely as ideas people. While their contributions may be used to gauge consumer expectations, it is highly unlikely that they are valued as innovators in their own right and are given the opportunity to contribute substantially to product development. Instead, users involved in innovation workshops and competitions tend to remain 'outsiders' who, while they might be generators of ideas, serve principally as objects of observation. The innovation processes themselves remain tightly controlled under highly controlled conditions (cf. Blutner 2010). It is unusual for them to be truly involved in dialogue and even less common that they might act as innovators or co-developers.

10.4 User knowledge in open innovation processes

Despite the limitations of user integration, open innovation processes are capable of generating interesting results, particularly with regard to the *diffusion* of sustainability innovations. By means of innovation workshops, competitions and the testing of toolkits (cf. Belz/Schrader in this volume), users identify their needs and requirements and thereby articulate (mostly implicit) *information about their preferences*. This information shows how they perceive and evaluate products and services and how future products might be accepted by consumers. In the workshops on transport, for instance, cleanliness and fair pricing emerged as key considerations. Users' preferences, which tend to be established reconstructively, are guided by their underlying beliefs and attitudes. In relation to transport, these beliefs could be summed up as "faster, more convenient and cheaper". In relation to housing, by contrast, users appeared to be more oriented towards the long term, preferring environmentally friendly construction methods and supply systems. Judging by the information gathered in our project, it appears that future sustainable transport solutions will need to take into account the still dominant mainstream preferences for optimal convenience, whereas in the areas of housing and nutrition sustainability issues are more widely accepted.

It was observed that workshop participants strongly focused on overcoming existing barriers to sustainability. They described how they had used existing products differently and often idiosyncratically; they made suggestions as to how existing products could be improved and new products developed. For example, they talked about how physically disabled people were negotiating the public transport system (especially the changes from one mode of transport to another); they talked about how ecologically-oriented builders were coping with insufficient information, with lossmaking and unsuitable products, while at the same time making inventive use of what was available to them. Generally, participants tended to shift the emphasis from the technical features of products to more intangible qualities, or from a naively optimistic vision geared towards steering user preferences (a common attitude in R&D departments), to the reality of persisting technophobic attitudes among certain members of the public. Such *pragmatic knowledge* may feed into the development of new, more user-friendly products and services. It can identify the potential usefulness of products that may not yet be available and thereby open up possible new markets. Conversely, it may prevent the futile development of unviable products.

Finally, open innovation programmes with user involvement tend to generate *knowledge about infrastructures*. If user-generated suggestions are fleshed out into whole scenarios (which we call "innovation regimes"), we might learn more about how sustainable innovations can be fitted to existing routines, current development approaches and previously identified development trends (cf. Beck/Kropp 2011; Kropp 2011). It is then possible to discern knock-on effects in other areas, path dependencies, long-term impacts and unintended side effects of innovations. If innovation is to succeed, early consideration of all these various interdependencies and potential side-effects is necessary since such infrastructural knowledge is crucial both for suppliers of the future products and services and for policy makers.

In addition to the different kinds of user knowledge described above, the user role referred to in section 10.3, namely Klaus Fichter's "marketing supporter", is of relevance in this context. Here, the role is to be understood not so much in the sense of opinion leader, pioneer and reference customer, but instead as a source of *diffusion knowledge*. Users can be regarded as scouts for those three levels, of which Rammert (2010) identifies analytically the strategic value for the diffusion of innovations: the "semantic level", where products and services are perceived, endowed with meaning and accepted; the "pragmatic level", where products and services are used in different ways and by changing user communities; the "institutional level", where products and services are institutionalised and regimes are formed (ibid., pp. 29 ff.). R&D departments as well as future customers have to consider whether the innovation is superior to what is currently available (*semantic level, preference knowledge*), whether it is viable and testable (*pragmatic level, pragmatic knowledge*), whether it is in demand, whether it conforms to legal requirements, and whether it is capable of being manufactured or 'rolled out' (*institutional level, infrastructural knowledge*). If suitable structures are in place, users can act as civil advocates: they can select which products to support, represent and manage the selection criteria and vote on development options. The role which Mr. L. describes in section 10.3 as "a bit of volunteering" corresponds to these scouting activities. Whether the scouts act as advocates for *sustainability* depends on how solidly anchored sustainability is in a particular product domain and how sustainability-oriented the scouts are. Their expertise could significantly contribute to the success of sustainability innovations – not so much in terms of generating ideas, but more in terms of putting an early halt to the development of unviable innovations. The basic dilemma of promoting a sustainable lifestyle is that – in a culture that is still largely materialistic in outlook – sustainable options are not regarded as desirable "innovations", but instead as "threatening deviations" from a convenient abundance of goods and services and conspicuous consumption.

10.5 From users to innovators

The hope that user involvement will lead to successful innovations – both in general terms and in the field of sustainability – derives from a number of successful real-world examples. They all share a common feature, namely that innovation is *initiated by users* rather than companies or their agents. In such cases, user involvement is not a procedure initiated by external actors, but instead arises from the social or personal needs of individuals who decide to collaborate on the solution of a perceived problem. Michael Ornetzeder and Harald Rohracher (2006; 2011) report on three sustainable innovations that have been successfully reproduced all over the world: wind farms in Denmark, self-build solar systems in Austria and car sharing schemes in Switzerland. Not only did these projects benefit from early practical user input, but development was initiated by individuals who had perceived a need and became proactive as a result.

Plans for the construction of nuclear power stations in Denmark in the early 1970s triggered a nationwide citizens' initiative whose objective was to prevent the construction of nuclear power stations and to come up with alternatives. According to Ornetzeder and Rohracher (2011), this anti-nuclear movement was a driving force behind the development of wind farms in Denmark. The developers of those first wind farms were able to draw on practical knowledge that had been accumulated and trials that had been undertaken since the 18th century. Private organisations and cooperatives were formed in order to take care of the technical development, the marketing and the operation of the wind farms. In 1996, 55,000 cooperative members owned a total of 2,150 wind farms (ibid., p. 179). As publicly funded research programmes became more and more involved in the promotion of wind farms, the environmental pioneers' involvement decreased somewhat. Even though larger companies have gradually replaced local firms and individual activists, the basic wind farm expertise in Denmark is still based on the technology that was developed and promoted by the original ecologically-oriented users. The principle of the double breaking system, for instance, pioneered in those early days, has been used to this day (ibid., p. 178).

Ornetzeder and Rohracher's second example is the "Austrian movement for self-build solar panels" (ibid.). The springboard for this development was a demand for solar water heating systems – triggered primarily by the oil crisis in the 1970s. The solar systems available at the time were both technologically flawed and expensive and therefore did not fulfil users' requirements. Demand grew nonetheless, but fell away once the tensions in the energy market had relaxed by 1980. As a consequence, many manufacturers withdrew from the sector. Parallel to these developments, some isolated attempts at constructing self-build solar panels in 1983 led to the coming together of a first group of builders who constructed a considerable number of solar panels. In the same year, two further, privately organised groups of builders were set up in the same local area. This early self-build movement was motivated by the prospect of "affordable solar systems", "increasing home comfort" and "protecting the environment" (ibid., p. 184). The movement was particularly successful in its diffusion strategy of the new technology: in the decade following their invention, self-build solar collectors covered a total of 400,000 square metres. A major driver in raising public awareness of the new technology was the tireless public advocacy over years and decades by committed campaigners. Commercial producers of solar panels learned from the cheaper self-build systems, but they regarded them as direct competition for their products. The very first practical trials for solar space heating systems originated within the self-build movement.

The third example cited by Ornetzeder and Rohracher concerns car sharing schemes in Switzerland. Individuals set up organisations with the objective of organising and managing car sharing to an extent that would not have been possible in private informal arrangements. They were motivated by the financial advantages in comparison to single car ownership (while still being able to use a car individually), by environmental considerations, and by a desire to reduce the need for car parking spaces in urban settings. The original networks set up by individual users soon developed into socially-oriented cooperatives and eventually attracted commercial players, including the current market leader in car sharing (ibid., p. 183).

Michael Ornetzeder and Harald Rohracher (2011) conclude that, in real life, users may take on far more active roles in innovation than in the workshops described in this article. The "defining of a space for problem solving" (Blutner 2010, p. 88) is a key term here. In all three scenarios described above, people were either driven to find alternatives to existing (or expected) inadequate products and services or to counter commercial ventures with ecologically and socially-oriented alternatives. The leaders of these movements could be described as "lead users" in terms of von Hippel's definition, but additionally, they were politically and socially engaged citizens. Although they liaised with commercial players in the field, they did not act under their direction.

In the above scenarios, innovative socio-technical relationships were established within a diverse network of players, similar to those networks that have been described in constructive technology assessments or studies in socio-technical change (Bijker et al. 1986; Bijker/Law 1994; Law 1991). From these perspectives, technological development is not inherently determined by technical principles and economic constraints, but – being linked to its social associations – is constantly co-created by evolving social institutions, as well as stakeholders' attitudes and interests. Usage seen in this sense, as 'co-creation' and 'appropriation', does not only influence the design and marketing, but the actual technological development of innovations. Seen in this light, the function and design, innovation and implementation, technical adaptation and modification all converge.

Ornetzeder and Rohracher rightly point out that "the emphasis should shift from the self-indulgent 'freedom' to appropriate products to being mindful of their embeddedness in the social and technical regimes (…)" (2011, p. 173; translated by C. Holzherr) – the power of technological 'scripts' and 'settings'. In the context of the possible citizen roles that users may adopt in innovation processes, it is important to stress that their *active* involvement, i.e. their "co-construction" (Oudshoorn/Pinch 2005), can contribute substantially to the success of sustainable innovations. A number of examples within technology development (ibid.) have demonstrated how users who have a personal interest in the development of an innovation, and who are socially and/or politically motivated to promote its development, can influence its creation, design and diffusion in a more socially responsible way.

"In such cases users' localised expertise about the requirements and interests of consumers combines with the necessary professional development skills, with target-oriented competences, with the experience gleaned from trials, with pilot applications and – most importantly – with far reaching socio-political perspectives" (Ornetzeder/Rohracher 2011, p. 187; translated by C. Holzherr).

10.6 Towards "open realisation"

The users involved in our workshops were able to generate three kinds of knowledge: knowledge about *user preferences*, i.e. how future products might be accepted by consumers; *pragmatic knowledge* about the demands that are made of products and services in different consumer contexts; *infrastructural knowledge*, essential for successfully establishing a product or a service on the market. Additionally, they contributed to the development of over forty interesting innovations, covering transport, energy-efficient housing and environmentally friendly packaging. Proposals included user-oriented marketing strategies (trial offers and loss-leaders), user-friendly technology and information services (portals, interfaces), novel and not so novel sustainable inventions (e.g. resealable and stackable bioplastics packaging, public transport solutions linking different transport services). Despite all the enthusiasm, productive energy and considerable output, it remains to be seen if any of the innovations will actually be developed and implemented.

As a final verdict on user involvement in sustainable innovation, we conclude that the focus should be less on the *generation of ideas* and more on the ways in which *innovations can be implemented* (cf. Klotz 2007). There seems to exist a certain "realisation gap", within both sustainability and other types of innovation (particularly in Germany), i.e. not enough innovation concepts reach market maturity (cf. Kriegesmann/ Kerka 2007; Bullinger/Klotz 2009).

Even optimistic assessments reckon with a mere ten percent take-up rate of innovations (Kriegesmann/Kerka 2007). The bottleneck in the development process does not appear to be in the generation of ideas, but in their development and diffusion; the obstacles have been identified as existing hierarchical innovation cultures (Bullinger/ Klotz 2009; Hauschildt 1993, pp. 89 ff.). Although researchers within innovation might hope otherwise, even products and services developed in open innovation processes can suffer from poor take-up.

Competitive innovations are thin on the ground not because there is a lack of ideas, but because, in the current innovation cultures, they are not successfully implemented (cf. Klotz 2007). Decision makers reject new innovations, not least because "new expertise tends to devalue existing knowledge and thereby shift the balance of power" (ibid., p. 185; translated by C. Holzherr). Klotz proposes to open up innovation by creating 'innovation bazaars', where "everyone communicates with everyone else before reaching their decisions" (2007, p. 188; translated by C. Holzherr). On the assumption that sustainability innovations will deviate substantially from a conventional understanding of the market and run counter to profit-maximizing decision-making, they will have to rely on methods of "open realisation". All the more so as sustainability innovations are less about incremental improvements to a given product than about transforming existing socio-ecological consumption regimes.

Ambitious open innovation programmes may well encourage the articulation of user needs and discussions around socio-political dimensions and sustainability concerns – and yet, they can succeed only if all participants are motivated and possess the necessary competencies to deliver sustainable innovations. That said, it may be more appropriate to select "lead users" not so much on the basis of their technical expertise (cf. Schrader/Belz in this volume), but of their motives and attitudes. Given the abundance of literature on the creativity, implementation and diffusion of open innovation, the attitudes and philosophies of users who contribute to innovations have so far been neglected. In our opinion, sustainability innovation programmes should aim to recruit users who are first and foremost committed to environmentally and socially beneficial causes and who are, or can be, trained to express competent judgments. If individuals are recruited whose attitudes are more mainstream, and who feel that sustainable innovations deviate from their beliefs and practices, then their contribution to social and ecological transformation will be relatively modest.

Endnotes

i   "Ein bisschen sehe ich das hier als ehrenamtliches Engagement, als ein sozialer Beitrag (…) zur Verbesserung des öffentlichen Verkehrs für alle. (lacht) Ich gebe meine Tipps, meine Kritik und die können das dann nutzen, denke ich, so umsetzen, dass der öffentliche Verkehr besser nutzbar ist, für mehr Leute nutzbar ist."

References

Arnold M. (2011): Methoden offener Innovationsprozesse. In: Belz F.-M., Schrader U., Arnold M. (eds.): Nachhaltigkeits-Innovationen durch Nutzerintegration. Marburg: Metropolis.

Beck G., Kropp C. (2011): Diffusionsszenarien: Verbreitung von Nachhaltigkeitsinnovationen durch Nutzerintegration? In: Belz F.-M., Schrader U., Arnold M. (eds.): Nachhaltigkeits-Innovationen durch Nutzerintegration. Marburg: Metropolis. 214–244.

Belz F.-M., Schrader U. (2011): Offene Innovationsprozesse. In: Belz F.-M., Schrader U., Arnold M. (eds.): Nachhaltigkeits-Innovationen durch Nutzerintegration. Marburg: Metropolis. 23–38.

Bijker W., Hughes Th., Pinch T. (eds.) (1986): The Social Construction of Technological Systems. New Directions in the Sociology and History of Technology. Cambridge, Massachusetts: MIT Press.

Bijker W., Law J. (eds.) (1994): Shaping Technology/Building Society. Studies in Sociotechnical Change. Cambridge, Massachusetts: MIT Press.

Blutner D. (2010): Vom Konsumenten zum Produzenten. In: Blättel-Mink B., Hellmann U. (eds.): Prosumer Revisited: Zur Aktualität einer Debatte. Wiesbaden: Verlag für Sozialwissenschaften. 83–95.

Bullinger H., Klotz U. (2009): Innovationsprozesse als Handlungsfeld von Gewerkschaften beim Übergang von der Industrie- zur Wissensgesellschaft. In: Bullinger H.-J., Spath D., Warnecke H.-J., Westkämper E. (eds.): Handbuch Unternehmensorganisation. Berlin: Springer. 71–87.

Clausen J., Fichter K., Winter W. (2011): Theoretische Grundlagen für die Erklärung von Diffusionsverläufen von Nachhaltigkeitsinnovationen. Berlin: Borderstep Institut für Innovation und Nachhaltigkeit.

Defila R., Di Giulio A., Kaufmann-Hayoz R., Winkelmann M. (in this volume): A landscape of research around sustainability and consumption.

Di Giulio A., Brohmann B., Clausen J., Defila R., Fuchs D., Kaufmann-Hayoz R., Koch A. (in this volume): Needs and consumption – a conceptual system and its meaning in the context of sustainability.

Fichter K. (2005): Modelle der Nutzerintegration in den Innovationsprozess. Möglichkeiten und Grenzen der Integration von Verbrauchern in Innovationsprozesse für nachhaltige Produkte und Produktnutzungen in der Internetökonomie. Berlin: IZT/IZT-Werkstattbericht Nr. 75.

Fischer D., Michelsen G., Blättel-Mink B., Di Giulio A. (in this volume): Sustainable consumption: how to evaluate sustainability in consumption acts.

Hauschildt J. (1993): Innovationsmanagement. München: Verlag Franz Vahlen.

Hellmann K.-U. (2010): Prosumer Revisited: Zur Aktualität einer Debatte. Eine Einführung. In: Blättel-Mink B., Hellman K.-U. (eds.): Prosumer Revisited. Wiesbaden: VS Verlag. 13–47.

Herstatt C. (1991): Anwender als Quelle für die Produktinnovation. Dissertation. Zürich.

Hoffmann E. (2007): Consumer integration in sustainable product development. In: Business Strategy and the Environment 16: 323–338.

Katz R., Allen T. (1982): Investigating the Not Invented Here (NIH) Syndrome: a look at the performance, tenure and communication patterns of 50 R&D project groups. In: R&D Management 12: 7–19.

Klotz U. (2007): Vom Taylorismus zur "Open Innovation" – Innovation als sozialer Prozess. In: Streich D., Wahl D. (eds.): Innovationsfähigkeit in einer modernen Arbeitswelt: Personalentwicklung – Organisationsentwicklung – Kompetenzentwicklung. Frankfurt/M.: Campus. 181–193.

Kriegesmann B., Kerka F. (2007): Innovationskulturen für den Aufbruch zu Neuem. Missverständnisse – Praktische Erfahrungen – Handlungsfelder des Innovationsmanagements. Wiesbaden: Gabler.

Kropp C. (2011): Erkennen und Gestalten – Lassen sich durch Szenarioprozesse Gestaltungsoptionen für ökosoziale Transformationen gewinnen? In: Elsen S. (ed.): Ökosoziale Transformation. Solidarische Ökonomie und die Gestaltung des Gemeinwesens. Neu-Ulm: AG SPAK. 154–180.

Ornetzeder M., Rohracher H. (2006): User-led innovations and participation processes: lessons from sustainable energy technologies. In: Energy Policy 34: 138–150.

Ornetzeder M., Rohracher H. (2011): Nutzerinnovation und Nachhaltigkeit: Soziale und technische Innovationen als zivilgesellschaftliches Engagement. In: Beck G., Kropp C. (Hrsg.): Gesellschaft innovativ – Wer sind die Akteure? Wiesbaden: VS Verlag. 171–190.

Oudhoorn N., Pinch T. (eds.) (2005): How Users Matter: The Co-Construction of Users and Technology. Massachusetts: MIT Press.

Paech N. (2005): Nachhaltiges Wirtschaften jenseits von Innovationsorientierung und Wachstum. Marburg: Metropolis.

Rammert W. (2010): Die Innovationen der Gesellschaft. In: Howaldt J., Jacobsen H. (eds.): Soziale Innovation. Auf dem Weg zu einem postindustriellen Innovationsparadigma. Wiesbaden: VS Verlag. 21–52.

Reichwald R., Piller F. (2006): Interaktive Wertschöpfung. Open Innovation, Individualisierung und neue Formen der Arbeitsteilung, Wiesbaden: Gabler.

Schrader U., Belz F.-M. (in this volume): Involving users in sustainability innovations.

Steiner S., Diehl B. (2011): Durchführung der Innovationsworkshops. In: Belz F.-M., Schrader U., Arnold M. (eds.): Nachhaltigkeits-Innovationen durch Nutzerintegration. Marburg: Metropolis. 80–97.

von Hippel E. (1986): Lead-Users. A source of novel product concepts. In: Management Science 32: 791–805.

von Hippel E. (1994). Sticky information and the locus of problem solving. In: Management Science 40 (4): 429–439.

von Hippel E. (2005): Democratizing innovation. Cambridge, Mass: MIT Press.

Zwick D., Bonsu S. K., Darmody, A. (2008): Putting consumers to work: 'co-creation' and new marketing governmentality. In: Journal of Consumer Culture 8: 163–196.

Birgit Blättel-Mink, Jens Clausen, Dirk Dalichau

11 Changing consumer roles and opportunities for sustainable consumption in online second-hand trading: the case of eBay

11.1 Introduction

Based on the observation that modern consumers increasingly involve themselves in the production of goods and services, the aim of this discussion is to assess the sustainability potential of online second-hand trading by private and (semi-)professional individuals. Electronic markets such as eBay offer their customers more than just convenient shopping. A clear shift has been noted away from the *object* of consumption towards the *act* of consuming (offering, negotiating, searching, selecting etc.). At the same time, there has been a dissolution of the traditional role division between consumers and producers. Exchange sites, auction platforms and other internet-based trading models – where users do not remain buyers, but simultaneously act as sellers – have brought about a distinct shift in traditional consumer roles.

Back in 1980, Alvin W. Toffler coined the term "prosumer" – a hybrid between a "producer" and a "consumer". The term denotes consumers who are either actively involved in the planning, organisation and creation of products and services, or who act simultaneously as sellers and thereby take on some of the traditional functions of producers. The increasing professionalisation of this form of trading (i.e. people earning a living from online trading) is a clear sign of the increase in prosuming activities.

The next two sections will explore the potentials for sustainability that the two new consumer roles outlined above may offer. Firstly, in order to identify the various participants' roles, current online trading trends will be explored. Secondly, a selection of findings from a survey of eBay users will be presented. Thirdly, the key outcomes of a life cycle assessment of selected products will serve to assess their sustainability potentials. Finally, a number of advantages and disadvantages of second-hand trading with respect to extending the life span of products will be discussed.[1] The article ends with a summary and some conclusive remarks.

11.2 The second-hand market

Unused goods often linger in private households even though they still retain some potential value. The reasons for this become clear upon closer inspection of the second-hand market. Prior to the advent of the internet, used goods had a very small and regionally restricted market. Products were sold at local flea markets and antique fairs, were given to second-hand dealers or passed on among families and friends. Finding specific or unusual items was time intensive and often unsuccessful. The arrival of the internet opened up a whole range of new opportunities for the trade of used goods. The online auction platform eBay was the first to use the new technology and probably remains the best-known online auction site. Selling unwanted goods has since become less complex and far more convenient. At a click of a mouse, goods can be put up for sale not just in Germany, but across the world. No need to hang around at the local flea market for hours only to sell one's products for a disappointing price or, conversely, pay too much for what is after all a second-hand item. Online trading allows sellers to locate buyers and buyers to find specific products. The life span of products is extended by selling them on to new owners who give them an new lease of life. The internet has contributed to a more transparent, convenient and manageable second-hand marketplace. Along with this development, second-hand goods have undergone a face-lift: marketed as vintage articles and rarities, they are no longer regarded as old, inferior and worthless, but modern, chic and fashionable. Initiated and successfully promoted by eBay, second-hand goods and products have known a true renaissance.

In line with the transformation of the second-hand market outlined above, the potential for sustainability may well change too. If existing products are used for longer, there will be less – or delayed – need for new production. Thus, (premature) degradation of the environment as a consequence of new production may be avoided by an extended use of old products. In principle, a longer life span of a product holds potential for sustainability, although the actual potential is highly dependent on the product in question. New and highly energy-efficient products may turn out to be more sustain-

---

1 The data referred to were collected as part of the project group "From Consumer to Prosumer – How changing consumer roles in the internet economy can hold potentials for sustainable consumption", conducted by the Institute for Futures Studies and Technology Assessment in Berlin, the Borderstep Institute for Innovation and Sustainability in Hannover and Berlin and the Goethe University in Frankfurt/Main.

able overall, compared to second-hand energy-guzzling products; refrigerators and washing machines from the nineties or – more recently – desktop computers are cases in point. By contrast, it is easier to assess the sustainability potentials for high quality goods with a long life span, such as furniture. If furniture is of outstanding quality (good craftsmanship etc.), its value merely depends on the preferences of the current owner. Apart from these two striking examples, online platforms offer a plethora of other products, each with its specific sustainability potential. In order to assess these diverse potentials, private eBay users who had bought or sold at least one new or second-hand product were interviewed. The survey revealed that practically all sold goods had been second-hand, whereas almost half of the bought products had been new. The survey and the life cycle assessment included second-hand products exclusively.

In the following section the online traders introduced above will be examined in more detail.

11.3 New players I: from consumers to "prosumers"

In step with a transforming second-hand market, players within this market have, over time, adopted different roles too. Online platforms such as eBay have enabled increasing numbers of consumers to put up for sale second-hand products, an activity which exceeds the remit of what we think of as activities of "normal" consumers. Back in the eighties, Toffler (1983) predicted the increasing significance of this type of individual 'freelance' work both in a general and in a specifically economic sense. As consumers have progressively contributed to the creation of value by companies, scholars from diverse disciplines have endeavoured to describe this phenomenon. From a perspective of industrial sociology, Günter G. Voß and Kerstin Rieder (2005) use the term "working customer" to describe this unofficial involvement of the consumer in the operations of a company. As an informal worker with skills tailored to the company, the prosumer blurs the boundaries between production and reproduction. Ralf Reichwald and Frank Piller (2009) consider the relationship between customers and companies from an economic perspective. They regard the interactive creation of value as a win-win situation, in which consumers – extrinsically or intrinsically motivated – contribute to innovation. Consumer involvement in the development of the internet is regarded by Heidemarie Hanekop, Andreas Tasch and Volker Wittke (2001) as a "new type of prosuming". Whereas Toffler (1983) largely regards the prosumer as a person who does physical or manual work, Hanekop et al. (2001) talk of the internet prosumer mainly as a performer of mental work. The new prosumer is required to manage new media, apply problem solving skills and – for eBay trading – also possess marketing and market research skills (presentation and pitching of products, plus effective timing). Referring to phenomena such as Wikipedia and Facebook, Alex Bruns (2008) talks about "produsage" as a hybrid form of *production* and *usage,* recognising that internet trading is less about production and more about usage. Individuals who engage in this hybrid activity are thus called produsers (Bruns 2010).

The various conceptual descriptions for essentially the same phenomenon outlined above resemble pieces of a jigsaw, each contributing to the overall picture, but remaining limited when seen in isolation. Raphael Menez and Daniel Kahnert (2010) urge us to try and overcome the contradictions that are an inevitable consequence of such diverse interpretations of the same phenomenon. "The differences in interpretation [...] can only be resolved if the different disciplines are willing to open up their boundaries to other disciplines in order to jointly conduct empirical investigations of the phenomena" (Menez/Kahnert 2010, p. 169; translated by C. Holzherr).[2]

The debate about the transformation of the online second-hand market is as complex and multifaceted as the transformation of more traditional consumer roles. eBay users exemplify the role of new prosumers insofar as they take pictures and devise descriptions of their products, place them on eBay's platform and track the bidding and payment transactions (Hanekop et al. 2001). They then package the sold products and send them for shipping, or they may refurbish the products before sending them off. In performing these activities, they fulfil the role of "traditional" prosumers as referred to by Toffler (1983). In their role as compilers and users of online catalogues they can be considered produsers (Bruns 2010): by adding to as well as buying from the eBay product range, they play an integral role in maintaining eBay trade – keeping themselves up to date in the process. They operate under the constraints of the platform, adhering to its rules and regulations. They exhibit the key characteristics of the "working customer" who is involved in the operations of the company by his or her own choice and is prepared to acquire the skills necessary for the task in hand. The interactive creation of value referred to by Reichwald and Piller (2009) is exemplified by the following scenarios: as working customers, eBay users "contribute" to the profit of the company in the sense that eBay obtains a proportion of the revenue from sold items as well as a further sum for fees.

The transformation "from consumer to prosumer" is thus not a simple matter of one change of role – it is multifaceted and dependent on point of view. For the purpose of our argument, the individuals involved in online second-hand trading will be referred to below in terms of Toffler's original coinage of prosumers.

---

2 For further discussion of the debate on user integration refer to the contributions of Kropp/Beck and Schrader/Belz in this volume.

As outlined above, second-hand trading per se offers considerable potential for sustainability. Further opportunities (Blättel-Mink 2010) arise when customers are involved in the production of goods and services. The shift from consumers to prosumers as part of online trading opens further avenues for sustainable consumption. The next section will take a closer look at the associated opportunities and risks.

11.4 New players II: from prosumers to "professionals"

While some users of online trading platforms such as eBay confine themselves to private trading, others use them as a stepping stone for setting up their own businesses. Online trading offers ideal opportunities for minimising start-up costs: with a computer and an email address a company can offer products to a potentially worldwide market – without having to rent premises and with practically no other overheads. In their report to the UN Global Compact (eBay 2010a, p. 6), eBay stresses the role that online platforms play in the development of young enterprises. "Our business enables hundreds of thousands of people to reach their personal aspirations of owning their own business, and during the last 15 years, we have trained hundreds of thousands of people on how to use eBay to grow their businesses. Additionally, in 2009, in collaboration with the Ewing Marion Kauffman Foundation – the world's largest foundation devoted to entrepreneurship – we launched the Sellers Challenge with the goal of empowering and inspiring entrepreneurs to start or grow businesses."

A survey in France, commissioned by eBay, shows that a large number of start-up companies were set up by people who were either unemployed, had physical handicaps or had little formal education. "eBay commissioned a study in France in 2009 and found that 26 percent of small business sellers have physical disabilities; and 49 percent do not have a diploma higher than high school. These numbers illustrate the striking opportunity that eBay represents for people who are historically disadvantaged to gain access to income and support their livelihoods" (eBay 2010a, p. 6).

The often voiced suspicions that the increasing professionalisation of online trading might lead to a desertion of city centres is put into perspective by the 2nd online business survey (eBay 2010b, p. 2): "The majority of traders use both online and offline approaches, thereby obtaining synergies that pay off. Online trade makes up a large part of turnover and thereby contributes to the survival of retailers. Nowadays, offline and online trading modes support and complement each other" (translated by C. Holzherr). Two thirds of online traders believe that a shop is more viable in the long-term if it also offers an online platform. 60 percent of online traders started their business as a normal shop. Of even greater interest is the fact that 28 percent of businesses started out trading online and subsequently acquired a shop as an addition to their business (eBay 2010b, p. 3). 22 percent of traders state that it is the large volumes of online turnover that makes it possible to keep shop premises (on average 58 percent of turnover). Online trading is particularly advantageous to traders in rural parts: 47 percent of online traders have their registered offices in communities of fewer than 20,000 inhabitants.

Overall, thanks to online trading platforms private consumer may develop into prosumers, semi-professional traders or even business owners with employees. The chances of success depend on the individual case. Whilst a few entrepreneurs go on to achieve revenues in the millions, the majority of semi-professional traders improve their income marginally and often only in the short term.

11.5 A new breed of actors: opportunities for and threats to sustainability

Second-hand trading offers ample opportunities for the protection of the environment. Environmental impact assessments have shown that extended usage of a product holds considerable sustainability potential (Erdmann 2011; cf. Behrendt et al. 2011). According to life cycle assessments undertaken as part of our study, this potential varies according to product and context of purchase (packaging, shipping, transaction mode online/offline etc.). "From an environmental perspective, second-hand online trade should be supported" (Erdmann 2011, p. 157; translated by C. Holzherr). Nonetheless, potential rebound effects should not be underestimated. For instance, money saved on second-hand goods might be spent on the acquisition of additional products.

An online survey[3] of 2,511 active eBay users, conducted in 2009, revealed that eBay users were not always environmentally motivated (Blättel-Mink et al. 2011). Generally, green attitudes appeared to be relatively common: 47 percent of the respondents agreed unreservedly with the statement "as citizens we can significantly contribute to the protection of the environment". By contrast, intended actions were relatively infrequent: only 27 percent of respondents agreed unreservedly with the statement that the protection of the environment was a motive for buying or selling via eBay. Significantly more people mentioned motives such as "practicality and convenience" (just under 76 percent agreed unreservedly), or "no need for relying on opening times" (just under 70 percent). A further motive appeared to be the fact that "unusual products" could be found on eBay (approx. 44 percent agreed unreservedly). Nonetheless, if this selection of respondents is compared to the overall population of Germany, the 27 percent of

---

3 For an outline of the methodology cf. Jaeger-Erben/Offenberger et al. in this volume.

environmentally motivated eBay traders compares with a mere 7 percent of "socially and ecologically minded" Germans overall (BMU 2010).

It appears that the respondents were quite willing to contribute to the protection of the environment through their actions. 64 percent were open to choose environmentally friendly shipping options, and a large proportion was also prepared to pay a small fee towards it. The potential for sustainability could thus be fully exploited by systematically following up such intentions.

These "potentials" for sustainability within eBay trading become even more evident upon closer inspection of eBay's customer base. To that end, a cluster analysis was applied to capture reported behaviours, motivations for eBay trading and willingness to act in environmentally responsible ways in the context of eBay trading.[4] Therefore, people with similar responses to the questionnaire questions were grouped together. The analysis revealed five groups, or "consumer types", of almost equal size (cf. Table 1): The first group consists of buyers who are, in the main, oriented towards saving money (the "price-conscious second-hand purchasers"); next come the "sceptics of second-hand trading", followed by the "online purchasers" who shop online frequently and specifically, but who are not particularly concerned with environmental matters. A fourth group, the "environment-oriented second-hand purchasers" can be characterised by not only being environment-oriented in an abstract sense, but by being willing to put their money where their mouth is. The members of this group tend to use both offline and online opportunities to act sustainably. However, a targeted re-sale of products with the intention of extending its life span is not particularly popular with this group. By contrast, the fifth group (23 percent) – the "prosumers" – were the most active online buyers and sellers within our survey. The potential for subsequent selling-on is an integral part of their consumer strategy, a characteristic that was not found in the other consumer types. Although not featuring at the top of their priority list, environmental impact is nevertheless taken into account by this group. Prosumers tend to be trend-conscious and use the opportunity for selling unwanted products as a means of acquiring more modern ones.[5] This does not, of course, preclude an immediate beneficial effect on the environment. By using products for a relatively short period before placing them back on the market, prosumers open up opportunities for buyers who could otherwise not afford such products. For instance, if a second-hand buyer opts for a high quality used (but nearly new) MP3-player instead of buying a lower quality new one, the production of new MP-3 players is delayed accordingly. Moreover, pro-

---

4 The cluster analysis was preceded by a factor analysis. The 21 factors which were identified were entered into the cluster analysis and thus characterise the five clusters or consumer types.

5 For an in-depth study of consumer types cf. Blättel-Mink et al. (2011).

Table 1: Overview of consumer types

Consumer type	Features	Percentage
Price-conscious second-hand purchasers	◆ price-consciousness as a motive ◆ low environmental orientation as a motive ◆ low trend-consciousness ◆ fairly strong environmental consciousness ◆ very low willingness to act environmentally responsibly ◆ high motivation to resell used products ◆ no willingness to take care of products and no wish to resell them ◆ intensive eBay trading ◆ low scepticism towards second-hand goods ◆ more online purchasing than selling ◆ more offline than online second-hand trading ◆ high behavioural change towards buying second-hand goods as a result of eBay (self perception)	20%
Sceptics of second-hand trading	◆ very high scepticism towards second-hand goods ◆ low second-hand trading, both online and offline (disposal of unwanted products) ◆ high barriers to selling, owing to "effort involved" ◆ highly trend-oriented ◆ fairly high willingness to act environmentally responsible ◆ fairly low actual environmentally friendly behaviour ◆ low eBay trading intensity (if at all, then new products) ◆ generally fairly low internet usage	20%
Online purchasers	◆ very high purchasing activity on eBay ◆ high barriers to selling, owing to "effort involved" ◆ more buying than selling ◆ high motivation to find rarities and collectors' items ◆ strong motivation for buying on eBay: relief from daily grind ◆ lowest price-consciousness ◆ very low environmental orientation as motive ◆ second-hand trading more offline than online ◆ second-hand trading more online than offline ◆ modest behavioural change towards buying second-hand goods as a result of eBay (self perception); eBay is rather used instead of offline purchasing opportunities used so far	15%
Environment-oriented second-hand purchasers	◆ strongest environmental consciousness as a motive ◆ strongest environmental consciousness of all types ◆ most pronounced actual environmentally friendly behaviour ◆ purchasers of green, sustainable, high-quality products with long life span ◆ low eBay trading intensity ◆ low scepticism towards second-hand goods ◆ very low willingness to take care of products in order to resell them ◆ frequent offline trader ◆ appreciate environmental potential of second-hand trading	22%

Prosumers	• highest motivation to sell products on • highest motivation to treat products with care with a view to selling them on • highest trend-consciousness • high online purchasing activity • greatest fun-factor motive • fairly low environmental orientation as a motive • highly positive attitude towards environmental issues • fairly low willingness to act environmentally friendly • fairly high trading intensity (purchase and sale) • far more online than offline second-hand trading	23%

The characterisation of the five consumer types is based on the 21 identified factors. Not all factors are relevant for each consumer type; therefore, only the most characteristic features are stated for each type in Table 1. Comparative terms (such as fairly etc.) always refer to features of one consumer type that differentiate it from other types.

duction of high quality durable products is promoted, which may in turn have positive consequences for sustainability. Consumption of high quality products by people who could not have afforded them first-hand also helps promote social inclusion and participation. However, the social advantage of a more equal distribution of goods may be in conflict with ecological sustainability: the net effect may well be a (further) acceleration of both production and consumption, which would, in turn, cancel out any desirable sustainability effects.

A key task will be to address the "new" players as ecologically and socially responsible consumers. If these individuals – who are already equipped with valuable competencies – were more aware of how to deal sustainably with (second-hand) products, the existing sustainability potential could be raised. For instance, if the "environment-oriented second-hand purchasers" (22 percent) could be motivated to resell their unwanted possessions, they could make a substantial contribution to the protection of the environment.

A further analysis of the survey indicated that the stage of life may impact on someone's second-hand trading habits – both online and offline. For instance, parents tend to sell high-quality children's clothing and toys online, while trading low-priced children's products offline. Retired people may use eBay when downsizing in retirement or buying products for a hobby. At the same time, they may use the opportunity to familiarise themselves – often reluctantly or at least sceptically – with the internet. Periods of unemployment also tend to be associated with specific behaviour patterns in second-hand trading. If it were possible to support these people in their online and offline trading, a further potential for sustainable resale could be realised.[6] Possible

---

6 Cf. Schäfer/Jaeger-Erben in this volume for general information on stages of life.

occasions might be young people's moving out of their parents' house, the birth of a child, relocation, weddings, divorces or deaths.

Another target group for tapping sustainability potentials are the growing number of semi-professional and professional second-hand online traders. They could exploit the hitherto neglected "treasures in people's attics". Even though the resale of such articles – when carried out by traders – might not result in substantial revenues for the original owners (they might not receive any money at the end of the day), at least the old unused items could be "rescued" and reused (Clausen et al. 2011) – an outcome that would undoubtedly meet with widespread approval, given that a respectable 52 percent of the respondents in the eBay survey stated that their used products could be of value to other people.

11.6 Conclusion

In this article, the example of eBay trading has been used to illustrate the potentials that two innovations in consumer practices may hold for sustainability: online second-hand trading and the involvement of consumers in a variety of roles in online second-hand trading. It has been shown that second-hand trading per se holds potentials for sustainability, which is enhanced by the active involvement of customers – be it private individuals, semi-professionals or start-up online businesses. It has further been demonstrated that people have diverse motives for buying and selling on eBay. Whilst green issues appear not to play a major role, they are not negligible either. "Prosumers" have been identified as a specific consumer type: consumers who buy goods with a view to selling them on and who therefore look after their products. However, given that this group is not particularly environmentally conscious, the concern is that their consumer behaviour may accelerate consumption and thereby counteract sustainability. The reverse is true for the "environment-oriented second-hand buyers" who like second-hand products, but are inclined to buy rather than sell products online. The task is to support both groups in their sustainable behaviours, i.e. fostering environmental responsibility in the prosumers and nudging the environmentally-oriented second-hand buyers to resell more of their unwanted products. A third task will be to use the opportunities that specific life stages present to promote the resale of unwanted products.

Regarding the professionalisation of online trading – and given that self-employment is often a consequence of unemployment – it is the social aspects that command attention. In addition, this group offers a doubly useful service of relieving the non-commercially-minded of their "hidden treasures in the attic" and making these available to people with low incomes. Generally, if there was a policy of associating second-hand trading not only with profitability, but with opportunities for protecting the environment, an as yet relatively unexplored potential for sustainability could be realised.

References

Behrendt S., Blättel-Mink B., Clausen J. (eds.) (2011): Wiederverkaufskultur im Internet: Chancen für einen nachhaltigen Konsum am Beispiel von eBay. Berlin, Heidelberg: Springer Verlag.

Blättel-Mink B. (2010): Prosuming im onlinegestützten Gebrauchtwarenhandel und Nachhaltigkeit. Das Beispiel eBay. In: Blättel-Mink B., Hellmann K.-U. (eds.): Prosumer Revisited. Zur Aktualität einer Debatte (Konsumsoziologie und Massenkultur). Wiesbaden: VS Verlag für Sozialwissenschaften. 117–130.

Blättel-Mink B., Bender S.-F., Dalichau D., Hattenhauer M. (2011): Nachhaltigkeit im online gestützten Gebrauchtwarenhandel: empirische Befunde auf der subjektiven Ebene. In: Behrendt S., Blättel-Mink B., Clausen J. (eds.): Wiederverkaufskultur im Internet: Chancen für einen nachhaltigen Konsum am Beispiel von eBay. Berlin, Heidelberg: Springer Verlag. 69–126.

BMU (2010): Umweltbewusstsein in Deutschland 2010. Ergebnisse einer repräsentativen Bevölkerungsumfrage. Berlin: Umweltbundesamt. [http://www.umweltdaten.de/publikationen/fpdf-l/4045.pdf; 27.06.2012]

Bruns A. (2008): Blogs, Wikipedia, Second Life, and Beyond: From Production to Produsage. New York: Publisher: Peter Lang.

Bruns A. (2010): Vom Prosumenten zum Produtzer. In: Blättel-Mink B., Hellmann K.-U. (eds.): Prosumer Revisited. Zur Aktualität einer Debatte (Konsumsoziologie und Massenkultur). Wiesbaden: VS Verlag für Sozialwissenschaften. 191–205.

Clausen J., Winter W., Behrendt S., Henseling C., Wölk M., Bierter W. (2011): Intensivierung des Gebrauchtwarenhandels: Neue Handelskulturen und Geschäftsmodelle. In: Behrendt S., Blättel-Mink B., Clausen J. (eds.): Wiederverkaufskultur im Internet: Chancen für einen nachhaltigen Konsum am Beispiel von eBay. Berlin, Heidelberg: Springer Verlag. 159–187.

eBay (2010a): UN Global Compact. eBay Inc. Communication on Progress. San Jose.

eBay (2010b): Online-Business Barometer. Eine Studie über Internet Händler in Deutschland. März 2010. [http://www.ebay.de; 27.06.2012]

Erdmann L. (2011): Quantifizierung der Umwelteffekte des privaten Gebrauchtwarenhandels am Beispiel von eBay. In: Behrendt S., Blättel-Mink B., Clausen J. (eds.): Wiederverkaufskultur im Internet: Chancen für einen nachhaltigen Konsum am Beispiel von eBay. Berlin, Heidelberg: Springer Verlag. 127–158.

Hanekop H., Tasch A., Wittke V. (2001): "New Economy" und Dienstleistungsqualität: Verschiebung der Produzenten- und Konsumentenrolle bei digitalen Dienstleistungen. In: SOFI-Mitteilungen Nr. 29/2001: 73–91. [http://webdoc.sub.gwdg.de/edoc/le/sofi/2001_29/hanekop-wittke.pdf; 27.06.2012]

Jaeger-Erben M., Offenberger U., Nentwich J., Schäfer M., Weller I. (in this volume): Gender in the focal topic "From Knowledge to Action – New Paths towards Sustainable Consumption": findings and perspectives.

Kropp C., Beck G. (in this volume): How open is open innovation? User roles and barriers to implementation.

Menez R., Kahnert D. (2010): Beyond Prosuming. Theoretische Perspektiven der Einbeziehung von Kunden und Nutzern in Arbeits- und Innovationsprozesse. In: Sozialwissenschaften und Berufspraxis (SuB) 33 (2): 153–173.

Reichwald R., Piller F. (2009): Interaktive Wertschöpfung. Open Innovation, Individualisierung und neue Formen der Arbeitsteilung. Wiesbaden: Gabler.

Schäfer M., Jaeger-Erben M. (in this volume): Life events as windows of opportunity for changing towards sustainable consumption patterns? The change in everyday routines in life-course transitions.

Schrader U., Belz F.-M. (in this volume): Involving users in sustainability innovations.

Toffler A. (1983): Die dritte Welle. Zukunftschance. Perspektiven für die Gesellschaft des 21. Jahrhunderts. München: Goldmann Sachbuch.

Voß G. G., Rieder K. (2005): Der arbeitende Kunde. Wenn Konsumenten zu unbezahlten Mitarbeitern werden. Frankfurt am Main: Campus.

Ulf Schrader, Frank-Martin Belz

12 Involving users in sustainability innovations

12.1 Introduction

Traditionally, consumer behaviour has been associated with the purchase, use and disposal of goods and services. Sustainable consumption defined in this sense is shared between, on the one hand, producers who develop and market sustainable products, and, on the other hand, consumers who buy, use and dispose of the products in the most environmentally friendly manner possible. For the purposes of the research project "Fostering Sustainable Consumption by Integrating Users into Sustainability Innovations" (short title "User Integration"), which is part of the social-ecological research programme, this division between producers and consumers has (at least partially) been dissolved. In a transdisciplinary research effort we attempted to find out how integrating consumers into innovation processes could promote the development and marketing of sustainable products and services. In this article we will present our approach and a selection of results.

Our focus was on which consumers to target and how best to involve them in innovation. In particular, our aim was to establish whether positive results in terms of sustainability innovation were entirely attributable to lead users, i.e. progressive users, or if non-lead users might also successfully contribute to such innovation processes.

This article begins by exploring the idea of opening innovation processes through user integration, and how sustainable consumption could be promoted in this context. Next, we will sketch out how we approached and implemented "open innovation" in our research project. The principal part of this article presents the quantitative evaluation of altogether twelve innovation workshops. The findings provide the basis for a successful integration of consumers into the generation of sustainability innovations. In the last section we will explore how the involvement of other users and stakeholders may optimise this process.

In contrast to the article by Cordula Kropp and Gerald Beck (in this volume), who report on the same project, but present results relating to the *diffusion* of innovations, our aim is to provide an overview of the project and focus on the *prerequisites* and *effects* of user integration.

12.2 Action for sustainable consumption through involving users in sustainability innovations

The "Open Innovation Paradigm" (Chesbrough 2003) can be understood as an alternative to traditional closed innovation processes, where product innovation and development up to market release is solely in the hands of company staff (e.g. in the R&D department). Potential users are consulted only as evaluators of the finished innovation (e.g. as part of traditional market research).

Opening up innovation processes can be beneficial for both consumers and companies (e.g. Hansen/Raabe 1991; von Hippel 2005; Reichwald/Piller 2006). By involving users in innovation processes, companies can tailor their products and services to customers' needs, thereby reducing the risk of a "flop" and increasing future users' willingness to pay for what is on offer. By employing creative methods for involving users, companies may not only obtain the explicit opinions of consumers, but they may gain access to implicit consumer knowledge (von Hippel 2005), so-called "sticky information". Open innovation can thus serve to pull together many different ideas and contribute to the development of a product or service. The 'contributing users' may benefit from open innovation by obtaining products and services that are optimally tailored to their needs. Additionally, they may experience joy, pride and recognition as a result of being part of a creative process (Steiner/Kehr 2011).

There are good reasons for involving users in generating sustainability innovations. On the basis of Fichter (2005), we define sustainability innovation as novel ways of problem-solving, capable of promoting viable, long-term styles of consumption. User integration and sustainability innovations are elements of sustainable consumption – a practice that is defined not in terms of consumers' *intentions*, but in terms of the *effects* of consumer behaviour (cf. Fischer et al. in this volume). People who would like to contribute to or make use of sustainability innovations need not necessarily have the explicit intention to contribute to sustainable development – although it may be beneficial if they are aware of sustainability issues.

In order to achieve the desired effect, sustainability innovations have to conquer the market. It is a fact that innovations with environmental and social benefits are often not perceived as sufficiently customer-oriented and therefore remain in a market niche. Open innovation can contribute in a number of ways to successful market penetration of such products and services. For instance, by taking into account customers' needs more consistently, and at an earlier stage in the development process, subsequent diffusion may be both facilitated and accelerated (Kropp/Beck in this volume). Furthermore, creative consumers who are aware of sustainability issues may come up with improvements that are socially or ecologically desirable and may have been overlooked by internal R&D departments. Furthermore, the sustainability of a product can often only be determined once it is actually used. The social and ecological benefits of a product depend on its actual usage and on any rebound effects it may cause. By involving users in the innovation process, these aspects can be assessed at an early stage. It should be clear from the above reflections that user integration into sustainability innovation is not only of benefit to participating consumers and companies, but can contribute to the promotion of sustainable consumption – which is not to say that user integration guarantees sustainable consumption.

12.3 "User Integration": the project group

Prior to the setting-up of the "User Integration" project, only a handful of schemes had applied open innovation methodology to sustainable consumption, one of which was the GELENA project (Social Learning and Sustainability), conducted by a group of junior researchers in Germany between 2002 and 2007 (e.g. Hoffmann 2007). Our project is unique in its wide-ranging scope: case studies involving lead users (LU) and non-lead users (NLU) were conducted across a total of eight sub-projects, spanning three industry sectors and involving five cross-sectional analyses (see Figure 1).

The case studies cover all stages of the innovation process. One idea that still requires further testing is "mobile mobility", i.e. transport services which link mobile communication technology with different transport providers. Packaging made of bioplastics has been technically feasible for some years. The task still remains to establish bioplastics on the market – if not immediately on the mass market, then initially as a niche product. By contrast, passive houses have been successfully launched on the market. The current challenge is to widen their appeal so that they can move onto the mass market. The various case studies included in the project were selected to cover three key ecological areas: housing, food and transport. Care was taken to select one field partner for each area for whom sustainability was part of the founding strategy and another one who is a (local) market leader in the field and who has – over time – embraced sustainable business practices. Within bioplastics a third field partner – a packaging producer – was also included.

Variety was also a criterion in the selection of participating users. Traditionally, user innovation focuses on "lead users" (von Hippel 1986). According to von Hippel, lead users are dissatisfied with what the market has to offer and would therefore benefit greatly from innovations, and by that token become involved in innovation processes. Genuine lead users are relatively rare in consumer goods markets, compared to capital goods markets. Furthermore, it was a basic hypothesis of the project that in order to

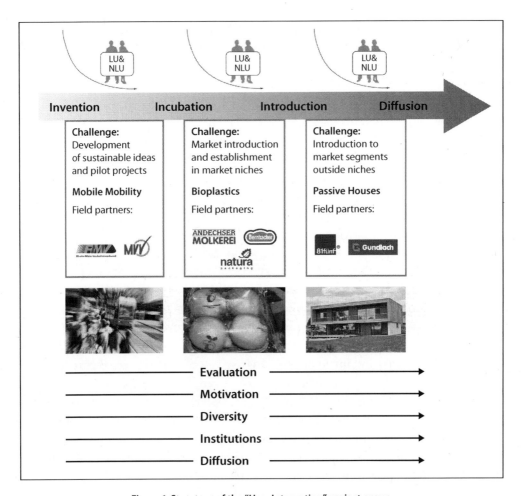

Figure 1: Structure of the "User Integration" project group

widen the distribution of sustainability innovations from a niche to a mass market, the contribution of "normal users", i.e. non-lead users, might be of special value. Therefore, both lead users and non-lead users were recruited. They were identified by means of a survey which elicited specific innovation-related characteristics in both user types (Ramakrishnan/Requardt 2011). A screening questionnaire was used that was based on Lüthje (2004) and Walcher (2007), who adapted von Hippel's lead-user characteristics for use within the consumer goods industry. The lead users and non-lead users selected for our project can be characterized by the extent to which they exhibited the following characteristics: dissatisfaction with what is currently available on the market, trend leadership, opinion leadership, involvement, expertise.

With each of the participating field partners (with the exception of the bioplastics producers) we conducted two innovation workshops: one with lead users and another with non-lead users. The twelve workshops with a total of 165 participants formed the basis for our empirical analyses. A theoretical framework based on motivational psychology was used in the design of the workshops: each one lasted 1.5 days and was characterized by a sequence of creative and evaluative elements (Steiner et al. 2011). The participants jointly developed a total of 38 innovation concepts. These are described in the case study reports for housing (Diel 2011), food (Requardt 2011) and transport (Ramakrishnan 2011).

The project consisted of eight sub-projects. In addition to the case studies, there were five further sub-projects, which evaluated the case studies by way of cross-sectional analyses and which covered the following areas:
- Evaluation of the environmental, social and economic impact of the concepts developed during the workshops (Cornet/Weber/Blaschke 2011)
- Motivation for participation and its development during the workshops (Steiner/Kehr 2011)
- Diversity of participants, particularly with regard to gender and age (Gebauer et al. 2011)
- Institutional factors within companies, sectors and society in general that are beneficial/detrimental to user integration (Backs/Siebenhüner 2011)
- Conditions for the diffusion of the concepts developed during the workshops (Beck/Kropp 2011; Kropp/Beck in this volume)

The focus of the analysis of the various case studies outlined below is on a) which users can most successfully be involved into sustainability innovation and b) how to optimally integrate these users.

12.4 Involving users in sustainability innovations – empirical insights

Since we generated our results especially within the twelve innovation workshops, the phases of workshop preparation, procedure and outcomes structure this section.

12.4.1 Workshop preparation
Alongside the development of a tailor-made approach (Steiner et al. 2011), the preparation mainly consisted of selecting and bringing on board suitable field partners and consumers. Contrary to von Hippel's (2005) optimism about the democratisation of knowledge and innovation, user involvement in (sustainable) innovation processes

is generally quite unfamiliar in Germany (Sichert/Siebenhüner 2011) and is not solidly anchored in institutions. This meant that initiating such an approach inevitably involved considerable effort and expense.

It became more and more obvious during the project that although it was difficult to identify lead users, once identified, they were easier to bring on board and they turned out to be more committed than non-lead users (Ramakrishnan/Requardt 2011). Our cross-sectional analysis of their motives in booking onto and participating in workshops indicated that lead users consistently outscored non-lead users on all measures (Steiner/Kehr 2011). The only area where lead users did not outscore non-lead users was in relation to extrinsic material rewards. Lead users' participation was clearly motivated by incentives such as increasing their knowledge, obtaining feedback on their skills and enjoying teamwork. Altruism and identification with the product and the company also played a role.

Given that altruism and identification with the product played a role in people's willingness to participate, it is important for future recruitment programmes not only to state the potential individual utility for users, but to stress the associated social and ecological benefits.

Extrinsic incentives (material rewards and career opportunities) were perceived as incidental by the participants in our project. This is consistent with findings of general lead-user research. In our project, neither lead nor non-lead users were particularly motivated by extrinsic rewards. Our participants were recompensed with an expense allowance of up to 100 euros for participating in the 1.5 day workshop. Our analyses could not clearly establish whether a higher "reward" would have incentivised people more to take part or whether this would have led to a "crowding out effect", i.e. to a reduction in non-material incentives in favour of material ones (Frey 1994).

A comparison of the three product sectors in our project indicated that willingness to participate also depended on the current relevance of the innovation to the individual consumer (Diehl 2011; Requardt 2011; Ramakrishnan 2011). Transport, for instance, is a crucial part of most people's lives. Unsurprisingly, recruitment was easiest for our mobile mobility case study, which involved local public transport users. In comparison, it was harder to enthuse people for our bioplastics project – probably because packaging matters less to people than transport. In contrast to passive housing, however, packaging is used on a daily basis and should therefore be relevant to most people's everyday lives. As expected, recruitment for the workshops on energy-efficient houses proved to be most difficult – especially in the case of non-lead users. This area is relevant only to very few people and at specific stages in their lives, i.e. immediately before they build or buy a house.

12.4.2 Workshop procedures

The measurements in Table 1 (cf. Steiner/Kehr 2011 for scales) confirm that the workshop design (Steiner et al. 2011) fulfilled its purpose of keeping motivation high throughout the 1.5 day workshop. This was especially true for the lead users who outscored non-lead users on motivation and wellbeing. Their levels of energy and perception of a smoothly running workshop tended to remain constant, while both decreased for non-lead users.

Table 1: Well-being and motivation during the workshop (mean and standard deviation)

		Point in time 1		Point in time 2		Point in time 3	
		M	SD	M	SD	M	SD
Hedonic tone	LU	5.65	0.86	5.77	0.95	5.68	0.88
	NLU	5.71	0.89	5.74	0.99	5.48	1.10
Energetic arousal	LU	5.53	1.01	5.80	0.95	5.58	1.27
	NLU	5.26	1.08	5.63	0.87	4.89	1.40
Tense arousal	LU	2.51	0.96	2.47	1.01	2.20	0.80
	NLU	2.58	1.01	2.46	0.89	2.32	1.06
Flow	LU			5.16	0.91	5.11	0.92
	NLU			4.96	0.81	4.62	0.98
Affective preferences	LU	5.91	0.91	6.04	0.85	6.05	0.74
	NLU	5.77	0.82	5.79	0.86	5.69	1.05
Cognitive preferences	LU	6.11	0.78	5.68	1.15	5.73	0.95
	NLU	5.80	0.87	5.27	0.91	5.37	1.09
Subjective skills	LU	5.47	1.00	5.90	0.80	5.81	0.68
	NLU	5.02	0.96	5.41	1.00	5.41	1.00

Source: Steiner/Kehr 2011 (scale from 1 = "totally true" to 7 = "not at all true")

Our results indicate that the challenge of involving non-lead users in innovation processes increases with the complexity of the applied methods. It is thus difficult to hold innovation workshops lasting several days with non-lead users.

In terms of diversity effects, qualitative interviews established that diversity among participants was consistently perceived as positive (Gebauer et al. 2011). Subjective perception of diversity appears to be more significant than actual objective differences between participants. For instance, even one single individual who is significantly different from other participants can cause a positive diversity effect. None of the respondents perceived too much or unproductive heterogeneity amongst the participants. This might be due the fact that the workshops were voluntary. Thus, a certain self-selection effect was to be expected, with participants sharing at least some common interest: people with no basic interest in the topic and with no desire for team work are unlikely to sign up for an innovation workshop, especially if there is no substantial material reward on offer.

12.4.3 Workshop outcomes

The workshop outcomes, i.e. the innovation concepts that were developed, were analysed in terms of both creativity and impact on sustainability.

Creativity was judged by a number of experts from the partner companies during a series of evaluation workshops – comprehensively documented in Diehl/Steiner (2011). Amabile's (1982) Consensual Assessment Technique (CAT) was used for the purpose, and the following dimensions of creativity were considered: originality, usefulness and elaboration (in terms of detailed elements of the concept). Overall, the workshop outcomes of lead users were perceived to be more creative (cf. Table 2).

Table 2: Evaluation of workshop outcomes (standardised overall evaluations)

Case study	Field partners	LU	NLU	d*
Passive houses	81fünf	10.11	8.44	2.27
	Gundlach	8.87	8.60	0.28
Bioplastics	Bernbacher	9.25	7.42	1.14
	Andechser	10.19	8.92	0.82
Mobile mobility	MVV	11.00	9.33	1.08
	RMV	10.00	8.33	0.79

* d = effect size; $d < 0.4$ = low; $0.4 > d < 0.8$ = medium; $d > 0.8$ = high

The overall average score for non-lead user workshops was less than 10 (potential maximum: 15). With regard to the lead user workshops, for 4 out of a total of 6 the average score was higher than 10, and for 2 slightly below 10. For the workshops held by the field partner Gundlach, the ratings of the innovation concepts of lead users and non-lead users were very similar. These workshops were the only ones where the aim was not to develop a product or service, but to work on diffusion approaches, i.e. on how to communicate and distribute passive houses. The results suggested that non-lead users may contribute in particularly effective ways at the later stages of innovation processes.

The fact that lead users are more creative is not only reflected in the average values in Table 2. A comparison of the separate innovations reveals that the five concepts with the highest creativity scores were developed in lead user workshops. However, three out of the 15 most creative innovation concepts were developed by non-lead users (cf. Table 3).

Table 3: Creativity ranking of top 15 final innovation concepts

Position	Final innovation concept	Field partners	K-Score	LU	NLU
1	Resealable packaging	Bernbacher	12.75	x	
2/3	Personal travel assistant	MVV	12.00	x	
	Intelligent packaging	Andechser	12.00	x	
4	Offer of the month	RMV	11.80	x	
5	Isar Flex	MVV	11.50	x	
6	Transfers between different modes of transport	MVV	11.25		x
7/8	Intelligent packaging	Andechser	11.00		x
	Intelligent packaging	Bernbacher	11.00	x	
9	Technology	81fünf	10.67	x	
10	Energy	81fünf	10.33	x	
11	Imminent prospect of greater comfort	MVV	10.25	x	
12/13	Organisation of an outing	RMV	10.00	x	
	Simple and robust packaging	Andechser	10.00	x	
14/15	Energy/resource conservation	Gundlach	9.80	x	
	Intelligent packaging	RMV	9.80		x

With regard to the different dimensions of creativity, lead users and non-lead users differed most in the extent to which they had elaborated their innovation concepts (Diehl 2011; Ramakrishnan 2011; Requardt 2011). This can be explained by the greater expertise and motivation of lead users.

The evaluation of ecological, economic and social impacts was carried out by researchers from the project team, who developed an "ecological traffic light" for the purpose (Cornet/Weber-Blaschke 2011). Given the abstract nature of many innovation concepts, the assessment of future applications and diffusion rates, as well as of side and after effects proved particularly challenging. The field partners were invited to contribute to this part of the evaluation. In the end, all innovations were evaluated in terms of pros and cons regarding sustainability. It was not possible to reach an overall consensus over whether lead users or non-lead users had come up with the most sustainable innovations. It was, however, observed that the innovations within energy-efficient housing (81fünf) differed considerably in terms of their content. Those developed by lead users were mostly concerned with the material aspects and the technical optimisation of the passive house as a product. By contrast, non-lead users concentrated less on the product per se, and focused instead on the living environment (Diehl 2011). It should be noted that women accounted for approximately 50 percent of the non-lead user workshop, whereas only one woman took part in the lead user workshop. Future research will show if the differences of focus were purely incidental or if they represent a trend, i.e. whether lead-users vs. non-lead users or gender differences were responsible for the differences in focus. Generally, it is safe to say that the effectiveness of sustainability innovations increases if not only the product itself, but the context in which it will be used is considered. In relation to houses, for instance, it would be advisable to take into account not just one house in isolation, but larger housing projects and their surroundings. In order for companies to modify their traditional frame of reference and embrace a bigger picture, improved organisational learning is often needed (Diehl 2011).

12.5 The outlook for open sustainability innovations

12.5.1 Differentiation within user integration

Our findings broadly tally with von Hippel's basic hypothesis that lead users tend to be best suited to the difficult task of developing a new product (von Hippel 1986). Whilst it is not easy to find lead users and bring them on board, once recruited, they tend to be more reliable, more motivated and more creative in their solutions than non-lead users. It has to be noted that our open innovation process was very time consuming and expensive and is unlikely (despite general appreciation and praise of the results) to be repeated by our field partners in the future without the support of public funding. Our verdict is therefore that either costs have to come down and/or the impact of the workshops has to be improved.

In order to improve impact, lead users could be selected more carefully. Proven innovative activity could be added to our selection criteria (dissatisfaction, trend leadership, opinion leadership, involvement, knowledge of the product and how to use it), so that lead users in the strict sense of the word – type I lead users – would be individuals who have actually developed products or services before (von Hippel 1986). Lead users who possess all the criteria mentioned above, but lack innovation experience would be type II lead users (cf. Figure 2).

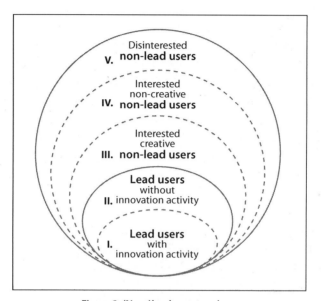

Figure 2: (Non-)lead user typology

One of the traditional lead user features is their expectation to draw a personal advantage from their participation (e.g. von Hippel 1986). In the context of sustainability innovations, where the aim is to arrive at solutions that benefit society at large, lead user selection criteria should therefore include a strong interest in sustainability issues (Belz et al. 2011). While it is no easy task to identify such a group of lead users who not only have experience in innovation, but are also sensitive to sustainability issues, the rewards of engaging such individuals may be worthwhile.

A subdivision within non-lead users may also enhance outcomes. Within the three groups of non-lead users, the individuals who are highly creative and motivated (III) despite not possessing lead user characteristics are of particular interest as recruits for user integration. By contrast, those individuals who are motivated and keen to participate in open innovation, but lack creative skills may weigh down open innovation processes (IV) and should therefore be excluded. The third non-lead user group are those individuals who show no interest in open innovation (V). A future challenge will be to develop recruitment instruments that are capable of excluding type IV individuals at the recruitment stage of the innovation process.

In order to cut down on expenses, it is conceivable to forego a targeted selection of participants altogether. An analysis of innovation workshops made up of randomly mixed lead and non-lead users was outside the scope of our project. Future empirical research projects could use a classification such as suggested in Figure 2 for analysing various combinations of groups in quasi-experimental designs.

12.5.2 From user integration to stakeholder integration

With regard to open sustainability innovation, there are more opportunities to cut costs and improve impact when, in addition to consumers, other stakeholders are involved.

Neyer et al. (2009) point out that open innovation need not be limited to involving end users. An innovation process that is originally restricted to the R&D department of a company can be opened up to employees from other departments (cf. Figure 3). Neyer et al. (2009) distinguish between "core inside innovators" (members of R&D), "peripheral inside innovators" (II: employees from other departments such as procurement, production and sales) and "outside innovators", such as the users who took part in our workshops. Employees participating in open innovation tend to be regarded as acting in their professional role only, i.e. the involvement into the innovation process of employees outside the R&D department tends to be justified by the fact that their experiences in different departments of the company can usefully feed into the innovation process (Neyer et al. 2009). It is often overlooked that the private lives of

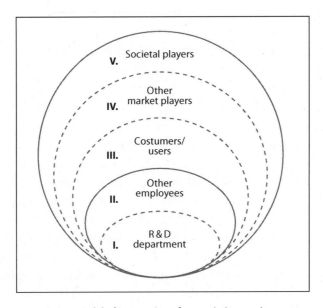

Figure 3: Onion model of integration of actors in innovation processes

employees (including those working in the R&D department) may hold potentials for innovation too. We would argue that a targeted involvement of employees as private consumers could be particularly promising for producers of consumer goods. Furthermore, it is often easier and cheaper for companies to involve their own staff instead of recruiting outsiders, and by using people who have an intimate knowledge of the company and its products or services, they can ensure that the innovations are always relevant. Finally, if people who are sustainability-conscious in their private lives are given the opportunity to express their convictions in their professional environment, they are highly likely to respond with job satisfaction and loyalty (Muster/Schrader 2011).

Additional external stakeholders include other market players (IV: e.g. retailers and suppliers) and societal players (V: e.g. environmental organisations, consumer organisations and public authorities).

It is crucial for sustainability marketing to enter into dialogue with these stakeholders (Belz/Peattie 2009; Schrader/Diehl 2010). The exchange with external stakeholders has long been discussed under the headings of "business dialogue" (Hansen et al. 1996) or "stakeholder dialogue" (e.g. Kaptein/Van Tulder 2003; O'Riordan/Fairbrass 2008), whereby the focus has mostly been on avoiding risks and increasing the legitimacy of companies (Arnold 2010). Stakeholder integration for the purpose of generating innovations is far less analysed (Dyllick et al. 1997). And yet, integrating people who are interested in and have knowledge of the general well-being of society may be particularly promising for innovations within sustainable consumption (Holmes/Smart 2009; Heiskanen/Lovio 2010).

Stakeholder dialogue tends to be framed as interaction with non-market players. However, the emergence of lead user methodology from within the business-to-business sector has demonstrated that intensive interaction with suppliers and institutional customers (e.g. retailers), i.e. with upstream and downstream value-adding partners, may also yield opportunities for the innovation sector. Whilst this practice is well established within open innovation generally, it has, as yet, been rarely adopted in a sustainability context. Interaction with market players who are at the same level in the value-added chain is yet another practice that could hold promise – as long as companies are able to collaborate as partners rather than competitors.

If innovation is to continue to open its boundaries and become a mainstream paradigm, it will not be long before the pool of willing and creative costumer participants will be exhausted and professional players will have to be drawn in. Zwick et al. (2008) describe in their critical paper "Putting Consumers to Work" how companies take advantage of users as badly paid workers in their open innovation schemes. Against this background, it might be advisable to limit intensive user integration to those innovations that benefit both the individual participants of open innovation programmes and society at large.

The extent to which user integration is capable of delivering sustainable results is very much dependent on the particular innovation in question. In our project, the focus was on the generation and evaluation of different approaches to innovation. The real sustainability credentials of a product or a service only emerge once it has been created, launched and established on the market – product development phases that were beyond the scope of our project. In order to thoroughly assess the potentials of user integration in the context of sustainable consumption, it would thus be necessary to include the implementation stage of a given innovation (Pobisch 2010). This is an area in which considerably more research is needed (cf. Kropp/Beck in this volume).

References

Amabile T. M. (1982): Social Psychology of Creativity: A Consensual Assessment Technique. In: Journal of Personality and Social Psychology 43 (5): 997–1013.

Arnold M. (2010): Stakeholder dialogues for sustaining cultural change. In: International Studies of Management and Organisation 40 (3): 61–77.

Beck G., Kropp C. (2011): Diffusionsszenarien: Verbreitung von Nachhaltigkeitsinnovationen durch Nutzerintegration? In: Belz F.-M., Schrader U., Arnold M. (eds.): Nachhaltigkeits-Innovationen durch Nutzerintegration. Marburg: Metropolis. 257–280.

Belz F.-M., Codita R., Moysidou K. (2011): Expected Social Benefits as a Novel Characteristic of Lead Users in the Context of Sustainability Innovations. Paper for the R&D Management Conference, Nörrkoping, Sweden, June 28–30, 2011.

Belz F.-M., Peattie K. (2009): Sustainability Marketing: A Global Perspective. Chichester: Wiley.

Belz F.-M., Schrader U., Arnold M. (eds.) (2011): Nachhaltigkeits-Innovationen durch Nutzerintegration. Marburg: Metropolis.

Chesbrough H. (2003): Open Innovation: The new imperative for creating and profiting from technology. Boston. Mass: Harvard Business School Press.

Cornet H., Weber-Blaschke G. (2011): Nachhaltigkeitsscreening innovativer Ideen und Konzepte. In: Belz F.-M., Schrader U., Arnold M. (eds.): Nachhaltigkeits-Innovationen durch Nutzerintegration. Marburg: Metropolis. 195–214.

Diehl B. (2011): Nachhaltigkeitsinnovationen im Bedarfsfeld Wohnen. In: Belz F.-M., Schrader U., Arnold M. (eds.): Nachhaltigkeits-Innovationen durch Nutzerintegration. Marburg: Metropolis. 121–144.

Diehl B., Steiner S. (2011): Kreativitätsbewertung. In: Belz F.-M., Schrader U., Arnold M. (eds.): Nachhaltigkeits-Innovationen durch Nutzerintegration. Marburg: Metropolis. 99–108.

Dyllick T., Belz F., Schneidewind U. (1997): Ökologie und Wettbewerbsfähigkeit von Unternehmen. München, Zürich: Hanser.

Fichter K. (2005): Interpreneurship. Nachhaltigkeitsinnovation in interaktiven Perspektiven eines vernetzenden Unternehmertums. Marburg: Metropolis.

Fischer D., Michelsen G., Blättel-Mink B., Di Giulio A. (in this volume): Sustainable consumption: how to evaluate sustainability in consumption acts.

Frey B. S. (1994): How Intrinsic Motivation is Crowded Out and In. In: Rationality & Society 6 (3): 334–352.

Gebauer S., Schütze L., Ihsen S. (2011): Diversity in Nachhaltigkeitsinnovationsprozessen. In: Belz F.-M., Schrader U., Arnold M. (eds.): Nachhaltigkeits-Innovationen durch Nutzerintegration. Marburg: Metropolis. 233–255.

Hansen U., Niedergesäß U., Rettberg B. (1996): Dialogische Kommunikationsverfahren zur Vorbeugung und Bewältigung von Umweltskandalen: Das Beispiel des Unternehmensdialoges. In: Bentele G., Steinmann H., Zerfaß A. (eds.): Dialogorientierte Unternehmenskommunikation. Grundlagen – Praxiserfahrungen – Perspektiven. Berlin: VISTAS. 307–331.

Hansen U., Raabe T. (1991): Konsumentenbeteiligung an der Produktentwicklung von Konsumgütern. In: Zeitschrift für Betriebswirtschaft 61: 171–194.

Heiskanen E., Lovio R. (2010): User-Producer Interactions in Housing Energy Innovations. In: Journal of Industrial Ecology 14 (1): 91–102.

Hoffmann. E. (2007): Consumer integration in sustainable product development. In: Business Strategy and the Environment 16: 323–338.

Holmes S., Smart P. (2009): Exploring open innovation practice in firm-nonprofit engagements: A corporate social responsibility perspective. In: R & D Management 39 (4): 394–409.

Kaptein M., Van Tulder R. (2003): Toward Effective Stakeholder Dialogue. In: Business and Society Review 108: 203–224.

Kropp C., Beck G. (in this volume): How open is open innovation? User roles and barriers to implementation.

Lüthje C. (2004): Characteristics of innovating users in a consumer goods field: An empirical study of sport-related product consumers. In: Technovation 24 (9): 683–695.

Muster V., Schrader U. (2011): Green Work-Life-Balance. A new perspective for Green HRM. In: German Journal of Research on Human Resource Management 25 (2): 140–156.

Neyer A.-K., Bullinger A. C., Moeslein K. M. (2009): Integrating inside and outside innovators: A sociotechnical systems perspective. In: R & D Management 39: 410–419.

O'Riordan L., Fairbrass J. (2008): Corporate Social Responsibility (CSR): Models and Theories in Stakeholder Dialogue. In: Journal of Business Ethics 83: 745–758.

Pobisch J. (2010): Konsumentenorientierte Produktinnovationen. Erfolgreiche Generierung, Verbreitung und Verwertung von Konsumentenwissen. Marburg: Metropolis.

Ramakrishnan S. (2011): Nachhaltigkeitsinnovationen im Bedarfsfeld Mobilität. In: Belz F.-M., Schrader U., Arnold M. (eds.): Nachhaltigkeits-Innovationen durch Nutzerintegration. Marburg: Metropolis. 169–193.

Ramakrishnan S., Requardt M. (2011): Auswahl der Teilnehmer der Innovationsworkshops. In: Belz F.-M., Schrader U., Arnold M. (eds.): Nachhaltigkeits-Innovationen durch Nutzerintegration. Marburg: Metropolis. 65–79.

Reichwald R., Piller. F. (2006): Interaktive Wertschöpfung. Open Innovation, Individualisierung und neue Formen der Arbeitsteilung. Wiesbaden: Gabler.

Requardt M. (2011): Nachhaltigkeitsinnovationen im Bedarfsfeld Ernährung. In: Belz F.-M., Schrader U., Arnold M. (eds.): Nachhaltigkeits-Innovationen durch Nutzerintegration. Marburg: Metropolis. 145–167.

Schrader U., Belz F.-M. (2011): Nutzerintegration in Nachhaltigkeitsinnovationen: Ein- und Ausblicke. In: Belz F.-M., Schrader U., Arnold M. (eds.): Nachhaltigkeits-Innovationen durch Nutzerintegration. Marburg: Metropolis. 331–346.

Schrader U., Diehl B. (2010): Nachhaltigkeitsmarketing durch Interaktion. In: Marketing Review St. Gallen 27 (5): 16–20.

Sichert D., Siebenhüner B. (2011): Institutionelle Rahmenbedingungen von Nutzerintegration. In: Belz F.-M., Schrader U., Arnold M. (eds.): Nachhaltigkeits-Innovationen durch Nutzerintegration. Marburg: Metropolis. 281–301.

Steiner S., Diehl B., Engeser S., Kehr H. (2011): Nachhaltigkeits-Innovationen durch Nutzerintegration. Implikationen für eine optimierte NutzerInnenansprache und Gestaltung von Innovationsworkshops. In: Zeitschrift für Umweltpsychologie 15 (1): 52–70.

Steiner S., Kehr H. (2011): Nutzermotivation. In: Belz F.-M., Schrader U., Arnold M. (eds.): Nachhaltigkeits-Innovationen durch Nutzerintegration. Marburg: Metropolis. 215–231.

von Hippel E. (1986): Lead-Users. A source of novel product concepts. In: Management Science 32: 791–805.

von Hippel E. (2005): Democratizing innovation. Cambridge, Mass.: MIT Press.

Walcher D. (2007): Der Ideenwettbewerb als Methode der aktiven Kundenintegration: Theorie, empirische Analyse und Implikationen für den Innovationsprozess. Wiesbaden: DUV.

Zwick D., Bonsu S. K., Darmody A. (2008): Putting consumers to work: 'co-creation' and new marketing govern-mentality. In: Journal of Consumer Culture 8: 163–196.

Section E

Design and efficacy of societal steering

Andreas Koch, Daniel Zech

13 Impact analysis of heat consumption – user behaviour and the consumption of heat energy

13.1 Introduction

Domestic heat consumption is influenced by a range of factors and can therefore vary considerably from one household to the next. From a technical point of view, energy use, often referred to as consumption, is determined by the properties of the building, such as the heat transfer coefficient of the building components, the airtightness of the building envelope and the efficacy of the heating system. In addition to these structural and technical aspects, energy use is dependent on user behaviour, i.e. the user can influence the heat consumption of a building to a considerable extent. As a result of different preferences for ventilation and room temperature, there is often a discrepancy between a theoretically calculated energy demand and the actual energy consumed in a given household. The objective of this article is to investigate the significance and the impact of different user behaviours on energy consumption. For this purpose, user behaviour – and in particular some key parameters for interactions between users and buildings – will be identified and quantitatively analysed. A range of building types will be compared, and the trend towards lower energy needs will be represented by the variation of the insulation standard for those building types.

It was one of the objectives of the parameter analysis carried out as part of the project group "consuming energy sustainably – sustainable energy consumption" to discriminate between different patterns of behaviour. This article will employ the conceptual system presented by Di Giulio et al. in this volume, according to which a comfortable and hygienic room temperature is seen as an objective need. The energy demand is based on the objective needs of the residents as well as the features of the building and its heating system. This objective demand is, so to speak, unavoidable, and is satisfied by the consumption of either fossil fuels or renewable energy sources. Any additional energy demand (e.g. number of heated rooms, room temperature, amount of hot water consumed) can be regarded as a subjective desire and could be contested on ethical grounds.

13.2 Energy use in residential buildings

Industry, transport and private households are the three largest energy consumption sectors in Germany. In recent years, they have been responsible for around a third of Germany's final energy use (Schoer et al. 2006). Within private households, 35 percent is attributable to private transport (fuel), and most of the remaining 65 percent to space and water heating (AGEB 2011). Apart from transport, savings on heating energy thus hold the highest potentials for promoting sustainable consumption. Figure 1 shows that space heating and hot water take up 85 percent of household energy use (transport was not included in this calculation).

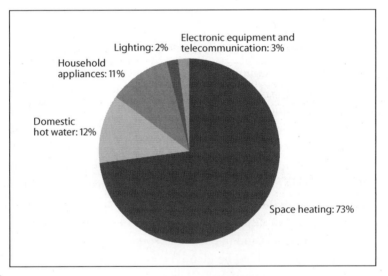

Figure 1: Final energy consumption by private households in 2008 according to category (BDEW 2010)

The main strategies for promoting a more sustainable use of energy are a) extensive renovation of buildings (aka "deep retrofit", resulting in a more efficient building envelope) and b) selection of renewable energy carriers (Bettgenhäuser/Boermans 2011). Both strategies require considerable financial outlay and their economic viability partly depends on the life cycle of the building and the efficacy of the heating system. Additional energy could, however, be saved if users were able to modify some of their behaviours in terms of energy consumption.

In order to describe that part of energy use that is associated with user behaviour, a distinction is made between *demand* and *use* of energy. "Demand" is to be understood as values calculated for heat and energy in terms of "fixed boundary conditions"

(DIN V 4108-6). The influence of individual user behaviour is explicitly excluded from these calculations. Heat demand represents the difference between *heat loss* through transmission and ventilation on the one hand, and internal and solar energy *heat gain* on the other. The discrepancy between heat loss and heat gain, plus the energy required for hot water, has to be supplied by a heating system. The *final energy demand* is the amount of energy required to meet the demand for space heating and hot water, after taking into account the losses resulting from heat distribution and hot water systems. The *primary energy demand* represents the amount of overall energy (production, conversion, transport) that is required to cover the *final energy demand* of a building.

Demand stands in contrast to *use*. Whereas demand represents an amount of energy calculated from pre-existing conditions, use denotes the amount of energy that has actually been *used*; this is a *measured* value. Energy use is often referred to as

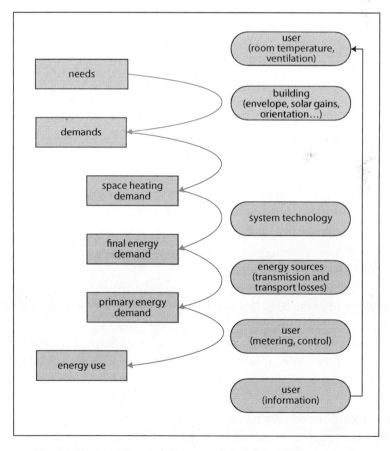

Figure 2: From needs to use – key parameters (after to Koch et al. 2008)

energy consumption. The German Energy Saving Ordinance of 2009 (EnEV2009), for instance, refers to "energy consumption" (Energieverbrauch) in their Energy Performance Certificates (§ 17, paragraph 1, sentence 1, EnEV 2009). A major problem in the identification of efficiency potentials is the fact that energy use and energy demand do not necessarily correspond. As a consequence of an inadequate operation of the heating system, actual energy use may be considerably higher than calculated demand. The operation of a heating system is considered inadequate if its output does not correspond to the requirements of the users. In these cases, the system is either badly adjusted or its functioning is somehow deficient. Whilst the actual energy use can theoretically be higher or lower than the calculated demand, in practice, use tends to outstrip demand. Higher or lower actual energy use is often referred to in everyday terms as being either "wasteful" or "economical", as the case may be. However, such broad judgments, which do not take into account specific household circumstances, tend to be unhelpful. A relatively high or low value may, for instance, reflect varying hot water demands as a result of different household sizes.

The stages from needs via demand to use are illustrated schematically in Figure 2. In the context of residential buildings, the consumption of heat energy can thus be divided into the following components. The largest component, the objective demand, arises from an objective need for a room temperature that is generally perceived as comfortable. Additional energy demand arises from subjective desires. The final portion of energy use arises from neither objective needs nor subjective desires as it is not related to a desired effect; it is, so to speak, 'unintentional' use (e.g. incorrect adjustment of heating system). Although it is possible to differentiate in principle between different components of energy use, issues of distributional fairness become relevant, as soon as individual cases are considered. It is thus a question of defining objective needs and subjective desires. Their combined effect results in a – partly justifiable – final energy demand.

13.3 Efficiency and consistency strategies for reducing primary energy demand

There are a number of ways of reducing the primary energy demand of residential buildings. Savings can first of all be achieved through efficient use of energy. This reduces the final energy demand, which in turn reduces the primary energy demand. Options include insulation of the building envelope or optimisation of the energy distribution and conversion system. Such measures are referred to as efficiency strategies. Additionally, the primary energy demand can be reduced by replacing fossil fuels with renewable energy sources; these measures are referred to as consistency strategies.

Examples include the use of solar radiation or biomass for heating purposes (e.g. wood pellets). The required energy efficiency standards for buildings in Germany are largely defined by the Energy Saving Ordinance (EnEV) which refers to the German Energy Savings Act (EnEG). The use of renewable energy sources for the purpose of heating buildings (i.e. consistency strategy) is defined in the German Renewable Energy Heat Act (EEWärmeG; in force since 01.01.2009).

The aim of the above strategies and the associated legal regulations is to reduce the non-renewable part of primary energy. On the assumption that renewable resources will continue to be available in the long term, a sustainable consumption of heat energy is possible only if it does not restrict the use by others (cf. Fischer et al. in this volume). Sustainable heat consumption is thus achievable through both consistency and efficiency strategies (Koch/Jenssen 2010).

A calculation geared towards this objective is contained in the EnEV and EEWärmeG for residential buildings (DIN 4108-6 and DIN 4701/10). The same calculation is used in our impact analysis. The primary energy demand (Q_P) is the result of the sum of the heat demand (Q_h) and the hot water demand (Q_{tw}), multiplied by the system expenditure (e_p). The latter denotes "the ratio of the primary energy fed into the heating system in relation to the system's output" (DIN 4701-10; translated by C. Holzherr).

$$Q_P = (Q_h + Q_{tw}) \times e_P$$

In accordance with the above equation, it is possible, within certain limits, to achieve a balance between *efficiency* and *consistency* in energy saving measures. It is also possible to describe those everyday behaviours that will lead to a reduction in the demand for heat energy, final energy and primary energy (cf. 13.4). In order to avoid an often unconsciously incurred waste of heating energy, there are, for instance, a number of adjustments that could be made to the heating system that would optimise the heating output and thus reduce energy use (Clausnitzer 2004).

Calculations such as described above always relate to energy demand in specific buildings. On the basis of calculations of the overall energy efficiency of individual buildings, a comparison can be drawn between different building types. Given that different buildings differ widely in their compactness and households within buildings show very different patterns in the use of energy, this method is not capable of capturing the energy efficiency of individual households.

13.4 Influence of user behaviour – results of impact analysis

The first objective in the analysis presented below is to apportion energy consumption to the various influencing factors. No judgement will be made as to which uses of energy are 'sustainably legitimate'. For the purpose of describing the relation of user behaviour and heat demand, a standardised Excel tool was developed, in which user behaviour was expressed as 'numerical' parameters. These were obtained by relating the various user profiles to descriptions of typical user behaviour. The intention was a) to identify those parameters with the most powerful impact, such as the desired room temperature or ventilation behaviour, and b) to quantify (by means of numerical parameters) the relevance of the various behaviours in relation to heat energy and primary energy demand. A comparable approach has been suggested by Loga et al. (2003). A further concern was to investigate whether the relevance of user behaviour to overall energy use is generally dependent on the overall quality of a building and whether it might become more relevant as buildings become more energy-efficient. The methodology is capable of delivering results that are consistent with measurement campaigns in low-energy and passive houses (cf. Diefenbach et al.; Ebel et al. 2003; Koch et al. 2008).

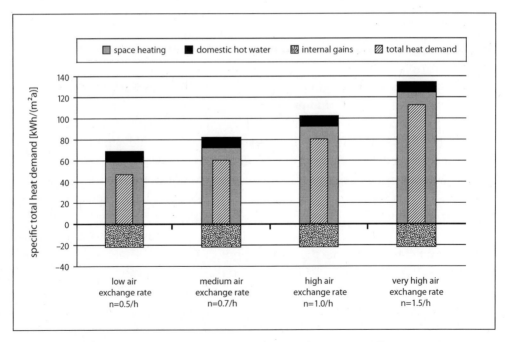

Figure 3: Composition of heat energy demand for a EnEV reference building – different air exchange rates

With regard to the heat demand of a building, the "air exchange rate" is the most important of the variables under investigation in our study. It is expressed as the ratio of fresh air coming into a room in one hour in relation to the overall air volume. A value of $0.5\,h^{-1}$ is regarded a "low air exchange rate"; it signifies that half the air in a room is replaced by fresh air within a given hour. As one would expect, a "high air exchange rate" ($1.5\,h^{-1}$) results in an additional consumption of heat energy of 66 kWh/(m^2a) (see Figure 3). Overall, optimal ventilation behaviour and high-quality insulation of the building envelope have a similar overall effect on energy savings. It should be noted that, in the context of our study, these variables were investigated only in relation to energy demand. In real life, the calculation of an optimal ventilation rate should additionally take into account the need to eliminate dampness in buildings (to prevent mould). The air exchange rates recommended by DIN 4108-6 (between 0.5 and 1; cf. DIN 4108-6; cf. Figure 3) correspond to an assumed minimum requirement for a hygienic internal air quality.

Overall, it can be said that user behaviour influences to a considerable degree the heating and hot water demand in a given building. Behaviour patterns range from "frugal usage" with a room temperature of 18 °C and modest consumption of hot water, to "energy-intensive usage" with a room temperature of 22 °C and heavy hot water

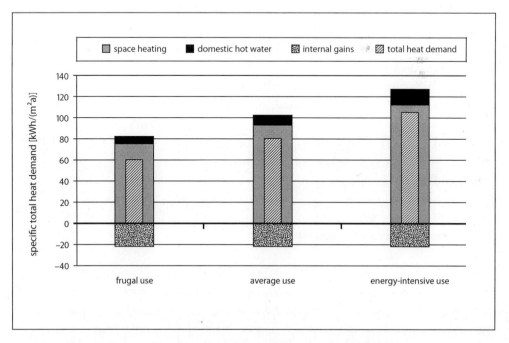

Figure 4: Heat energy demand of an EnEV reference building – different user types

consumption. Such differences in energy demand result in values deviating from plus 31 percent to minus 25 percent from a calculated average value for a reference building as calculated according to EnEV (see Figure 4).

Therefore, more attention should be paid to user behaviour, both in the EnEV's calculations of average energy demand and in the general planning of buildings and heating systems. Heating systems could then be built to match individual user behaviour patterns, and an efficient use of heating energy, as well as an adequate level of comfort (for everyday living) could be assured. Calculations of heat demand with individual parameter adjustment could be deployed in explaining to residents why their actual energy use might be deviating from estimated energy demand. Even if it is difficult to establish individually tailored, detailed parameters, indicators such as household size (number of people living in a household) could be used for calculating hot water demand. Such an indicator would be more meaningful than the current EnEV value, which is established purely on the basis of the surface area of a residence. Lastly, installers of heating and ventilation systems should instruct residents how to operate and regulate their systems to optimal effect.

Our investigations have shown that, in absolute terms, the largest energy savings can be achieved in poorly insulated buildings with a high heat demand. Although the share of the heat demand that the user is able to influence is proportionally smaller in these buildings than in well-insulated buildings, behavioural changes can reduce resource consumption considerably. Even in well-insulated buildings with efficient heating systems there is scope for energy savings through intelligent user behaviour. Although in relative terms, the amount of heat demand that can be influenced by the user is considerably higher in these buildings, the absolute savings are considerably lower than in poorly insulated buildings. In high performance buildings, user behaviour is a last 'lever' for optimisation of energy use. In summary, the higher the heat demand of a given building, the higher the absolute savings potentials through optimal user behaviour, but the lower the proportion of overall heat demand that can be directly influenced by user behaviour.

13.5 Limitations of standardised assessment

A comparison of different types of households illustrates how various objective needs and subjective desires of the household lead to considerable variation in energy use. The actual use is contrasted with a standardised value for demand. It is calculated on the basis of individual user parameters, relating to room temperature, air exchange rate, hot water consumption and solar as well as internal gains (Koch et al. 2008). In principle, standardised calculations such as those prescribed in the German Energy Saving Ordinance can generate technically acceptable values capable of representing user behaviours – even when behaviour deviates from the norm. The only exception is the calculation for the demand of hot water of households, which uses surface area rather than a user-related reference value and is therefore incapable of taking into account household size and other socially relevant factors.

Buildings- and systems-related energy savings as well as those arising from changes in user routines have been discussed above. A consideration of objective user needs and subjective user desires is capable of giving further information about individual household requirements. These needs and desires will be discussed below with reference to the utilisation of living space and to a typology of residential buildings.

For the purpose of comparing energy demand in private households, the specific energy demand relating to the surface area (kWh /(m^2a)) is a widely accepted measure. Given that this value does not depend on user behaviour, it is possible to compare the energy performance of different building types when using the heating needs. When referring to final energy demand, also the efficiency of the heating systems is considered. An analysis of private use of natural resources between 1995 and 2004 by the German Federal Statistical Office (Schoer et al. 2006) showed that implemented saving measures led to a 9.1 percent reduction in energy use for space heating. However, during the same period, a 13.1 percent increase in the usage of residential living space more than offset those savings (Figure 5). This combined effect is reflected in the overall energy use of households, but not in specific energy demand, which reflects merely the increase in efficiency.

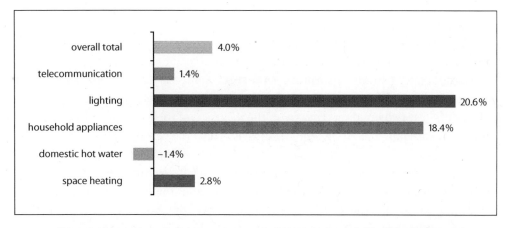

Figure 5: Changes in overall energy consumption of private households from 1995–2004 (Schoer et al. 2006, p. 20)

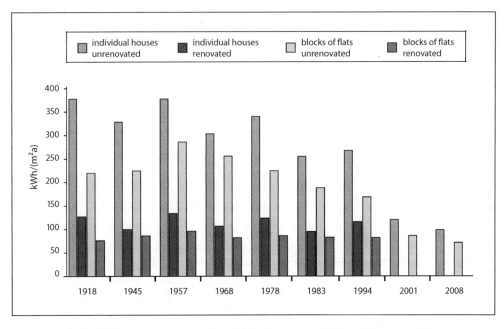

Figure 6: Specific final energy demand in individual houses and blocks of flats according to year of construction, renovation measures according to EnEV (calculations by authors, based on the German building typology, IWU 2003)

In addition to the factors discussed in the parameter analysis, a comprehensive description of objective needs requires the elaboration of further criteria relating to individual households or to individuals themselves. In addition to the (primary) energy demand of buildings (as referred to in the current regulation), a description of the (primary) energy demand of a household would thus not only cover the building itself, but also the intensity with which a specific household uses energy. This would be a better basis for evaluating sustainable consumption.

Another factor that directly influences energy demand is the type of residence. For instance, individual houses generally use more energy than blocks of flats. The comparison of these two building types (with the inclusion of their year of construction) in Figure 6 is based on a building typology developed by the German Institute for Housing and the Environment (IWU 2003; also compare Friedrich et al. 2007).

Figure 6 compares the respective energy demand of old, unrenovated individual houses and old blocks of flats to their potential consumption in a renovated condition (according to recommended EnEV renovation standards). The lower energy demand of blocks of flats can be attributed primarily to the fact that they have a higher compactness. This ratio of the outer surface of the building, through which heat transmission losses occur, and the internal volume is advantageous for blocks so that the specific energy demand is lower compared to individual buildings. The desire to live in individual houses can thus be seen as a subjective desire that has an impact on the energy use of a household. As the specific primary energy demand is used as a key performance indicator in the current German regulation, the requirements are higher for less compact buildings and thus partly counter-balance the above described effect.

13.6 Conclusion

The parameter analysis has demonstrated the importance of taking into account user behaviour in the assessment of private household energy use. Our results are consistent with comparable studies of measuring energy use in low energy buildings (Loga et al. 2003). We conclude that in addition to an overall strategy of reducing energy use in private households based on renovating the building envelope and the heating system, behavioural measures aimed at more efficient use of energy (e.g. optimising the operation of the heating systems) should be considered. The latter could be included in the recommendations provided in the Energy Performance Certificates.

Our analytical framework has drawn a distinction between objective needs and subjective desires of users. Energy use that exceeds needs and desires, and thus produces no genuine benefit to users, can be classified as wasted energy. A standardised calculation of energy demand based on average user behaviour could be regarded as a first attempt to reflect the objective needs and subjective desires of users. The German Energy Savings Ordinance stipulates that, as a means of reducing energy use, both efficiency strategies and consistency strategies are acceptable. According to our investigation as presented in this article, the fulfilment of subjective desires can be compatible with sustainable (energy) consumption so long as fewer resources are exploited to fulfil people's objective needs. It is suggested here that, in addition to the criteria applied in our parameter analysis, further indirect influences such as intensity of use (e.g. people per household) and type of residence (individual houses vs. blocks of flats) should be considered in future energy saving strategies.

The aspects of energy use in households discussed in this article could be applied in a wider context of sustainable consumption, involving estimations on what amount of energy resources is justifiable for each human being. The question then arises whether current standards as set out in the regulations do in fact reflect the demand that has been calculated purely from objective and legitimate needs (i.e. excluding those subjective desires that are ethically questionable), and if there might thus be leeway for curbing consumption, i.e. employing a strategy of eco-sufficiency (Linz et al. 2002;

Linz-Scherhorn 2010). If the basis of calculation were to be the (primary) energy demand of households, the outcome would not necessarily be a limitation of energy use. Instead, it would be an opportunity to look at subjective desires in terms of their compatibility with sustainable consumption.

References

AGEB (2011): Auswertungstabellen zur Energiebilanz für die Bundesrepublik Deutschland 1990 bis 2009. Berlin: Arbeitsgemeinschaft Energiebilanzen.

BDEW (2010): Endenergieverbrauch der privaten Haushalte 2008 nach Anwendungsarten. In: Arbeitsgemeinschaft Energiebilanzen BDEW PGr Nutzenergiebilanzen, DIW (ed.). 2011 (18.7.2011).

Bettgenhäuser K., Boermans T. (2011): Umweltwirkung von Heizungssystemen in Deutschland. In: Umweltbundesamt (ed.): Climate Change. 2. Dessau-Roßlau: Ecofys Germany GmbH.

Clausnitzer K.-D. (2004): Marktrecherche "Zentrale Einzelraum-Temperaturregler" für Gebäude. Bremen: Bremer-Energie-Institut.

Diefenbach N., Loga T., Born R. (2005): Wärmeversorgung für Niedrigenergiehäuser – Erfahrungen und Perspektiven. Darmstadt: Institut für Wohnen und Umwelt GmbH.

Di Giulio A., Brohmann B., Clausen J., Defila R., Fuchs D., Kaufmann-Hayoz R., Koch A. (in this volume): Needs and consumption – a conceptual system and its meaning in the context of sustainability.

Ebel W., Großklos M., Knissel J., Loga T., Müller K. (2003): Wohnen in Passiv-und Niedrigenergiehäusern: Eine vergleichende Analyse der Nutzerfaktoren am Beispiel der 'Gartenhofsiedlung Lummerlund' in Wiesbaden-Dotzheim. Teilbericht Bauprojekt, messtechnische Auswertung, Energiebilanzen und Analyse des Nutzereinflusses. Darmstadt: Institut für Wohnen und Umwelt.

Fischer D., Michelsen G., Blättel-Mink B., Di Giulio A. (in this volume): Sustainable consumption: how to evaluate sustainability in consumption acts.

Friedrich M., Becker D., Grondey A., Laskowski F., Erhorn H., Erhorn-Kluttig H., Hauser G., Sager C., Weber H. (2007): CO_2 Gebäudereport 2007. In: Bundesministerium für Verkehr, Bau und Stadtentwicklung (ed.): Berlin: co2online gemeinnützige GmbH, Fraunhofer-Institut für Bauphysik.

IWU (2003): Deutsche Gebäudetypologie – Systematik und Datensätze. Darmstadt: Institut Wohnen und Umwelt GmbH (IWU).

Koch A., Huber A., Avci N. (2008): Behaviour Oriented Optimisation Strategies for Energy Efficiency in the Residential Sector. Proceedings of the 8th International Conference for Enhanced Building Operations, Berlin, Germany, October 20–22. [http://repository.tamu.edu/bitstream/handle/1969.1/90834/ESL-IC-08-10-72.pdf?sequence=1; 14.09.2011]

Koch A., Jenssen T. (2010): Effiziente und konsistente Strukturen – Rahmenbedingungen für die Nutzung von Wärmeenergie in Privathaushalten. Stuttgart: Institut für Sozialwissenschaften, Abt. für Technik- und Umweltsoziologie, Universität Stuttgart.

Linz M., Bartelmus P., Hennicke P., Jungkeit R., Sachs W., Scherhorn G., Wilke G., Winterfeld U. v. (2002): Von nichts zu viel – Suffizienz gehört zur Zukunftsfähigkeit. Über ein Arbeitsvorhaben des Wuppertal Instituts. Wuppertal Papers. 125: Wuppertal: Wuppertal Institut für Klima, Umwelt, Energie GmbH.

Linz M., Scherhorn G. (2011): Für eine Politik der Energie-Suffizienz: Impulse zur "WachstumsWende". Wuppertal: Wuppertal Institut für Klima, Umwelt, Energie GmbH.

Loga T., Großklos M., Knissel J. (2003): Der Einfluss des Gebäudestandards und des Nutzerverhaltens auf die Heizkosten – Konsequenzen für die verbrauchsabhängige Abrechnung. Darmstadt: Institut für Wohnen und Umwelt GmbH.

Schoer K., Buyny S., Flachmann C., Mayer H. (2006): Die Nutzung von Umweltressourcen durch die Konsumaktivitäten der privaten Haushalte: Umweltökonomische Gesamtrechnungen (UGR). Wiesbaden: Statistisches Bundesamt.

Georg Sunderer, Konrad Götz, Sebastian Gölz

14 The evaluation of feedback instruments in the context of electricity consumption

14.1 Introduction

Already in the 1980s, Hans-Joachim Fietkau and Hans Kessel had argued that certain preconditions had to be in place in order for society to embrace a greener lifestyle (Fietkau/Kessel 1981; Fietkau 1984). Topics under discussion included knowledge, attitudes, values, infrastructural opportunities, incentives for behavioural change, and the importance of feedback in relation to consumers' green and not so green behaviour, and the consequences thereof. The last two factors, feedback on behaviour and feedback on consequences of behaviour, remain a neglected issue to this day.

One area in which feedback has recently experienced a surge in research interest is that of energy consumption. A good number of studies have demonstrated that feedback is capable of reducing energy consumption (cf. Fischer 2008; Darby 2006; Abrahamse et al. 2005), although conclusions vary, with predictions of savings ranging between 0 and 27 percent (cf. Büttner et al. 2008). Judging by pilot trials so far conducted, feedback on electricity consumption appears to be a useful strategy for reducing household energy consumption under certain circumstances.

The above findings were used as a starting point by the "Intelliekon" project group. In a field trial, two feedback instruments were tested as to their effectiveness in reducing electricity consumption. One was a web portal; the other was paper-based feedback, i.e. a letter containing information on individual electricity consumption. However, the aim of the project was not only to measure the effect of feedback instruments on electricity consumption, but also to obtain users' evaluations of the instruments. We believe that positive user evaluation is an important prerequisite for future acceptance and dissemination of feedback instruments.

The main objective of this article is the evaluation of the feedback instruments used in "Intelliekon". In this context we will address the following issues:
- How are the feedback instruments evaluated by the users?
- Do users' evaluations of feedback change over time?

- To what extent is the evaluation influenced by general attitudes towards saving energy?

As a first step, we will introduce those aspects of feedback instruments that were evaluated by users (section 14.2). In section 14.3, we will focus on possible relationships between general attitudes towards saving energy and the evaluation of the feedback instruments. Sections 14.4 and 14.5 are devoted to the methodology employed in our field trial, followed by empirical analyses of the three research questions in section 14.6. In section 14.7, we will outline two further issues that were investigated by the "Intelliekon" project, namely the electricity savings that were achieved by the use of feedback instruments, and the frequency with which the web portal was consulted. Our findings will be discussed in section 14.8.

14.2 Evaluation of various aspects of feedback instruments

In this section we will briefly introduce those aspects that were evaluated. The first was the *design of the feedback instruments*. The importance of good design is frequently emphasised in the literature on feedback instruments (cf. Darby 2000 and Fischer 2008). In practical terms, they should a) be understood by users, b) provide the information requested by users, and c) be user-friendly and attractively designed. Thus, these aspects of design must be taken into account in the analysis of the evaluation. A further aspect included in our analysis (relating to the acceptance and distribution of feedback instruments) was the *overall usefulness that is attributed by users to feedback instruments*. A qualitative pilot study (Birzle-Harder et al. 2008) with 76 structured interviews, conducted as part of the "Intelliekon" project, revealed that feedback instruments were not universally regarded as useful. It became clear in the course of our interviews that some respondents felt that they had already fully exploited all opportunities to save electricity and that feedback would therefore not yield any additional savings. Others simply did not believe that feedback instruments would contribute to saving energy.

A further aim was to investigate whether *feedback instruments could increase consumers' interest in electricity consumption* and whether *people enjoyed engaging with electricity consumption by using the feedback instrument*. These aspects assume particular significance against the background of the fact that many people pay little attention to their electricity use and are unaware of the extent of their consumption (Fischer 2008; Abrahamse et al. 2005). Consistent with this phenomenon is the fact that, in a marketing context, energy products tend to be regarded as "low-involvement products" (Felser 2007). The barriers to saving energy that arise from such a lack of interest in one's individual consumption could possibly be lowered by re-engaging people by means of feedback instruments – especially if the feedback meters were capable of generating a certain "enjoyment" (Burmester et al. 2002).

Finally, two potential drawbacks of feedback instruments were analysed. *The first relates to concerns about a lack of data protection and privacy protection.* The above-mentioned pilot study confirmed that even some people who, in principle, were favourably disposed towards feedback instruments had certain privacy concerns (Birzle-Harder et al. 2008). The second barrier relates to pro-social behaviour and has been a subject of research in social psychology. It has been found that external pressure can undermine people's internal motivation to behave altruistically (e.g. Schwartz/Howard 1982, p. 344). In terms of feedback instruments, this could mean that the presence of feedback instruments might put pressure on consumers and cause them to react defensively. It was thus our intention to establish *if participants in our trial felt pressurised by the presence of feedback instruments.*

14.3 Attitudes to saving energy as influencing factors on the evaluation of feedback instruments

For the purposes of analysing people's evaluations of feedback systems, we are drawing on the sociological concept of framing (e.g. Kühnel/Bamberg 1998; Kroneberg 2005). When someone is confronted with a feedback instrument, he or she forms a "subjective definition of this situation" (Esser 1996). The subjective definition of a situation is a situational frame that determines which conditions for decisions are perceived in a certain situation (e.g. the perceived alternatives of action and criteria for decision-making). A person's evaluations of different aspects of a feedback instrument can thus be regarded as elements of his or her definition of the particular situation.

The framing of a situation is influenced by a person's attitudes (cf. Kühnel/Bamberg 1998; Bamberg/Braun 2001). Starting from this result of previous research, we assume that attitudes also have an effect on the framing of the situation if someone is confronted with a feedback instrument. We tested this hypothesis by examining to what extent three different attitudes towards saving energy had an impact on the evaluation of feedback instruments (regarded as an element of the situational frame):

- The financial saving attitude: the view that one needs to save electricity for personal financial reasons, i.e. in order to save money.
- The climate-related saving attitude: the view that one should save electricity in order to protect the climate.

- The third attitude combines two aspects: a lack of interest in saving electricity and the opinion that saving electricity requires too much effort (referred to hereafter as 'lack of interest/too much effort').

Given that the first two attitudes each provide an explanation why feedback instruments could be of interest, we proceeded on the assumption that people with those attitudes would be more likely to rate feedback instruments positively. By contrast, people with the third attitude would be more likely to give negative evaluations of feedback instruments.

14.4 The feedback instruments used in the field trial

One of the objectives of the qualitative pilot study (cf. Birzle-Harder et al. 2008) was the development of the two feedback instruments for the field trial: the web portal and the paper-based feedback. Overall, the web portal offered more features and more

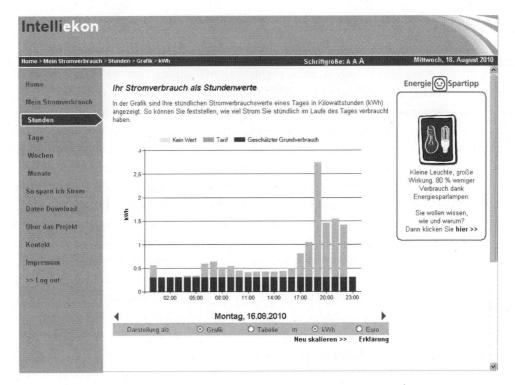

Figure 1: Screenshot of web portal

options for analysis than the paper-based feedback (cf. Figure 1 for web portal). The respective features of the two instruments are detailed below:

- The *web portal* gave users the opportunity to view and compare their electricity consumption in monthly, weekly, daily and hourly values. Additionally, there was a choice between viewing the data as either bar charts and tables, or in numerical terms of energy usage (kWh) and costs (euros). All representations of the data included an estimate of base load, i.e. the consumption of standby and cooling equipment, calculated as a proportion of the overall consumption within a given time period. The portal also offered practical tips on saving energy, links to related information and a download function.
- The *paper-based feedback* consisted of a two-page document containing multi-coloured, graphically represented information on the amount of electricity consumed by the household over the preceding months, weeks and days. The same design features were used for both web portal and paper-based feedback. The paper document was sent to households on a monthly basis.

14.5 Design of the field trial and the data base for the empirical analyses

The field trial, which was part of the "Intelliekon" project, was conducted in eight German cities (Celle, Hassfurt, Kaiserslautern, Krefeld, Münster, Oelde, Schwerte, Ulm) and one Austrian city (Linz). First, the local utility providers made available lists of potential participants. These were randomly assigned to a pilot group (who would receive a feedback instrument) and a control group (who would receive no feedback instrument). In the pilot group, 2,348 households agreed to take part; in the control group, 1,324 households agreed to participate. The households in the pilot group were free to choose either the web portal or the paper-based feedback. 1,241 households opted for the web portal and 1,107 households opted for paper-based feedback.

The field trial in all cities except Münster and Linz began between May and July 2009. The trial in Münster and Linz started in November and December 2009 respectively. All trials ended in November 2010. As part of the field trial, the electricity consumption of households and the frequency of use of the internet portal (number of clicks for each option within the portal) were recorded. The field trial was accompanied by a panel survey (standardised telephone interviews). The interview was conducted with the household member that was most familiar with the subject of electricity consumption. The first interview took place at the beginning of the trial, once the households had familiarised themselves with the feedback instruments. A second interview was conducted about six months later. In the empirical analyses below we will refer to these two waves of interviews. The first wave resulted in a total of 1,114 exploitable interviews with the participants from the pilot group. (Those households who had moved home or had never used the portal were excluded.) The wave of interviews conducted six month later yielded a total of 853 exploitable interviews.

The two subgroups of the pilot group, the web portal group (web) and the group with paper-based feedback (post), differed in their socio-demographic composition. In the web portal group there was a higher proportion of men than in the group with paper-based feedback (proportion of men for web: 74 percent, for post: 52 percent). Additionally, the web portal participants were younger (average age for web 46, for post 54) and tended to be better educated (proportion of university entrance diploma for web: 50 percent, for post: 27 percent). Furthermore, in the case of web portal participants, the proportion of households with three or more persons was higher and the proportion of one- and two-person households was lower (one- and two-person households for web: 52 percent, for post: 67 percent). Given that participants were free to choose their preferred feedback instrument, the socio-demographic composition indicated that men, younger individuals, individuals with higher academic qualifications, and those who lived in households with three or more persons tended to opt for the web portal more frequently. The reverse tendencies were observed for the paper-based feedback.

In order to analyse different aspects of evaluation in each wave of the panel survey, two question blocks were used. In the first block, respondents were asked to rate four aspects concerning design (informational content, user friendliness, clarity and appearance) and the overall usefulness of their feedback instrument. This was done on a scale of 1 to 6, whereby 1 was the highest and 6 was the lowest approval rating. In the second question block, respondents were asked to indicate their (dis)agreement with a number of statements about feedback instruments on a scale of 1 to 4 ranging from "totally agree" (1) to "totally disagree" (4). Overall usefulness (learning effect, waste of time) was covered by two statements; the issues interest, fun, concerns about data protection and pressure to save energy were each covered by one statement, which the participants were requested to rate.

The financial saving attitude and the climate-related saving attitude were represented by one statement each.[1] The four statements[2] used to measure the 'too much effort/lack of interest' attitude were aggregated to an additive index. The internal consistency of these statements is acceptable (Cronbach's Alpha = 0.70).

---

1 The items were "I have to save electricity for financial reasons" and "With regard to climate protection I am willing to think about saving electricity".

2 For instance, one item was "Reducing my electricity consumptions is too time-consuming".

14.6 Empirical findings concerning the evaluation of feedback instruments

14.6.1 Descriptive results from the first and second survey

Figure 2 presents the responses to the initial survey; it shows the distribution of responses of the question block where rating of different aspects of the feedback instruments was required. Both feedback instruments were rated either highly or very highly by the majority of respondents. Poor ratings (4–6) were given only by a relatively small number of respondents. Compared to the web portal, the paper-based feedback was rated somewhat lower on all criteria (with the exception of appearance). The differences between means within each category are statistically significant at the 5 percent level of significance – with the exception of the item on informational content. In numerical terms, they are, however, relatively small.

The results of the second question block (rating statements on a scale of 1 to 4) in the initial survey are illustrated in Table 1. A clear majority of participants reported that their interest in electricity consumption was stimulated by the feedback instrument, believed in learning effects and enjoyed using it. Too much pressure to save energy as a result of the feedback instrument or the sentiment that engagement with the feedback was a waste of time were reported by a relatively small proportion of respondents.

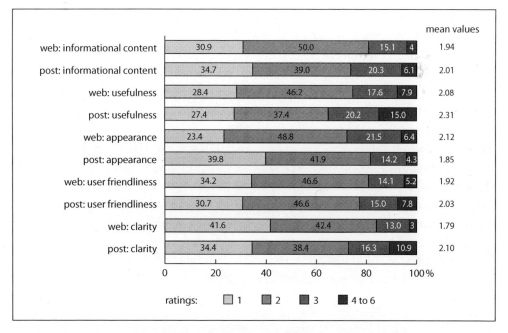

Figure 2: Evaluation of feedback instruments by means of ratings
(proportions in percentages on the basis of all participants in the initial survey)

Table 1: Statements about feedback instruments – level of agreement and mean values

	Web portal		Paper-based feedback		Two-tailed t-test web/post; significance
	agreement in %*	mean	agreement in %	mean	
Learning effect: The feedback instrument (FI) has taught me to use electricity effectively.	77.1	1.98	84.3	1.74	0.000
Interest: The FI has awakened an interest in electricity consumption.	84.1	1.83	84.4	1.76	0.123
Fun: I enjoy engaging with electricity consumption using the FI.	76.4	2.00	71.7	2.03	0.625
Waste of time: I regard engagement with the FI as a waste of time.	10.4	3.58	13.9	3.48	0.027
Pressure: I feel too pressurised to save electricity by the presence of the FI.	5.5	3.69	11.5	3.50	0.000
Data protection: I feel too exposed by using an FI.	23.5	3.20	25.7	3.17	0.679

*Values are collapsed for "I totally agree" (1) and "I partly agree" (2); basis: all participants in initial wave of survey.

Concerns about privacy on the other hand were expressed somewhat more frequently, although still only by a minority (25 percent). A comparison of the results for the two feedback instruments revealed statistically significant differences between means for three items (Table 1, last column): the respondents using the paper-based feedback rather tended to associate learning effects with the feedback. The respondents using the web portal were somewhat less likely to regard engagement with the feedback system as a waste of time or to feel pressurised by its presence. Nonetheless, the overall evaluations of the web portal and paper-based feedback were very similar.

In the comparison of the differences in evaluations between the initial survey and the second survey, only data from those households were included who had also taken part in the second survey.[3] The averages of the evaluations of the households in the first survey remained more or less constant, after excluding those households that had not participated in the second survey. The comparison of the two surveys showed that with

---

3 In our analysis, only the aggregate level was considered. As regards differences in individual evaluations, only few participants differed considerably in their rating between the initial and the second interview. In the first question block, for example, 83% to 95% of respondents gave either the same or a very similar rating to all items (plus or minus 1 band) in both surveys.

regard to the first question block both feedback instruments were rated almost identically on informational content, appearance, user friendliness and clarity in both waves. This result was expected, given that the feedback instruments remained unchanged during the field trial. In terms of overall usefulness, some minor differences were found, which, nonetheless, were statistically significant: both feedback instruments were deemed somewhat more useful in the second survey. The proportion of high and very high ratings rose by 4 percent for the web portal and by 7 percent for the paper-based feedback. The ratings of 4 and below dropped by 4 and 6 percent respectively. With regard to the statement section of the survey for both feedback instruments, evaluations of three statements differed significantly between the two surveys. In the second survey, the statements concerning the 'interest' and 'fun' aspects were each less often agreed with and using the instrument was more often regarded as a waste of time, although the differences were small. Depending on the item, the proportion of agreement rose or fell by 3 to 9 percent.

In summary, both the web portal and the paper-based feedback met with overall approval in both surveys. Given that the novelty of any new 'gadget' tends to wear off, the slightly inferior ratings for 'fun' and 'interest' in the second survey came as no surprise. The changes in the ratings for 'overall usefulness' point to a learning effect, i.e. that people gradually learnt how to use the feedback instrument. In apparent contradiction to this result was the finding that the proportion of people who regarded the feedback instrument as a waste of time increased with time. This could be explained by the fact that some people evaluated feedback instruments as basically useful in both waves, but regarded a *further* use of feedback instruments as a waste of time at the point of the second wave.

Regarding the comparison between the paper-based feedback and the web portal, the better ratings of the web portal in terms of informational content, usefulness, user friendliness and clarity was predictable, given that the web portal offered more functions and opportunities for analysis. The stronger agreement for the 'pressure' aspect in the case of the paper-based feedback can be explained by the fact that people might well feel more uncomfortable receiving an official-looking letter detailing individual electricity consumption than they would in having the information available online and having the choice between engaging with it or ignoring it. Regarding the 'appearance' aspect, it is likely that the participants used different benchmarks: the web portal might have been compared to a vast range of different, often very attractively designed websites, whereas the paper-based feedback might have been compared to conventional utility bills containing consumer information. A comparison between the two feedback groups in relation to this particular aspect might therefore be considered unhelpful. It is difficult to find an explanation for the difference in evaluation for the

Table 2: Influences on evaluation of feedback instruments

	evaluation in terms of ratings				
	usefulness	informational content	appearance	user friendliness	clarity
age (reference: over 60)					
18 to 30	−0.213	0.139	0.659**	0.165	−0.318
31 to 45	0.041	−0.068	0.244	−0.067	−0.219
46 to 60	0.075	−0.062	0.178	0.110	−0.148
gender (ref.: male)	0.162	−0.084	−0.439**	−0.042	0.180
education (ref.: university entrance diploma)					
basic certificate of secondary education	−0.416**	−0.566***	−0.445**	−0.526***	−0.380**
general certificate of secondary education	−0.425**	−0.466**	−0.323*	−0.278	−0.234
household size (ref.: 3+ people)					
one-person household	−0.141	−0.202	0.337*	0.187	−0.040
two-person household	−0.056	0.076	0.114	0.010	0.139
feedback group (ref.: web portal)	0.464***	0.272*	−0.429**	0.367**	0.462***
too much effort/lack of interest	−0.046*	−0.003	−0.017	0.001	−0.018
financial saving attitude	0.060	−0.077	−0.020	0.052	−0.049
climate-related saving attitude	0.273***	0.259***	0.129	0.199**	0.194**
Nagelkerke's pseudo-R2 (overall model)	0.051	0.038	0.072	0.034	0.038
Nagelkerke's pseudo-R2 (model with neither control variables nor feedback group variable)	0.024	0.014	0.013	0.010	0.008
number of cases	1,042	1,054	1,055	1,049	1,057

The non-standardised regression coefficients as well as the level of significance are shown for the explanatory variables in each model (* $p < 0.05$; ** $p < 0.01$; *** $p < 0.001$). Values of the variables: All evaluations according to ratings: 1 = score 1, 2 = score 2, 3 = score 3, 4 = score 4 to 6. All statements: 1 = totally agree, 2 = partly agree, 3 = partly disagree, 4 = totally disagree.

'learning effect' aspect. We believe that it may be due to the different composition of groups (cf. section 14.5). Group composition might well have been responsible for some of the other differences too. With this in mind, the purpose of the next section is also to test whether the differences in evaluation between the two feedback instruments still exist when socio-demographic variables and attitudes towards saving energy are controlled for.

Table 2 continued

		statements			
learning effect	interest	fun	waste of time	pressure	concerns relating to data protection
0.181	−0.078	0.109	0.565*	0.159	−0.300
0.100	−0.225	−0.141	0.480*	0.050	−0.136
−0.043	−0.169	−0.163	0.657**	−0.042	−0.338*
−0.077	−0.134	0.293*	−0.117	−0.055	0.315*
−0.639***	−0.311*	−0.591***	−0.071	−0.590***	−0.222
−0.386*	0.175	−0.275	0.021	−0.507*	−0.275
−0.159	0.073	−0.148	0.143	0.252	0.118
−0.069	−0.004	−0.132	0.277	0.287	0.190
−0.300*	−0.086	0.135	−0.064	−0.408**	−0.015
−0.035	−0.060**	−0.089***	0.261***	0.231***	0.112***
0.156**	0.133*	0.167**	0.079	0.130	0.157*
0.479***	0.450***	0.340***	−0.167*	−0.038	−0.021
0.125	0.087	0.089	0.144	0.138	0.056
0.087	0.073	0.061	0.128	0.108	0.039
1,064	1,067	1,067	1,068	1,069	1,067

Explanatory variables: too much effort/lack of interest: lower values indicate that the two attitudes are more pronounced. Financial saving attitude and climate-related saving attitude: high values represent lower saving attitude. For coding of all other variables refer to the description of the variable in the table.

14.6.2 Attitudes towards saving energy as factors influencing the evaluation

The influence of attitudes to saving energy was investigated by means of ordered logit models. In each model the dependent variable was one of the ratings or statements. The models are based on data collected in the first survey and include the participants of both feedback groups. Control variables were gender, age, level of education and household size. In addition, all models included as a variable the specific feedback group.

Table 2 provides an overview of the models. The results show that – with the exception of appearance – all aspects of evaluation show statistically significant relationships with at least one attitude towards energy saving. With one exception[4], the direction of the significant effects is as expected. Lack of interest and the attitude that saving electricity requires too much effort result in more negative evaluations, whereas the other two factors, i.e. the financial saving attitude and the climate-related saving attitude, boost evaluations in a positive direction. If the evaluation of the feedback instrument is regarded as part of a subjective definition of the situation, the influence of attitudes towards saving energy on the framing of the situation is confirmed. However, as is shown by Nagelkerke's pseudo R^2, attitudes have only small to middling explanatory power. The 'waste of time' aspect has the greatest explanatory power. By contrast, aspects relating to the design of feedback instruments have practically no explanatory power. Thus, the framing effect does not affect all evaluation aspects to the same degree. Nonetheless, it can be shown that a person whose opinion is in favour of saving energy will tend to give higher ratings to feedback instruments. Negative attitudes towards saving energy have the opposite effect.

The socio-demographic variables also exhibit significant effects. Persons with lower academic qualifications tend to give better ratings in terms of design (appearance, informational content etc.), usefulness, learning effect, enjoyment and interest. At the same time, these individuals feel more often pressurised by the presence of the feedback instruments. Three significant effects have been detected for the variable gender: in comparison to men, women tend to give higher ratings to appearance, are less likely to enjoy engaging with the feedback and tend to have fewer reservations about data protection. In terms of the variable age, younger people tend to give lower ratings to appearance. People over 60 are more likely than other age groups to regard feedback instruments as a waste of time. Additionally, over 60s appear to be least concerned about data protection.[5] No significant effects were found for household size – with the exception of one-person households, who gave lower ratings to appearance. The reasons behind this phenomenon were difficult to fathom. The overall explanatory power of the socio-demographic variables is low (as reflected in the pseudo R^2 values).

With the help of the variable 'feedback group', it is now possible to see that the differences in evaluation reported above (with the exception of waste of time) persist even when the socio-demographic variables and the different attitudes have been controlled for. This supports the hypothesis that the differences between the feedback instru-

---

4 Concerns about data protection seem to increase in line with the desire to save money. We suspect that this effect is caused by variables that have not been included in our model (e.g. traditional values).

5 The differences relating to age groups "18 to 30" and "31 to 45" are, however, not statistically significant.

ments were responsible for the differences in evaluation. We are, however, still doubtful about the difference in the evaluation of the 'learning effect' aspect since we see no theoretical explanation for this phenomenon.

14.7 Saving effect and frequency of use of the web portal

In this section we will present some further findings of the "Intelliekon" project, relating to the potential of feedback instruments to bring about energy savings, and to the frequency of use of the web portal.

Electricity savings were calculated by means of a multiple regression analysis (for details cf. Schleich et al. 2011). It was shown that the provision of feedback instruments had led to average savings of 125 kWh, equating to an average 3.7 percent. Compared to savings identified in other studies (cf. 14.1), this is a relatively modest saving. Nonetheless, such a saving is not negligible in an ecological context. In terms of the overall electricity consumption in Germany, 3.7 percent corresponds to about 5 TWh, which is the approximate annual output of the (now decommissioned) nuclear power plant Neckarwestheim I, or the six-month output of more modern power plants such as Grundremmingen B and C (cf. Federal Ministry of Economics and Technology 2011). Such a comparison shows that, despite relatively modest savings, feedback instruments can appreciably contribute to overcome dependency on nuclear and fossil energy.

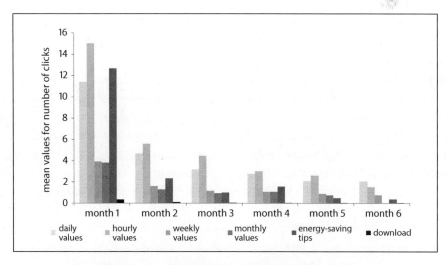

Figure 3: Average number of clicks for the options of the web portal during the first six months of the trial

In order to investigate the use of the web portal during the field trial, we monitored when and how often the various options in the web portal were clicked by each participant, and calculated an average number of clicks for each portal option. Figure 3 shows the average number of clicks during the first six months of the trial for the following options: hourly, daily, weekly and monthly values, energy-saving tips and the download function.

As expected, the web portal was used most intensively during the first month of the trial. The hourly and the daily values, plus the energy-saving tips were the most frequently accessed options. During the second month, usage dropped by 50 percent and continued to decline over the course of the subsequent months. The options hourly values, daily values and energy-saving tips remained the most popular. Further analyses show that one reason for the considerable drop in usage after one month was that half the participants did not access the web portal at all between month 2 and month 6. Continuous usage (at least one log-in per month for the six month period) was registered for 7 percent of participants.

14.8 Discussion of results

Our results indicate that, judging by participants' evaluations, feedback instruments may well have a future in German households. Overall, the two feedback instruments investigated as part of our project were widely regarded as informative, useful, well designed, user friendly and clear. Therefore, it can be assumed that there is great potential for acceptance for such instruments.

Despite the general positive evaluation of the web portal, only a small number of participants used it on a regular basis. This lack of interest cannot be explained by the small proportion of participants (around 10 percent) who regarded the web portal as not useful. Suspecting that other factors might be responsible for this lack of interest, we asked participants in the second survey why they had stopped (or almost stopped) using the portal. It appears that the effort involved in engaging with the feedback represented the main barrier. Additionally, it was mentioned relatively often that whilst it was interesting to be able to view one's electricity consumption at the beginning, once people had established how much they consumed, they saw no point in tracking it any further.

Similarly, the electricity savings of 3.7 percent achieved as a result of the use of our feedback instruments were – in comparison with savings achieved in other trials – somewhat disappointing. Our results thus indicated that a broadly positive evaluation did not necessarily lead to great savings. This finding recalls the "gap between attitude and behaviour" which is confirmed by numerous studies within the social sciences (cf. Eckes/Six 1994). Furthermore, this result indicates that the extent of savings might depend on other variables. One such variable could have been the extra effort involved in engaging with feedback. On this assumption, a straightforward feedback instrument, such as a permanent easy-to-install display might increase savings. Another means to generate savings could come from external data interpretation coupled with householders interpreting the feedback data together with an expert who gives them personalised advice in their own homes. Moreover, there is some evidence from previous studies that an alarm function (e.g. a signal when electricity consumption rises above a certain level) or metering devices for individual appliances may increase savings (cf. Büttner et al. 2008; Darby 2006; Fischer 2008). Given that existing research into the factors that are likely to increase the saving effect of feedback so far remains somewhat inconclusive, we suggest that future research efforts should be concentrated in this direction.

References

Abrahamse W., Steg L., Vlek C., Rothengatter T. (2005): A review of intervention studies aimed at household energy conservation. In: Journal of Environmental Psychology 25: 273–291.

Bamberg S., Braun A. (2001): Umweltbewusstsein – ein Ansatz zur Vermarktung von Ökostrom? In: Umweltpsychologie 5: 88–105.

Birzle-Harder B., Deffner J., Götz K. (2008): Lust am Sparen oder totale Kontrolle? Akzeptanz von Stromverbrauchs-Feedback. Ergebnisse einer explorativen Studie in vier Pilotgebieten im Rahmen des Projektes Intelliekon. Intelliekon-Berichte. Frankfurt am Main.

Bundesministerium für Wirtschaft und Technologie (2011): Zahlen und Fakten: Energiedaten. [http://www.bmwi.de/BMWi/Redaktion/Binaer/energie-daten-gesamt,property=blob,bereich= bmwi,sprache=de,rwb=true.xls; 27.06.2012]

Burmester M., Hassenzahl M., Koller F. (2002): Usability ist nicht alles – Wege zu attraktiven Produkten. In: i-com 1: 32–40.

Büttner M., Sothmann D. D., Schäffler H. (2008): Internationale Studien zu Demand-Response-Programmen. Projektbericht Fraunhofer-Institut für Solare Energiesysteme (ISE).

Darby S. (2000): Making it obvious: designing feedback into energy consumption. Proceedings of the 2nd International Conference on Energy Efficiency in Household Appliances and Lighting. Italian Association of Energy Economists/EC-SAVE programme.

Darby S. (2006): The effectiveness of feedback on energy consumption. A review for DEFRA of the literature on metering, billing and direct displays. Environmental Change Institute, University of Oxford.

Eckes T., Six B. (1994): Fakten und Fiktionen in der Einstellungs-Verhaltens-Forschung: Eine Meta-Analyse. In: Zeitschrift für Sozialpsychologie 25: 253–271.

Esser H. (1996): Die Definition der Situation. In: Kölner Zeitschrift für Soziologie und Sozialpsychologie 48: 1–34.
Felser G. (2007): Werbe- und Konsumentenpsychologie. Heidelberg: Schäffer-Poeschel.
Fietkau H.-J. (1984): Bedingungen ökologischen Handelns. Weinheim, Basel: Beltz.
Fietkau H.-J., Kessel H. (1981): Umweltlernen. Königstein, Taunus: Hain.
Fischer C. (2008): Feedback on household electricity consumption: a tool for saving energy? Energy Efficiency 1 (1): 79–103.
Kroneberg C. (2005): Die Definition der Situation und die variable Rationalität der Akteure. In: Zeitschrift für Soziologie 34: 344–363.
Kühnel S., Bamberg S. (1998): Überzeugungssysteme in einem zweistufigen Modell rationaler Handlungen: Das Beispiel umweltgerechteren Verkehrsverhaltens. In: Zeitschrift für Soziologie 27: 256–269.
Schleich J., Klobasa M., Brunner M., Gölz S., Götz K., Sunderer G. (2011): Smart metering in Germany and Austria – results of providing feedback information in a field trial. Working Paper Sustainability and Innovation. Fraunhofer ISI.
Schwartz S. H., Howard J. A. (1982): Helping and cooperation: A self-based motivational model. In: Derlega V. J., Gozelak J. (ed.): Cooperation and helping behavior: Theories and research. New York: Academic Press. 327–353.

Andreas Klesse, Joachim Müller, Ralf-Dieter Person

15 Achieving and measuring energy savings through behavioural changes: the challenge of measurability in the actual operation of university buildings

15.1 Introduction

Energy savings from behavioural changes, rather than from refurbishments or improvements to technical installations, can be achieved at lower investment cost. It is estimated that behaviour-related savings on electricity and heating in public buildings amount to up to 15 percent of overall consumption (cf. Matthies/Thomas in this volume). The aim of the "Change" project ("Changing sustainability-relevant routines in organisations") was to investigate user behaviour in university buildings that were used in similar ways to offices. The present article focuses on certain aspects of the "Change" project and its findings and discusses the measurability of energy savings. On the basis of information gathered from other sources, including the HIS GmbH (Hochschul-Informations-System), this article will focus particularly on how to assess energy savings in university settings. Given that the study excluded students' behaviour, as well as lecture theatres and classrooms, the results could conceivably be transferred to other premises similar to universities, such as administration departments (Matthies/Thomas in this volume).

The general aim of this article is to give an account of challenges encountered by projects that attempt to measure the effects of behavioural changes. Specifically, it will be considered how needs can be quantified in the context of energy consumption within buildings. Furthermore, the significance of reference values and the issues around collecting and analysing data on energy consumption in university settings will be discussed. The assessment will focus on final energy consumption of heat energy and electricity consumed in office environments (including lighting and electrical devices such as PCs, but excluding cooling devices). The study exclusively focuses on effects that can be influenced by user behaviour.

15.2 Background

The energy consumption of buildings is affected by four main variables: weather conditions, building envelope, technical installations, and how the building is used – including user behaviour (Wagner/Rudolph 2008). Adopting the same approach as the "Change" project, office-type environments will be used as a starting point, focusing on "individual user behaviour", given that this variable alone has been shown to strongly influence energy consumption. In view of the fact that energy consumers in public buildings remain largely anonymous, it is practically impossible to match up individual users with individual consumption. In order to apportion energy consumption in terms of the above four variables and to assess the impact of behavioural changes, it is necessary to start with an analysis of objective data – both on consumption and demand for energy. To that end, all available data on consumption (and demand) has to be collected, and subsequently classified into energy demand arising from legitimate user needs and additional amounts that are used 'inadvertently' (e.g. energy used for stand-by functions).

The ideal technique for measuring consumption and observing changes within energy usage is a metering device that is capable of monitoring individual buildings or even individual areas within buildings (e.g. institutes, chairs). A more detailed evaluation of behavioural changes should ideally include observations and questionnaires.

In the analysis of consumption data a distinction is made between heat, electricity and water consumption (the latter was not included in this experiment). If single effects of *individual* behavioural changes are to be assessed, it is necessary to either keep the other three main variables (weather conditions, building envelope and technical installation) constant or to neutralise their effects in the calculations. Furthermore, given that the choice of reference value dictates the extent of the observed change, it is crucial to state this value in the presentation of the results. Reference values can be obtained from the analysis of previously recorded consumption data or from benchmark data.

15.3 Quantification of needs

In this context, the term consumption is to be understood as the usage or utilisation of goods for the purpose of fulfilling human needs (for a definition of terms, see Di Giulio et al. in this volume). In this case, the goods are heat energy and electricity consumed for the purpose of heating (or cooling) office spaces and the operation of buildings and office equipment ("objective needs"). An issue frequently discussed in this context is the extent of potential savings that can be achieved as a result of changing user behaviour in office environments. This question in turn raises two further important issues, which need to be evaluated individually. First, the users of public office-type buildings (as opposed to households or petrol stations) remain largely anonymous. They are not direct participants in the 'energy market' and therefore cannot be influenced by economic (market-oriented) values (Klesse/Wagner 2011). Whilst it may be possible to map energy consumption on to particular spatial or functional areas, a precise allocation to individual users is not feasible in the majority of cases – neither would this be desirable from a data protection perspective. Secondly, in order to determine potential savings achieved through behavioural changes, it is essential to take account of not only that part of energy consumption that arises from objective user needs, but also the share of energy consumption that is a result of other factors. In the workplace, the need to perform tasks smoothly and without hindrance is of paramount importance. Any proposals for behavioural changes should therefore not infringe on the comfort and well-being of staff. In the context of the project, the terms 'comfort' and 'well-being' do not refer to standardised values of interior air quality (cf. DIN EN 13779), but to the situation as we found it in the buildings that were investigated, which was thus used as a reference value for proposed changes in user behaviour. This consideration is beyond the comfort assessment in ISO 7730. These levels may go beyond the assessment of comfort as prescribed in ISO 7730.

Additionally, workplaces have to meet the statutory requirements (e.g. workplace regulations) for minimum air exchange rate, minimum lighting levels and minimum room temperature. Any energy consumption over and above that is, by definition, not needed for the fulfilment of the above criteria; neither is it necessarily caused by individual user behaviour. In this respect, the consumption of energy differs from other consumer behaviours (e.g. reflected purchase decisions). A further peculiarity of energy consumption is the fact that, apart from its functional use, it is imbued with very little symbolic, social or emotional significance – at least not in those situations where a reduction of energy consumption is perceived as a restriction (cf. Fischer et al. and Kaufmann-Hayoz/Bamberg et al. in this volume).

In order to quantify that amount of energy consumption that exceeds the above-mentioned objective needs, a building-specific analysis of energy consumption is required. From there it is possible to identify potential savings that could be achieved through changes in user behaviour. Whilst individual user behaviour is largely determined by subjective criteria of well-being and is characterised by established behavioural patterns and habits, it can nevertheless vary from situation to situation. This is why a statistical approach (e.g. DIN 18599) is inappropriate. For the purposes of a dynamic building simulation, used in academic research, the energy simulation software package Trnsys has proven successful (Klesse et al. 2010). Figure 1 illustrates the methodology adopted in the quantification of potential energy savings resulting from modified user behaviour.

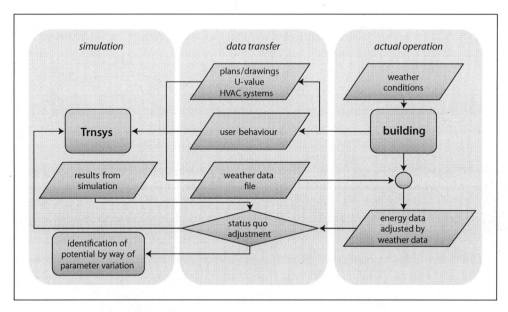

Figure 1: Methodology of calculation of potentials (Klesse et al. 2010, p. 10)

The aim is to quantify the effects that are the result of behavioural changes (e.g. opening and closing windows) in real, often less than optimally insulated and equipped buildings. For that purpose, it is necessary to isolate user behaviour from the other three main variables. A simulation of a standardised energy demand (as in DIN V18599), intended for comparing theoretical values of demand with values from real usage, was not intended.

As a first step, the collected data on energy consumption have to be prepared for calibrating the simulation model. The raw data collected in situ and adjusted for variations in weather conditions are used as reference values in aligning heat energy and electricity consumption with the results from the simulations. Once the currently dominant user behaviour has been established through observations and questionnaires, it is integrated into the model as an independent dynamic process (Klesse et al. 2010). As a next step, the individual needs are fed into the model. It is now possible to determine the maximum potential that a change in user behaviour can achieve. In this context, needs are to be understood as general well-being. Physical, psychological and social well-being does not only promote physical health, but can enhance performance (Willems et al. 2006). It is principally the requirements of thermal comfort that define the limits of behavioural change in connection with heat energy consumption (cf. ISO 7730). In order to simulate a fully operative building, the data on weather conditions, the data relating to the building and the technical installations together with the data on dynamic user behaviour is fed into the model. As a next step – by varying the parameters of user behaviour in the simulation – potential energy savings are determined.

15.4 Selection of reference values

Behavioural changes in the context of buildings can be evaluated using a wide range of methods, including questionnaires, observations and actual measurements of energy usage. It is the quantifiable energy consumption data that is of most interest to buildings operators or those who receive the energy bills.

The calculation of the energy savings depends to a significant extent on an appropriate choice of reference values. Annual energy consumption values cannot be readily compared because of the different boundary conditions. The proportion of heat consumption due to variations in weather conditions can be discounted, and the final heat consumption adjusted accordingly. In Germany, these adjustments are made by taking into account heating degree days and degree days. These two measures are defined by the Association of German Engineers (VDI guideline 3807 and 2067). Both adjustments are based on the standardisation of heat energy consumption (HEC), illustrated in the following equation by means of appropriate indicators.

$$HEC_{standardised} = HEC_{measured} \times \frac{G_{norm}}{G_{spec}}$$

The climate correction factor G_{stand}/G_{spec} represents the ratio of the average of degree days (measured over several years), the standardised size G_{norm}, to the degree days of the period of measurement, G_{spez}. G can be calculated in terms of degree days according to VDI 2067 and, in terms of heating degree days, according to VDI 3807. In the analysis stage of the "Change" project, calculation methodology, standardised size and weather data were combined in a number of different ways. The time series analysis showed that it was not crucial *how* the data adjustment for variations in weather conditions was calculated, but merely that it *was* calculated.

In order for the result not to be dependent on an accidentally high or low consumption value in one particular year, it is necessary to determine a constant value, i.e. a reference value. Options include a single data point or the analysis of a time series. Whilst single data points are generally easy to establish, they do not make for optimal reference values: firstly, they do not reveal temporal progression, and secondly, they can be strongly influenced by organisational variables such as variations in meter readings taken at different times.

In order to minimise these effects, an average value is often assumed – currently this tends to be an arithmetically calculated average value. The IFEU (Institute for Energy and Environmental Research) in Heidelberg (2004) proposes a consumption average covering the last three years that has been adjusted for variations in weather conditions. This simple procedure of estimating quasi-constant values has proved more suitable than opting for a single value from the preceding year.

If the data is not uniform over the years, it makes more sense to use trends. In the measurements of heat energy this method is technically and analytically limited. Given that comparable values come up only once a year, the number of data points was limited. Furthermore, in view of the legal requirement for energy meters to be replaced every five years (German calibration directive 2007) and the periodic renovations of buildings and technical installations, a comparison over more than five years is best avoided.

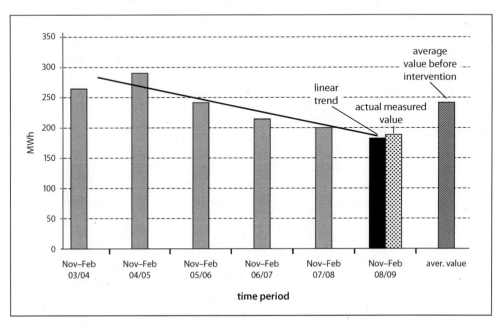

Figure 2: Different ways of generating reference values

It has been shown in the evaluation of an energy saving campaign that energy consumption in a given building is best measured by an expected value based on a linear extrapolation from the preceding years. This allows for the inclusion of changes in the use and operation of the building and operational management (Klesse et al. 2010). Figure 2 illustrates different ways of generating reference values.

Figure 2 illustrates that, when the measured values do not progress consistently, neither the value of the preceding year nor the mean represents a satisfactory reference value. With a downward trend, the mean overestimates the expected value; with an upward trend, the mean underestimates the expected value. A linear approach, however, remains an approximation and cannot reliably be continued indefinitely. It is possible to reduce energy consumption up to a certain point by optimising the technical installations (e.g. heating systems). However, once a ceiling is reached, further reductions in energy consumption can be achieved only by means of major refurbishment.

15.5 Infrastructure for measuring and metering energy consumption in German universities

A crucial prerequisite for the evaluation of savings achieved through interventions is their measurability. Regardless of whether energy-saving interventions are of a technical, organisational or behavioural nature, organisations generally wish to evaluate their savings, or are required to justify their investment costs.

15.5.1 Current situation

In practice, universities tend to use a variety of methods for collecting data on energy consumption: they range from 'manual readings' of separate meters by university staff to automated data logging by interconnected metering systems capable of generating real-time data.

Energy costs tend to be allocated to appropriate budget headings or cost centres. A differentiation within energy costs or an allocation to different user groups tends to be rare. Occasionally, costs might be assigned to different departments (e.g. institutes, chairs), but this tends to be mainly for information purposes. To date, there are a very few universities – one of which is the University of Göttingen – that systematically allocate energy costs to separate user groups.

A conflict frequently arises when new or refurbished buildings are fitted with meters. Requests made by prospective buildings operators for sophisticated systems capable of differentiating between various departments are often rejected for cost reasons.

15.5.2 Prerequisites

In an ideal situation, different areas of energy usage should be distinguished and electricity and heat energy consumption should be monitored and analysed separately for each area. This would mean that space heating, lighting, general building technology, etc. would be monitored separately from experimental facilities and laboratories. The values for the experimental facilities and laboratories should not flow into the overall calculation, because they do not lend themselves to the same energy savings measures as the office-type buildings. The boundaries are, however, not always that clear. Ventilation in a laboratory, for instance, is highly dependent on the experiments undertaken in the laboratory and does therefore not conform to generally recommended ventilation standards for buildings. Here, industry-specific regulations (e.g. accident prevention regulations) stipulating minimum air exchange rates have to be adhered to. In contrast to energy-intensive experiments, which should not be included in an overall assessment because they would skew the overall data, there is nonetheless some scope for organisational and behavioural changes in the ventilation of laboratories. These savings do belong in an overall calculation of energy savings.

In the face of the difficulties involved in separating the different areas for measurement, many projects – not least the "Change" project – limit themselves to monitoring office-type spaces, where conditions do not depend on additional energy spent on research or teaching activities. When addressing the overall energy consumption of a university, and the reality that laboratories can be very energy-intensive, such limitations are less than ideal (Birnbaum et al. 2007).

One way of differentiating between different spatial/functional areas and of guaranteeing the generation of user-specific data would be the installation of a comprehensive metering system across the university – an option that is, however, associated with a considerable level of investment. The following calculation will illustrate this point. In order to simplify matters, only two metering devices will be considered here: an electricity meter and a heat energy meter. The current minimum requirement in German universities is more or less one meter (of each type) per building. A typical university comprises an average of 100 buildings and spends around 6 million euros on energy. According to the Johannes Gutenberg University in Mainz, the costs for the installation and replacement of a meter that can be read remotely via an M-bus[1] interface amount to 600 euros for an electricity meter and 650 euros for a heat energy meter, resulting in an investment of 125,000 euros for 100 buildings – with no immediate direct benefit. Such a high financial outlay is difficult to justify in a university con-

---

[1] Meter-Bus, European standard for the remote reading of gas or electricity meters, developed at the University of Paderborn [http://www.m-bus.com].

text. This cost will rise even further if more sophisticated systems (i.e. more than one meter per building) or calibrated meters (this tends to be a requirement in Germany) are required.

15.5.3 Practical solutions

A cheaper solution is the use of data loggers. They are capable of recording data and are particularly suited to time-limited interventions. Even though the devices are inexpensive and can be installed relatively quickly, there is still an installation cost, plus the overhead for the evaluation of the data. In the absence of a metering infrastructure, the Institute of Nuclear Technology and Energy Systems (IKE) at the University of Stuttgart used data loggers as early as 1999 (REUSE 1999; Schmidt et al. 2005). Other universities followed suit with similar approaches.

For comprehensive metering infrastructures to be financially feasible, a number of prerequisites have to be in place. The Georg-August-University Göttingen (Knöfel 2010) fulfils such preconditions:

- As a foundation university of the Federal German state of Lower Saxony, the University of Göttingen is responsible for its own budgeting in relation to buildings and therefore is in a position to invest in metering systems and energy-saving intervention measures.
- Rising energy prices in the context of a frozen overall budget made a detailed cost breakdown essential.
- The buildings management team was supported by the university directorate in its effort to curb energy usage and save costs.

If an institution such as a university is to be fitted with metering devices, these should ideally not be restricted to buildings, but be installed as part of any large technical installations. In the case of modern monitoring and control systems, they are simple to install or, as the case may be, to retrofit.

A summary of recommendations relating to the installation of metering systems is available from the brochure "EnMess", produced by the Mechanical and Electrical Engineering Working Group for State and Local Governments (AMEV 2001).

15.6 Data collection and analysis in a university context

Given the high costs involved in collecting data on energy consumption in universities, only those energy usages that can potentially be influenced should be taken into consideration. Energy data tends to be analysed by the universities themselves and should ideally form the basis for so-called "energy controlling".

15.6.1 Energy data collection

Universities are often already equipped with energy meters, but these have to be read manually at regular intervals. Predictably, this comes with considerable overheads. In order to collect data more efficiently and more frequently, for example every few minutes, it is necessary to invest in automated metering systems. A more flexible collection of data allows for a range of evaluation approaches.

A monthly cycle of data collection is particularly useful for monitoring trends. By comparing the data with previous years, energy consumption targets can be established. If the targets are not achieved, causes can be investigated and action can be taken. On the basis of discrepancies between existing and newly generated data, flaws in technical installations or in the buildings themselves can be detected at an early stage.

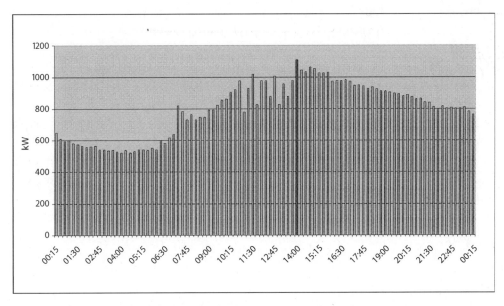

Figure 3: Example of a recorded electricity load profile of a building (15-minute measuring intervals; daily maximum value is highlighted)

Daily average values may be appropriate for analysis at individual user level (determination of base load and peak loads). In order to monitor particular load patterns, measuring intervals can be reduced to 15 minutes (cf. Figure 3).[2] This allows a detailed analysis and may also be useful for accounting purposes. Specific data can also illustrate the effects of particular energy saving measures – including behavioural changes. The assessment intervals must then be aligned with the starting dates for the introduction of new technical, organisational and behavioural measures.

15.6.2 Energy data analysis

The analysis of the data is part of the "energy controlling" process, which in turn is part of the energy management of an organisation. Energy data are often used to determine KPIs (key performance indicators); data on energy consumption are analysed in terms of a base number, such as surface area or number of staff. The measured values are used in a variety of contexts. Annual averages are, for instance, frequently documented in environmental reports (University of Osnabrück 2008) to illustrate long-term energy consumption or effects of particular measures. Universities who use energy management systems (e.g. EMAS[3], DIN EN ISO 14001, DIN EN 16001) can immediately see from their values if their energy-saving measures have had an impact.

In recent years, energy data has increasingly been analysed against the background of climate change. Paralleling initiatives in the transport and business sector, universities participate in climate change projects (carbon neutral universities[4]) that are based on the collection of energy data. Where projects are initiated by individual departments within an institution, appropriate metering systems are required for collecting those specific data. Other projects collect data on a much larger scale. One example is the German Federal State of Hessen, which included all its universities in a comprehensive, state-wide CO_2-balancing exercise (Müller/Person in 2010).

Detailed load profiles can be detected by measuring electricity consumption at more frequent and flexible intervals. For instance, the electricity consumption outside the normal usage periods (e.g. at night) will give a figure for base load, or fluctuations can be traced when facility manager switch on all the lights and use their cleaning equip-

---

[2] Current data and examples of analyses are available from the University of Stuttgart (University of Stuttgart 2011).

[3] Eco-Management and Audit Scheme (EMAS) is a certification system developed by the European Union; it is an environmental management and organisational audit system based on the EMAS regulation.

[4] Two examples are the Leuphana University [http://www.leuphana.de/nachhaltigkeitsportal/klima-neutrale-universitaet.html; 10.07.2012] and the University of Marburg [http://www.uni-marburg.de/aktuelles/projekte/co2; 10.07.2012].

ment in the morning, and when various electrical devices are being used during the day. Additionally, the settings of heating, cooling and ventilation systems can be evaluated by monitoring the times when they switch themselves on and off.

Another aspect of the evaluation of energy data is the possibility of feedback to the user. This can be a key motivator for permanent behavioural changes. Some universities are aware of the importance of feedback and share with students and staff the outcomes of energy-saving strategies. The Leuphana University Lüneburg, for instance, displays data in the canteen, and the University of Dresden (TU Dresden) is planning a display in the foyer of one of its main buildings. In addition, a range of research projects on automated, state-of-the-art user feedback and visualising systems are currently in development (e.g. "Mustercampus Energie" by the Saarland University).

Meter data can be analysed with the help of purpose-built energy management systems. The universities of Göttingen[5], Mainz[6] and Münster[7] all use such systems. A wide range of different systems involving spreadsheets and computer databases are currently in use. An overview of the scope of data collection and analysis is provided in the "Energy 2010" brochure published by AMEV (AMEV 2010).

Expert interviews conducted by HIS GmbH as part of the "Change" project revealed that among the universities who took part in their survey, some had started to put in place the prerequisites for "energy controlling", while others were further advanced and had actually employed specialist staff for the express purpose of energy management (from full-time posts down to 0.25, depending on size of university). Although smaller universities tend to have less scope for employing specialist staff, they may well be in a position to install meters in each of their buildings – purely because their campuses comprise fewer buildings. This means that they will at least have in place the technical prerequisites for detailed data analysis. In the municipal sector, there is already a widespread willingness to invest in energy data collection systems (examples include the cities of Aachen, Frankfurt and Munich)[8].

---

5 [http://www.uni-goettingen.de/energie; 10.07.2012]

6 [http://www.his.de/publikation/seminar/Praxisseminar_Energie_062008/vortrag_liers.pdf; 10.07.2012]

7 [http://www.uni-muenster.de/WWUmwelt/energiemanagement/index.html; 10.07.2012]

8 [http://www.aachen.de/de/stadt_buerger/planen_bauen/gebaeudemanagement/ABTEILUNGEN/3_Technisches_GM/energiemanagement.html; 10.07.2012] and [http://www.energiemanagement.stadt-frankfurt.de; 10.07.2012] and [http://www.muenchen.de/Rathaus/bau/en_bau/115678/index.html; 10.07.2012]

15.7 Summary and future prospects

Universities use energy for research, for teaching purposes and for a number of support systems. For the purpose of the experiment reported in this article, a distinction was made between electricity and heat energy. Energy consumption in buildings is principally influenced by four factors: weather conditions, building envelope, technical installations such as HVAC and ICT systems, and usage or user behaviour. It is known (from various assessments and reports from universities) that opportunities exist for reducing the consumption of energy and that organisational measures and changes in individual user behaviour are effective for that purpose. In order to assess the opportunities and record the effects of behaviour-related savings, a quantitative data analysis of energy consumption is required.

With the help of simulation models, the energy that is required to meet the needs of users can be compared to the actual energy consumed in the building. The discrepancy corresponds to unnecessary energy consumption. On that basis, it can be estimated if and where savings are possible, and targeted measures can be put in place. The data so collected can also be used as proof of energy savings for accounting purposes.

It has been demonstrated in this article that, for the purpose of identifying the potential savings, the optimal reference value is the mean of a number of values collected over a period of time. The difference between expected value and measured value is useful for both accounting purposes and for maintaining the motivation to save energy.

In order to analyse consumption data and provide evidence of savings, a comprehensive metering system is essential. Ideally, meters should be installed in separate spatial and functional areas. In order to save costs, it is important a) to monitor only those areas of energy consumption where changes can realistically be implemented, and b) to avoid focusing on very small areas of consumption. Once the data from the various areas are measured at monthly or more frequent intervals, they can be analysed accordingly.

Results from data monitoring have shown that concerted efforts to change user behaviour are an effective way of saving energy. It is therefore important to extend and promote energy management efforts in general, and "energy controlling" in particular. As well as installing a comprehensive network of metering devices and an overall energy control system, it will also be necessary to employ staff who are specifically responsible for energy management. The focus of energy management should always be on specific usage areas (spatial or functional). Advances in measuring techniques and building management systems may lead to data collection that is both cheaper and finer-grained.

Discussions with experts in universities have shown that some institutions already carry out "energy controlling", for instance by using energy management systems. With a view to promoting this development and facilitating targeted communication between universities, an information and communication portal has been developed. The objective is to promote good practice and to give individual support to universities. The "Higher Education Information System" (HIS) portal [http://www.his.de/energie] has so far proved to be an effective tool in the dissemination of relevant information.

References

AMEV (2001): Arbeitskreis Maschinen und Elektrotechnik staatlicher und kommunaler Verwaltungen (AMEV): Messgeräte für Energie und Medien (EnMess 2001). Berlin: AMEV.

AMEV (2010): Arbeitskreis Maschinen und Elektrotechnik staatlicher und kommunaler Verwaltungen (AMEV): Hinweise zum Energiemanagement in öffentlichen Gebäuden (Energie 2010). Berlin: AMEV.

Birnbaum U., Gabrysch K., Schiller H., Göttsche J., Kauert B., Kleemann M., Schwarzer K. (2007): Leitfaden für die energetische Sanierung von Laboratorien. Jülich: Forschungszentrum Jülich GmbH.

Di Giulio A., Brohmann B., Clausen J., Defila R., Fuchs D., Kaufmann-Hayoz R., Koch A. (in this volume): Needs and consumption – a conceptual system and its meaning in the context of sustainability.

DIN EN 13779 (2007): Lüftung von Nichtwohngebäuden – Allgemeine Grundlagen und Anforderungen für Lüftungs- und Klimaanlagen und Raumkühlsysteme. Berlin: Beuth Verlag.

DIN EN 16001 (2009): Energiemanagementsysteme – Anforderungen mit Anleitung zur Anwendung. Berlin: Beuth Verlag.

DIN EN ISO 7730 (2006): Ergonomie der thermischen Umgebung – Analytische Bestimmung und Interpretation der thermischen Behaglichkeit durch Berechnung des PMV- und des PPD-Indexes und Kriterien der lokalen thermischen Behaglichkeit. Berlin: Beuth Verlag.

DIN EN ISO 14001 (2009): Umweltmanagementsysteme – Anforderungen mit Anleitung zur Anwendung. Berlin: Beuth Verlag.

DIN V 18599 (2007): Energetische Bewertung von Gebäuden – Berechnung des Nutz-, End- und Primärenergiebedarfs für Heizung, Kühlung, Lüftung, Trinkwarmwasser und Beleuchtung. Berlin: Beuth Verlag.

Eichordnung (2007): Eichordnung vom 12. August 1988 (BGBl. I S. 1657), last amended in Article 1 of the Regulation of June 6, 2011 (BGBl. I S. 1035).

Ifeu-Institut Heidelberg (2004): Auswertung der Budget- und Anreizsysteme zur Energieeinsparung an hessischen Schulen – Leitfaden. Wiesbaden: Hessisches Ministerium für Wirtschaft, Verkehr und Landesentwicklung (publ.).

Kaltschmitt M. (2001): Energie aus Biomasse. Berlin: Springer-Verlag.

Kaufmann-Hayoz R., Bamberg S., Defila R., Dehmel C., Di Giulio A., Jaeger-Erben M., Matthies E., Sunderer G., Zundel S. (in this volume): Theoretical perspectives on consumer behaviour – attempt at establishing an order to the theories.

Klesse A., Hansmeier N., Zielinski J., Wagner H.-J., Matthies E. (2010): Energiesparen ohne Investitionen – ein Feldtest in öffentlichen Liegenschaften. In: Energiewirtschaftliche Tagesfragen 60 (4): 8–12.

Klesse A., Wagner H.-J. (2011): Energiesparen und dann? Anonyme Energieverbraucher in Energiemärkten? 7. IEWT, Energieversorgung 2011. Märkte um des Marktes Willen? 16.–18.02.2011 TU Vienna, Austria.

Knöfel H. (2010): Klimaschutzbericht. Gebäudemanagement Georg-August-Universität Göttingen. Technisches Gebäudemanagement, GM 32 Energiecontroller. Göttingen [http://www.uni-goettingen.de/de/128028.html; 10.07.2012]

Matthies E., Thomas D. (in this volume): Sustainability-related routines in the workplace – prerequisites for successful change.

Müller J., Person R.-D. (2010): CO_2-Bilanz für Hochschulen – In: HIS-Magazin (4): 10.

REUSE (1999): Rational Use of Energy at the University of Stuttgart Building Environment. Vorhaben BU 343/94 DE. IKE-Bericht, Juli 1999.

Schmidt M., Stergiaropoulos K., Schmidt F. (2005): CAMPUS – Energie- und Gebäudemanagement im Campus Pfaffenwald und seine Auswirkungen auf die Effizienz der Energieerzeugung (report LHR-1-05). Stuttgart: Lehrstuhl für Heiz- und Raumlufttechnik der Universität Stuttgart. [http://www.kenwo.de/downloads/Uni_CAMPUS_Bericht.pdf; 10.07.2012]

Universität Osnabrück (2008): Umweltbericht der Universität Osnabrück. [http://www.uni-osnabrueck.de/UmweltmanagementDokumente/Umweltbericht_03._September.pdf; 10.07.2012]

Universität Stuttgart (2011): Energiemanagement Server. [http://www.energie.uni-stuttgart.de/Energieserver/index.html, 10.07.2012]

VDI 2067 (2000): Wirtschaftlichkeit gebäudetechnischer Anlagen – Grundlagen und Kostenberechnung (Blatt 1). Berlin: Beuth Verlag.

VDI 3807 (2007): Energie- und Wasserverbrauchskennwerte für Gebäude – Grundlagen (Blatt 1). Berlin: Beuth Verlag.

Wagner U., Rudolph M. (2008): Energieanwendungstechnik. Berlin, Heidelberg: Springer-Verlag.

Willems W. M., Dinter S., Schild K. (2006): Handbuch Bauphysik. Teil 1 – Wärme- und Feuchteschutz, Behaglichkeit, Lüftung. Wiesbaden: Friedrich Vieweg & Sohn Verlag.

Bettina Brohmann, Veit Bürger, Christian Dehmel, Doris Fuchs, Ulrich Hamenstädt, Dörthe Krömker, Volker Schneider, Wilma Mert, Kerstin Tews

16 Sustainable electricity consumption in German households – framework conditions for political interventions

16.1 Introduction

Private households account for around a quarter of total electricity consumption in Germany. Despite increasing efficiency of appliances in recent years, consumption in this sector is still not falling (AGEB 2011). This is partly a result of the rising number and size of appliances in homes and the increase of single households. At the same time, there is great scope for potential savings, which have not yet been exhausted. If efficient use of electricity is to be the key to the transition to non-nuclear electricity production in future, these potential savings in private households must be more effectively addressed in the political arena. In doing so, it is important to understand how consumers can be motivated to change their behaviour to save more energy and how to remove barriers that are beyond consumer control. This means that it is a case of how to achieve absolute reductions in private electricity consumption, for example through more informed consumer decisions on the one hand, and the reduction of market failures, for example as a result of existing asymmetries in information, on the other hand.

For some time now, when dealing with targeted political interventions, economic instruments have been discussed against the backdrop of political paradigms that have been focused on the market and on voluntary consumer actions. The primary question is the extent to which the desired behaviour can be encouraged by making it cheaper or providing subsidies or benefits, and how undesired behaviour can be reduced by making it more expensive by adding additional taxes, for instance. In addition to a thorough investigation of the framework conditions for political interventions in the interests of more efficient use of electricity by private households in Germany, this poses the question of if and how economic instruments could promote electricity savings in private households in Germany (see also Kaufmann-Hayoz/Brohmann et al., Brohmann/Dehmel et al. and Koch/Zech in this volume).

Various perspectives must be considered to extensively investigate the underlying framework conditions for political interventions to promote sustainable electricity consumption in private households. The first task is to identify the areas with the greatest potential for savings in terms of electricity consumption in private households. Second, reports must be compiled on the effectiveness of various economic instruments to address such potentials for savings, in particular the economic and psychosocial determinants of the actions of German consumers. At the same time, the political instruments used in other countries should be examined, along with the extent to which these can significantly influence electricity consumption when compared with other determinants. Based on an investigation of these various aspects, this paper provides a basis for considerations on selected promising political interventions in the field of energy savings in private households in Germany. Furthermore, the results of the analyses of the various aspects form the basis of more detailed investigations of selected political instruments, particularly progressive electricity tariffs and bonus schemes for electricity-efficient home appliances by the "Transpose" project group (see Brohmann/Dehmel et al. in this volume).

The paper is structured as follows: The next section specifies the greatest potentials for savings in electricity consumption in private households in Germany. The third section assesses the price sensitivity of consumers to get an idea of the potential effects on the price of focused intervention strategies. The fourth section discusses psychosocial determinants of consumer behaviour in private households in Germany, before the fifth section presents the findings on the existence and effectiveness of relevant political instruments outside Germany on the basis of a comparative statistical analysis. The final section of the paper summarises the results of the various studies, drawing conclusions for policies and research.

16.2 Key potential savings

With regard to the framework conditions for more efficient use of electricity in German households to be considered, the first question is where the greatest potentials for reducing electricity consumption in private households lie, so that they can then be properly addressed. In doing so, a distinction must be made between technical potentials, which can be realised through investments, and usage-based potentials.

In theory, the maximum potential for saving electricity, which could be achieved through creating more efficient household appliances or by exchanging thermal storage heaters and water heaters run on electricity (i.e. through investment), amounts to around 90 TWh (Bürger 2009; 2010). This corresponds to around 60 percent of the current electricity consumption of all private households in Germany.[1] In the area of household appliances, particularly for cooling/freezing (and cooking/baking), there certainly is a great potential for saving electricity that can be achieved relatively easily. These are followed by the areas of lighting and entertainment. However, TV appliances, for example, are currently only responsible for a relatively small portion of electricity consumption in comparison to cooling appliances. Yet, the current consumer tendency towards larger and larger screens is a significant trend in this area.

The overall potential for electricity savings, which could be achieved in theory through changed user behaviour, is around 30 TWh, which is equivalent to around 20 percent of the total electricity consumption in German households (Bürger 2009; 2010). Classical household appliances (fridge/freezer, oven/cooker, washer/drier) represent the greatest potential areas for savings. From a purely economic perspective, achieving this potential is always profitable as it is not associated with monetary expenditures.[2]

Priority areas for action in political interventions can be identified using these figures. Reducing the electricity consumption of private households by investing in more efficient cooling and TV appliances (while simultaneously attempting to buck the trend of making appliances bigger and bigger) as well as changed user behaviour and the replacement of heaters running on electricity all represent particularly interesting areas of activity. An analysis of the economic framework conditions, particularly the price sensitivity of consumers and the psycho-social determinants of consumers, can help determine what types of intervention in these areas appear particularly promising.

16.3 Price sensitivity in electricity consumption

To what extent can economic instruments effectively promote a reduction of electricity consumption in private households? To what extent do consumers react to changes to the price of electricity, and how does it influence the demand for electricity-saving household appliances? Given the existing potential savings and trends in electricity consumption outlined above, it stands to reason that, in addition to the general price elasticity of electricity consumption, the willingness of households to pay for efficient cooling and TV appliances should be taken into account.

---

[1] Electric heaters and electric water heaters contribute almost half of the total theoretical potential for savings. However, both areas come with considerable obstacles to achieving the potentials (costs, "unpleasantness" of the building work needed).

[2] However, opportunity costs in terms of the time required to perform the work must be considered here.

An experiment which helped simulate purchase decisions for efficient cooling and TV appliances given changing electricity prices came to the finding that the price sensitivity in private households is relatively low in the short run (Hamenstädt 2009).[3] This finding is further supported by the existing scientific literature (including OECD 2008). Therefore, in order to reduce private electricity consumption through pricing signals, the price would have to be increased significantly (and not just marginally), or the pricing signals would need to be accompanied by other political instruments. A slight increase in the price of electricity alone would not be sufficient to produce the desired result.

In addition to the price, further determinants of consumer behaviour would have to be considered in order to identify relevant accompanying measures. In connection with the experiment, for example, the income of the households, the age of the people, whether the property was rented or owned, gender and the education of the people were identified as important factors for influencing decisions to purchase cooling and TV appliances (Hamenstädt 2009). Information plays a key role here. Therefore, one of the experiment's findings was that people were far more willing to pay for energy-saving cooling appliances than they were for energy-saving TVs. One reason for this, in addition to the question of different consumer requirements for both types of appliance,[4] may have been the EU's efficiency label for cooling appliances, which was easily accessible throughout the experiment both in terms of logistics and understanding. At the time of the experiment, there was no such label for TV appliances. This indicates that information could be an important additional measure during price interventions.

16.4 Environmental psychological determinants of consumer behaviour

Further information about determinants of consumer behaviour in economic contexts and therefore about promising political intervention strategies in terms of electricity consumption in private households can be gleaned through approaches used in environmental psychology. Which characteristics of consumers should political instruments address in order to effectively promote more sustainable electricity consumption in the areas identified as particularly relevant for potential electricity savings? A representative telephone survey (N=1,000) among German households analysed

---

3 Price sensitivity is understood here as the so-called short-term cross price elasticity of electricity consumption and purchase decisions for energy-saving household appliances (for a more detailed outline, see Hamenstädt 2008; 2009; 2011).

4 In this way, the television is also characterised by a high symbolic effect.

this question in terms of four target behaviours, which resulted from the aforementioned potentials for savings: investments in more efficient cooling and TV appliances, replacing electric heaters, changed user behaviour, as well as the use of switchable connection plug boards. The survey revealed that specific combinations of psycho-social influences are relevant for individually desired target behaviours, such as the early substitution of energy-efficient cooling appliances. These influential variables are partly object-related (problem perception), subject-related (self-image, e.g. materialism) and action-related factors (e.g. moral obligation, feasibility), but also emotions (e.g. feelings of guilt), habits and particular reasons for certain behaviour.[5] Consequently, specific political measures must be developed for individual target behaviours, considering these influences (Krömker/Dehmel 2010).[6]

Cooling appliances
For cooling appliances it was apparent that the consumption of 70 percent of those surveyed was above 400 kWh per year, a value which can be accepted as a potential standard for a desirable level of consumption for households (this value was calculated by the operation of two 300 l cooling appliances of efficiency class A++). The actual consumption increases with subjective demand. The subjective perception of the need for more cooling also leads to increased electricity consumption due to the use of second or third appliances. In terms of the energy efficiency of cooling appliances, it was found that the average electricity consumption used for cooling increased in households with larger surface areas and more property. Furthermore, the widespread (94 percent) intention to invest in an energy-efficient appliance in the next purchase is particularly fostered through convictions regarding the climate, innovative self-concepts and the identification of one's own behavioural control possibilities. However, the survey also found that a majority of households asked (78 percent) are not planning to purchase a new cooling appliance in the near future.

These relationships provide further information in terms of the relevance of different intervention strategies. It therefore becomes apparent that information and advertisement regarding the effects cooling appliances have on the climate, stressing innovative technical aspects and promising the possibility to control options all support purchase decisions in favour of more efficient cooling appliances. These may be additional measures to assist economic interventions. However, as a large proportion of those asked

---

[5] The respective influence variables were formed from the constructs stated in brackets or their usual methods and designs.

[6] See Kaufmann-Hayoz/Bamberg et al. in this volume for more information on behavioural models as a basis for such surveys.

stated that they wanted to invest in an efficient appliance on their next purchase,[7] an intervention to bring the time of the purchase closer is all the more important. Given this, providing economic incentives, for example via bonus schemes, may be an interesting possibility for intervention. Even if insufficient financial resources were not identified as a hindrance to the purchase of more efficient cooling appliances in the survey (in contrast to the experiment), financial incentives may represent additional motivation, as recent examples, such as the scrapping premium for cars ('cash for clunkers'), have shown. However, when utilising bonus schemes, the problem of second and third appliances would need to be considered so that the inefficient old appliance does not continue to run in addition to the subsidised, energy-efficient new appliance.

TV appliances

TV appliances currently still represent a low proportion of electricity consumption. However, a trend of significant growth has also been identified here (cf. section 16.2). The majority of households still use conventional cathode ray tube appliances, which consume less electricity by comparison. With regard to the use of TV appliances, the survey discovered that the current electricity consumption was promoted by materialistic self-perceptions and the availability of special offers at retail, as well as higher income and the subjectively formulated need for more appliances (Krömker/Dehmel 2010). The vast majority of those asked firmly intended to make their next TV purchase an energy-efficient appliance although that purchase is not planned in the near future.[8] This intention is greatly strengthened by factors oriented around protecting the environment, such as an ecological self-image and the feeling of a moral obligation. It is partly hindered by the perception that energy-saving devices are not available.

Again, it is also possible to address an individual's ecological self-image and moral obligation, as well publicising the greater availability of appliances through information and advertising in order to further facilitate the purchase of energy-efficient appliances. Corresponding financial incentives can bring the time of the investment forward, as, according to the statements, there would be a tendency to purchase an energy-saving device. However, this intention must be questioned given the observed trend towards purchasing larger appliances in certain product groups, as the energy efficiency gains achieved through technical innovations, for example with TVs, may

---

7 Of course, the problem of the common tendency towards providing socially desired answers when responding to surveys must be considered.

8 See footnote 2.

be overcompensated by larger screens.[9] There is also the problem of second and third appliances as already seen with cooling appliances. The provision of better information on the availability of energy-saving appliances on the market would be less controversial. Efficiency labels for TVs developed by the EU will also contribute to this improved information transfer.

Electric heaters
With regard to the replacement of electric heaters, it has been observed that this area cannot be approached without considering the age structure of homeowners. Only a minority of those asked (6 percent) want to exchange their electric heaters in the near future. The older the person surveyed, the less they can and want to change their heater, but rather see it as the responsibility of the next generation. Younger people and those asked who felt a greater moral obligation to save electricity and those who saw the financial feasibility of the necessary investments were more likely to intend to replace their electric heater with a more eco-friendly alternative. In this case, the conclusion of the survey is that instruments need to be used depending on the specific audience targeted. For older homeowners, neither economic incentives nor the targeted provision of information will change much. In contrast, for younger homeowners, help with investments could mean that the desired remedial action would be possible (earlier). The group in the middle in terms of age would probably be of most interest as they are still undecided about whether it is worthwhile to tackle the financial and organisational efforts involved in replacing electric heaters. Economic incentives, accompanied by the relevant information, could make a particular difference.

Switchable connection plug boards
Switchable connection plug boards were found in 93 percent of households of those asked, with almost half of households owning four or more. Regular use of switchable multi-plug adaptors was reported by a vast majority of those asked (70 percent). A self-image of being innovative, and the self-confidence to actually think about switching it off, promotes the use of switchable devices. However, the greatest influence came from habits which prevent use, and from the conviction that this is accompanied with sacrificing comfort (e.g. the necessity to reprogram devices regularly). This last target behaviour examined in the survey therefore showed the boundaries of political intervention possibilities in general and economic instruments in particular. Habits and perceptions of sacrificing comfort are difficult to change, if this is at all possible. At best,

---

9   At the same time, the selection of a plasma or LCD screen and additional features plays a significant role.

they are changed in the course of targeted interventions in which a number of communicative instruments are used (cf. Schäfer/Jaeger-Erben, Matthies/Thomas and Sunderer et al. in this volume). Information on the electricity consumption of appliances which appear to be turned off, coupled with higher electricity prices, would only provide another incentive to the continuous use of multi-plug adaptors. However, an obligation of manufacturers to lower the electricity consumption of appliances not in use, which has also been established on European level in the meantime, is surely more effective.

16.5 Experiences from outside Germany

Finally, in addition to better knowledge of the framework conditions, experiences from outside Germany can help gain a better understanding of how to use potential economic instruments to promote energy-saving in private households. A quantitative analysis of various political instruments and their effects in a macro-environment was performed to compare countries (Mayer et al. 2011; Schneider et al. 2011). Due to the variety of instruments used and the numerous context factors that can affect developments in electricity consumption in private households, such an investigation does not come without problems. Despite this, it can yield some interesting insights.

First and foremost among them is the already large number and scope of instruments used in private households in OECD and developing countries in the area of electricity consumption. This number has risen rapidly, particularly over the course of the last two decades. At the same time, economic development (measured by GDP) and climatic conditions explain many of the differences in the level of per capita electricity consumption and how this has changed over the last 14 years. However, electricity prices had an equally great impact. Finally, the cross-section regressions with various instrument groups revealed statistically significant savings for the policy instrument "Consumer information" in all models. Connections between the establishment of energy efficiency agencies and economic measures were equally significant. However, there was positive correlation between the instruments and electricity consumption, which either indicates reverse causality (in particular countries with rising electricity consumption utilise such instruments) or "rebound effects". The existence of reversed causal relationships is also suggested by the findings of the analyses about the connection between the growth in electricity consumption and the number of measures used. Countries with higher electricity consumption growth intervened more through political instruments.

A qualitative, comparative study on the use and effectiveness of the political instruments used in countries outside Germany for saving energy in private households showed the great number and scope of the measures. On the other hand, it underlined the effectiveness of communicative and economic instruments combined with cross-section instruments. This supports the findings outlined above that only using economic instruments, without sufficient supporting information, would not sufficiently lead to the desired effect.

16.6 Conclusions

The efficient use of electricity in households is of great importance for climate-friendly policies and the energy transition. However, the significance of electrical energy efficiency has been severely neglected in the political discussions regarding measures to be taken to expand production capacities and power lines as required (cf. Vorholz 2011). This poses the question of which measures, or which packages of measures, would be particularly promising possibilities for intervention in Germany. This paper particularly emphasised economic instruments as these are given particular attention in political debates.

To identify promising potential interventions, however, fundamental contextual factors for political interventions in private energy consumption must be clarified first. On the one hand, this includes the greatest potential areas for saving energy, the economic context with regard to the price sensitivity of consumers, and on the other hand the environmental psychological determinants of consumer behaviour. Ultimately, this chapter also builds on the findings of a study on the utilisation and effectiveness of political instruments in the area of electricity consumption outside Germany.

With regard to the relevant potential savings in electricity, the chapter showed that reducing electricity consumption in private households by investing in more efficient cooling and TV appliances, changed user behaviour and the replacement of electric heaters represent particularly interesting areas of action. The investigations of the economic and psycho-social framework conditions and the analyses of the use of relevant political instruments outside Germany also showed that individual instruments would not be sufficiently effective to achieve this energy-saving potential. Instead, packages of measures, particularly the combination of economic and communicative instruments, would be desirable (see also Kaufmann-Hayoz/Brohmann et al. and Brohmann/Dehmel et al. in this volume).

In this way, the experiment on the price elasticity of electricity and electricity-saving household appliances, as well as the representative survey of consumers, highlighted the importance of accompanying economic incentives in the form of price increases or bonuses with information. In doing so, it became clear that controlling the electricity price alone would not be sufficient to achieve the desired goal, at least as long as the increases are not substantial. Even if studies (including OECD 2008) indicate that the price sensitivity of private households is increasing over time, additional informative strategies would be needed to achieve significant energy reductions in the short-term (see also Brohmann/Dehmel et al. in this volume). The potentials which could be realised ad hoc include in particular those which address the behaviour of electricity consumers. The environmental psychological analysis of consumer behaviour highlighted the key role played by consumer attitudes when consuming electricity. Given these analyses, it is the accompanying communicative instruments, which target the subjective consumer norms and enable a consumer to reconcile their own behaviour with a specific self-image, that appear promising.

As a whole, the investigative approaches presented here paint a complex picture of consumer behaviour and the consumer environment. The analysis built on a number of scientific approaches and methods which shared a common factor in that they slightly open the "black box" of electricity consumption. Thus, this paper also shows that the combination of different analytical perspectives and approaches promises complementarities and added value in terms of quality, which can be achieved through inter-disciplinary research.

References

AGEB (AG Energiebilanzen e.V.) (2011): Energieverbrauch in Deutschland im Jahr 2010. Berlin.

Brohmann B., Dehmel C., Fuchs D., Mert W., Schreuer A., Tews K. (in this volume): Bonus schemes and progressive electricity tariffs as instruments to promote sustainable electricity consumption in private households.

Bürger V. (2009): Identifikation, Quantifizierung und Systematisierung technischer und verhaltensbedingter Stromeinsparpotenziale privater Haushalte. TRANSPOSE Working Paper No.3. Freiburg. [http://www.uni-muenster.de/Transpose/en/publikationen/index.html; 02.07.2012]

Bürger V. (2010): Quantifizierung und Systematisierung der technischen und verhaltensbedingten Stromeinsparpotenziale der deutschen Privathaushalte. In: Zeitschrift für Energiewirtschaft 34 (1): 47–59.

Fuchs D. (1995): Incentive Based Approaches in Environmental Policy – "Little We Know". Background Report "Clear Vision, Clean Skies" project, Southern California Edison, The Claremont Graduate School. Claremont.

Hamenstädt U. (2008): Bestimmung der Preiselastizität für Strom. Münster: Westfälische Wilhelms-Universität.

Hamenstädt U. (2009): Stromsparen über den Preis? Ein Experiment. TRANSPOSE Working Paper No 4. Münster. [http://www.uni-muenster.de/Transpose/en/publikationen/index.html; 02.07.2012]

Hamenstädt U. (2011): Teaching Experimental Political Science. Experiences from a seminar on methods. European Political Science – Advance Online Publication: 24.06.2011.

Helm D. (ed.) (1990): Economic Policy Towards the Environment. Cambridge: Cambridge University Press.

Kaufmann-Hayoz R., Bamberg S., Defila R., Dehmel C., Di Giulio A., Jaeger-Erben M., Matthies E., Sunderer G., Zundel S. (in this volume): Theoretical perspectives on consumer behaviour – attempt at establishing an order to the theories.

Kaufmann-Hayoz R., Brohmann B., Defila R., Di Giulio A., Dunkelberg E., Erdmann L., Fuchs D., Gölz S., Homburg A., Matthies E., Nachreiner M., Tews K., Weiß J. (in this volume): Societal steering of consumption towards sustainability.

Koch A., Zech D. (in this volume): Impact analysis of heat consumption – user behaviour and the consumption of heat energy.

Krömker D., Dehmel C. (2010): Einflussgrößen auf das Stromsparen in Haushalten aus psychologischer Perspektive. Transpose Working Paper No 6. Münster. [http://www.uni-muenster.de/Transpose/en/publikationen/index.html; 02.07.2012]

Matthies E., Thomas D. (in this volume): Sustainability-related routines in the workplace – prerequisites for successful change.

Mayer I., Schneider V., Wagemann C. (2011): Energieeffizienz in privaten Haushalten im internationalen Vergleich. Eine Policy-Wirkungsanalyse mit QCA. In: Politische Vierteljahresschrift 52 (3): 399–423.

Michaelis P. (1996): Ökonomische Instrumente in der Umweltpolitik. Heidelberg: Physica-Verlag.

OECD (2008): Household Behaviour and the Environment. Reviewing the evidence. OECD. [http://www.oecd.org/dataoecd/19/22/42183878.pdf; 02.07.2012]

Schäfer M., Jaeger-Erben M. (in this volume): Life events as windows of opportunity for changing towards sustainable consumption patterns? The change in everyday routines in life-course transitions.

Schneider V., Schorowsky N., Tenbücken M. (2011): Die Wirkung politischer Maßnahmen auf die Senkung des Stromverbrauchs in privaten Haushalten: Ein makroquantitativer internationaler Vergleich. In: Fuchs D. (ed.): Die politische Förderung des Stromsparens in Privathaushalten. Herausforderungen und Möglichkeiten. Berlin: Logos. 46–68.

Sunderer G., Götz K., Gölz S. (in this volume): The evaluation of feedback instruments in the context of electricity consumption.

Vorholz F. (2011): Ihr wollt gar nicht sparen! In: Zeit online (07.07.2011). [http://www.zeit.de/2011/28/Energiewende; 02.07.2012]

Bettina Brohmann, Christian Dehmel, Doris Fuchs, Wilma Mert, Anna Schreuer, Kerstin Tews

17 Bonus schemes and progressive electricity tariffs as instruments to promote sustainable electricity consumption in private households

17.1 Introduction

There are a number of different instruments available to support the greater sustainability of electricity consumption in private households (see Kaufmann-Hayoz/Brohmann et al. in this volume). Among them, it is the economic instruments which have attracted much attention as early as the 1980s, as they appear particularly compatible with a socio-political paradigm focused on the market and competition (Fuchs 1995; Helm 1991; Brohmann/Bürger et al. in this volume). Two types of economic instruments, which appear particularly noteworthy both in terms of experiences gleaned from outside Germany and against the backdrop of the gains in efficiency necessary for the energy transition in Germany, are bonus schemes for energy-efficient household appliances and progressive electricity tariffs. Pursuant to one of the central research interests of the "Transpose" project group, this paper examines the design and conditions necessary for the success of these instruments using selected case studies from outside Germany and builds on this by offering considerations regarding their application to Germany.

17.2 Bonus schemes

Bonus schemes should incite the earlier and/or better quality replacement of old household appliances with new, more energy-efficient ones by offering a monetary incentive in form of a bonus payment.[1] For decades, they have been the instrument of choice in

---

[1] From an energy perspective, it could be recommended to replace cooling appliances older than five years (Rüdenauer et al. 2007).

terms of steering purchasers towards a specific, for example particularly energy-efficient, appliance. In the USA, bonus payments for white goods (predominantly fridges) have been used as early as the 1980s, with the support of the regulatory authority at the time, which also helped manage electricity consumption during peak hours (least cost planning). While the specific efficiency standards in the USA led to significant success in terms of energy savings, the size, equipment and number of typical household refrigerators also rose. Therefore, it is expected that the energy consumption, which had decreased in the past decade, will be overcompensated by new appliances in the 2010s (Deumling 2004).

National bonus schemes were also implemented by energy efficiency funds and agencies (Elsparfonden, NOVEM) in Denmark and the Netherlands for various household appliances between 1999 and 2005 in order to increase the market penetration of the most efficient models. In 2009, the purchase of cooling appliances given the top efficiency rating of A++ by Umweltforum Haushalt (UFH)[2] was encouraged along with the return of old appliances.[3] The three approaches display interesting differences and similarities, which are described below (see also Dehmel 2010).

Both the introduction of bonus schemes in the three countries and the differences in design can be put down to different interests represented by the key parties involved. In Austria, it was the economic interest represented by the UFH, who wanted to improve sales figures for new appliances in a period with lower sales as well as making more use of the group's own recycling plant. In contrast, the environmental motivation and the objective of prompting a transformation of the market were decisive in Denmark and the Netherlands.

There were also significant differences in terms of how the programmes were financed. Austria, or the UFH, had the advantage that funds were already available, having resulted from the recycling costs for cooling appliances which had been paid by consumers under earlier legislation and could now be "passed back" to them.[4] In Denmark and the Netherlands, the bonus programmes were financed through environmental levies in form of energy taxes and possible tax reductions for traders who paid the bonuses.

---

2 The UFH is an interest group in the electrical and electronics industry in Austria.

3 The bonus scheme was therefore given the name "Trennungsprämie" (replacement bonus).

4 Between 1993 and 2005, part of the recycling costs had to be paid when purchasing a cooling appliance in Austria; for this payment customers received a so-called "Kühlschrankpickerl". Due to the changes in EU law, the disposal of household appliances has been free since 2005. The money paid for the "Kühlschrankpickerln" could be reclaimed, but only a small fraction of consumers did so. The project was managed by the UFH, which is organised as a private foundation, so it was able to operate relatively freely with the remaining funds.

At the same time, common conditions for success in terms of the introduction and effectiveness of the bonus schemes were identified. One such condition appears to be the transfer of responsibility to a small, independent organisation whith good contacts to other relevant actors, which means that transaction costs remain low and trust between the parties involved is high (cf. Dehmel 2010). The provision of information regarding the bonus schemes to consumers, retailers and industry via as many channels as possible is regarded as another condition for success. For example, the website of the bonus scheme in Denmark, which contained information on the best and most price-effective cooling appliances, had a great influence (Noergaard et al. 2007). In terms of the duration of the bonus schemes, it had been assumed in the past that placing time constraints on the programmes was beneficial. However, aside from the financial expenditure required, there were no disadvantages identified in the longer-running Dutch programme.

All three bonus schemes were regarded as successful (cf. Tanzer 2010; Thomas 2007; Noergaard et al. 2007). Essentially, the long-term support both in Denmark and the Netherlands triggered a significant market transformation. In Denmark, the market share of A-class cooling appliances rose from 14 percent in 1999 to 74 percent in 2003. There was major growth in the Netherlands with the market share rising from 23 percent in 1999 to 98 percent in 2003. In Austria, the market share of A++ appliances rose from 8 percent to 26 percent in the short time in which the bonus scheme was in effect.

Yet in terms of energy savings or the reduction of CO_2 emissions per annum, the results were more mixed. For Austria, 8,299,000 kWh/a energy savings or a reduction of 1,344 tonnes of CO_2 emissions per annum were calculated for the difference between the new cooling appliances and the old devices surrendered (Tanzer 2010). In Denmark and the Netherlands, however, there are few comparable figures regarding the energy savings and reduced CO_2 emissions achieved.[5] However, estimates would also need to be revised for the fact that there was no obligation to surrender an old cooling appliance when using the bonus scheme in the countries, leading to a suspicion that the old appliances were not replaced and simply continued to run.[6] To that extent, the obligation to return the old appliance in Austria is an important design difference.

---

[5] The only estimates available relate to the electricity saving potential and CO_2 emissions of all promoted household goods (see Rambøll Management 2004; Thomas 2007).

[6] Consumers in focus groups on the Austrian "Trennungsprämie" (disconnection bonus) stated that as long as the devices were still working, they would have kept old appliances as a secondary appliance or given them as a present to someone else if they had not been obliged to dispose of them (Mert/ Schreuer 2010).

However, in terms of a potential "rebound effect", it is interesting that none of the programmes examined considered the potential detrimental effects on the environment of a trend towards "greater comfort", i.e. larger fridges. Quite the contrary – in some cases, even higher premiums were paid for larger appliances. This aspect should certainly be considered when designing any potential future bonus schemes.

From an economic point of view, it is ultimately important that the bonus schemes are designed in such a way to keep the free-rider effect as small as possible. This finding has been evident since the first programmes were introduced in the USA. To avoid the free-rider effect observed through purchaser bonuses, following the assessment of first experiences, credits were provided to manufacturers who produced above-average sales figures for efficient appliances. When evaluating the Dutch bonus scheme, it was also established that there was a possibility that subsidising the purchase of appliances through purchaser bonuses may lead to free-rider effects (Herling/Brohmann 2011) Therefore, appropriate framing, or the combined use of instrument packages, which include various stakeholders, i.e. including producers and traders, was needed to tackle the free-rider effect. One example here is the "green carrot" incentive for manufactures of LED lights and fridge/freezers, which led to faster market penetration of efficient appliances in the USA (Grießhammer 2011).

The market for TV appliances is currently being rationalised, due to the binding introduction of an EU efficiency label. Accordingly, the introduction of a bonus scheme will have to wait (Öko-Institut 2009). The introduction of the standards can trigger and support the replacement of appliances with highly efficient ones, with no additional effect created by the incorporation of a bonus scheme (Rausch/Timpe 2010). This is largely because of the dynamics of a market introduction of new appliances through the introduction of the label. For the new generation of A+ televisions, electricity consumption is so low that simply the visible differences in electricity consumption will make the replacement of old appliances sooner sensible for consumers and no additional incentive appears necessary. In the past, simply the introduction of the energy label for white goods also led to a relatively quick readjustment of the market.

The analyses performed under the "Transpose" project and similar projects (Rausch/Timpe 2010; Brohmann et al. 2010) show that the instrument of purchaser bonuses can essentially provide adequate support for the introduction of more energy-efficient household appliances. This view holds if an actual saving target is aimed at, only for certain groups of appliances with specific design options and as long as they are complemented by other political instruments such as information and consulting or standards. Alternatively, changing from a purchaser bonus to a seller/manufacturer reward may be sensible for certain situations in terms of the market and regulation.

17.3 Progressive electricity tariffs

Under progressive electricity tariffs, the price per kWh of electricity consumed rises along with growing consumption, providing households with a financial incentive to keep their electricity consumption low. On a national level, progressive electricity tariffs for private households in the EU are only used in Italy (Dehmel 2011). Outside the EU, California also represents an interesting case that displays similar historical developments and a few significant differences (Dehmel 2011; Tews 2011).

Progressive electricity tariffs were introduced in both Italy and California as early as the middle of the 1970s as a response to the oil crisis.[7] In addition to the political target of ensuring the overall energy supply with decreased private electricity consumption, social objectives also played a part as an attempt was made to reduce the financial burdens placed on private households as a result of the rising electricity prices. In both cases, the idea was to make the first 50 to 70 percent of the average electricity consumption relatively cheaper as a result of the progression (Hennessy et al. 1989; AEEG 2008). In Italy, additionally a capacity limit was set, which also contained a progressive element. The price of a contract with the limit of 3 kW output rose progressively to a contract with a limit of 6 kW.[8] The introduction was decided in strongly regulated and vertically integrated electricity markets by the authorities responsible for regulation and tariff design, with a strong influence coming from the respective governments and trade unions in Italy and environmental and consumer organisations in California (Prontera 2010; Hennessy et al. 1989).

In Italy, the tariff design was adapted in the course of the European market deregulation in the 1990s. Thus, today only the proportion of the total electricity price which is not subject to competition is progressive, i.e. network fees, system costs and taxes.[9] The price per kWh of electricity used by all households in Italy is rising in block structures of the proportion of the total electricity price made up of network fees, system costs and electricity tax, which are specified by the regulatory authority (AEEG). At the same time, the progression has remained in place with the capacity limitation.[10]

---

7 In California, all statements refer to the proportion of households (approx. 75 percent) receiving their electricity from investor owned utilities (IOU) (in contrast to publicly owned utilities (POUs)), which have been the subject of all of the large state regulation schemes since the 1970s.

8 The limit affects the parallel use of electrical appliances. This way, given a 3 kW contract, appliances with a higher power uptake, such as dishwashers and washing machines, cannot be used simultaneously.

9 For some of the households, there is also a progression in the price of the services considered in the proportion of the production costs.

10 Over 90 percent of all households in Italy still have the cheaper 3 kW contract (AEEG 2009).

The initially planned abolishment of the progressive electricity tariff has not yet been realized for social and increasingly environmental political reasons, as this would massively increase the electricity bill for almost all households.

A new progressive tariff structure was also introduced in California in the course of tackling the energy crisis in 2000/2001.[11] However, the progression here was in the proportion of the total electricity price constituted by production and supply costs, which are calculated by IOUs and regulated by the California Public Utilities Commission (CPUC). A binding 5-block tariff structure was introduced for IOU customers, in which rising electricity consumption is reflected in rising production prices. Although the fixed 5-block structure is currently being relaxed, the basic progressive structure of the electricity price is to be maintained.

This means that various components of the total electricity price can be altered towards progressivity. In this way, the Italian model is better suited to a liberalised electricity market, as it was adapted in the context of the liberalisation of the European electricity market so that the progression only affects the pricing components which are not subject to competition (network fees and tax). However, the *binding nature* of the progressive tariff is always an essential prerequisite for activating potentials for electricity savings through tariffs. As customer movements cannot be prevented in the liberalised electricity market, large consumers would change their provider or tariff if the progressive tariff were only an additional and not a binding tariff structure. However, the actual effect of progressive electricity tariffs on electricity consumption in private households is not easy to calculate as electricity consumption is influenced by a variety of factors. Yet, the results of individual studies have concluded that progressive electricity tariffs have a positive influence on the reduction of total electricity consumption in private households. In this way, a corresponding influence of various price adjustments in the progressive tariff structure in Italy on total electricity consumption was established (FNLE/CGIL 1989), while analytical models by American researchers have calculated the potential for lowering the average annual electricity consumption of Californian households as lying between 6 and 10 percent (Reiss/White 2004; Faruqui 2008).[12]

---

11 In the course of the deregulation of the electricity provision in the 1990s, there were severe disruptions on the electricity market in California (Dormady/Maggioni 2009). To restabilise the electricity market and avoid the competition of several electricity providers, the current tariff structure was introduced and the right to change electricity providers for IOU customers suspended simultaneously.

12 In the long-term, the electricity savings contributed up to 20 percent, while a combination of progressive tariffs with time-variable tariffs and complementary measures in terms of demand would provide the optimal support for the savings (Faruqui 2008, pp. 22 and 27). At the same time, the simulations highlighted the central role of the tariff design, for example the definition of sufficient price differences between the steps, to induce significant price signals (op. cit. p. 24).

17.4 Considerations on implementing the measures in Germany

So what can Germany learn from the experiences from other countries? Not everything that works in other countries can be implemented and effective in Germany. Therefore, in order to assess how these instruments can be transferred to Germany, the conditions for the introduction of these instruments and the central details of the instrument's design for the German context must be examined.

Essentially, the bonus schemes should transfer well to Germany. Such programmes have already been and are being used by various (communal) providers (Herling/Brohmann 2011). Premium bonuses are issued for fridges and freezers, washing machines and driers. The net savings actually achieved this way depend largely on specific user behaviour in households and only estimates have been produced up until now. At the same time, the aforementioned analyses have shown that the bonus schemes need to be designed in such a way that there is an obligation to return old appliances, it does not facilitate a trend towards larger appliances, and a free-rider effect can be excluded to the greatest possible extent. In terms of the party responsible for the scheme, it is recommended to tend towards placing greater trust in public actors (Mert/Schreuer 2010). Furthermore, in times where the state's coffers are empty, indirect financing possibilities, such as the potential to reduce tax burdens, must be considered. These also affect the state budget, but are easier to represent politically. Pay-as-you-go financing would also be conceivable, with a mark-up to be paid for less efficient appliances, which could be used accordingly as a reward for more efficient fridges.

Progressive tariffs have been repeatedly discussed in Germany since the 1980s. There was also little disagreement with regard to the objective of correcting the degression of the electricity price, from an ecological perspective. The two-part tariff structure of the basic price, which is not based on consumption, and the work price, which is based on consumption, is undesired in that regard. However, there was and is no consensus regarding the specifics of how to implement such a political instrument. Introducing a progressive tariff structure in Germany comes with both substantial political challenges and the necessity for more detailed considerations on how the tariffs are designed in terms of a number of problem areas (Dehmel 2011; Tews 2011).

First, the necessary binding nature of a progressive tariff structure must be considered a political obstacle to realising desired potential savings. This would require a change of the Energy Management Act (EnWG), but runs contrary to the longstanding objectives of the majority of those involved in terms of state interventions into the design of the electricity price. However, the potentials for savings attributed to this tariff model cannot be achieved through additional and optional progressive tariffs. That was the finding of our focus groups, which showed a limited willingness of consumers to choose a progressive electricity tariff (Mert/Schreuer 2011). Without the possibility of attractive offers for customers, a market-oriented approach would also not generate demand for such tariffs – neither for load-variable and time-variable tariffs nor progressive tariffs. Second, the prerequisites in terms of measurement technology are lacking for individual balancing. A demand-oriented, competitive approach of introducing this measurement technology over a wide area failed due to the high costs in relation to low uptake in private households (cf. Ecofys et al. 2009). Third, with regard to design variants, it seems natural to use the same method as the Italian model and embed the progression in network fees and taxes. However, the price-setting model for network fees in Germany is fundamentally different to the Italian model. In Germany, end consumers do not pay a uniform price for the network fees. This is partly because of the structural characteristic of the German electricity market – a large number of actors in the network operation – which creates different network fees. An optimal starting point for integrating progressive elements into the network fee would be uniform network fees for end consumers. There was and is also a politically articulated demand for making the network fees uniform, which could be built upon (see Ruhebaum 2010; Tews 2011, pp. 37 ff.). There are also some administrative and legal barriers for progressive design of the electricity tax, although these can be overcome. Moreover, the proportion of annual total electricity costs of a sample household made up by the electricity tax is only around 9 percent, which is hardly likely to have a noticeable steering effect on consumer behaviour. This means that significant leverage could only be generated by spreading the interest rate from zero to a suitably higher maximum price. Given this, the possibility of a progressive design of other elements of the total electricity price specified by the state, particularly the EEG reallocation charge (for renewable energies), would also be worthy of assessment.

17.5 Outlook

The pressure to manage the energy transition and reduce greenhouse gas emissions is immense and creates new opportunities for the necessary regulatory and administrative reforms. It is also possible that bonus schemes and progressive tariffs could be identified as a strategic option, despite empty state coffers, political opposition and the necessary legal adjustments. Electricity efficiency and savings are the bridge to the renewable energy provision of the future. Building this bridge requires an ambitious mix of policies. In addition to the economic instruments discussed here, which focus on consumer reactions, the policy mix must naturally also do more to engage the providers, network operators and appliance manufacturers. At the same time, despite the expected resistance and challenges, the use of potentially promising intervention strategies, such as combining bonus schemes and progressive tariffs, appears worthy of consideration given the current problems in terms of energy policy.

References

Brohmann B., Brunn C., Fritsche U., Hünecke K., Rausch L., Schönherr N., Teufel J. (2010): Bundling of Sustainable Consumption Instruments and Hypotheses on the Effects of Instrument Bundles for Food and Housing. Draft Report Dec. 2010 of D 4.1 Part 2, EUPOPP. Darmstadt.

Brohmann B., Bürger V., Dehmel C., Fuchs D., Hamenstädt U., Krömker D., Schneider V., Tews K. (in this volume): Sustainable electricity consumption in German households – framework conditions for political interventions.

Dehmel C. (2010): Austausch von Kühlgeräten durch effiziente Neugeräte in privaten Haushalten – Die Trennungsprämie in Österreich im Vergleich zu ähnlichen Programmen in Dänemark und den Niederlanden. TRANSPOSE Working Paper No. 9. Münster. [http://www.uni-muenster.de/Transpose/en/publikationen/index.html; 02.07.2012]

Dehmel C. (2011): Der Einfluss von progressiven Tarifen auf den Stromkonsum in privaten Haushalten in Italien und Kalifornien. TRANSPOSE Working Paper No. 10. Münster. [http://www.uni-muenster.de/Transpose/en/publikationen/ index.html; 02.07.2012]

Deumling R. (2004): Thinking outside the Refrigerator: Shutting down Power Plants with NAECA? Berkeley. [http://www.eceee.org/conference_proceedings/ACEEE_buildings/2004/Panel_11/p11_2/paper; 11.07.2012]

Dormady N., Maggioni E. (2009): Climate Change Mitigation Policy and Energy Markets: Cooperation and Competition in Integrating Renewables into Deregulated Markets. [http://www.polsoz.fu-berlin.de/en/polwiss/forschung/systeme/ffu/studium/promotion/events/participiants/paper/dormady_maggioni.pdf; 11.07.2012]

ECOFYS, EnCT, BBH (2009): Einführung von lastvariablen und zeitvariablen Tarifen. [http://www.bundesnetzagentur.de/cae/servlet/contentblob/153298/publicationFile/6483/EcosysLastvariableZeitvariableTarife19042010.pdf; 02.07.2012]

Faruqui A. (2008): Inclining Toward Efficiency. Is Electricity Price-elastic Enough for Rate Designs to Matter? In: Public Utilities Fortnightly (August): 22–27.

FNLE, CGIL (1989): Tariffe elettriche: Uno strumento di governo della politica energetica, 3–4. Roma.

Fuchs D. (1995): Incentive Based Approaches in Environmental Policy – "Little We Know". Background Report "Clear Vision, Clean Skies" project, Southern California Edison/The Claremont Graduate School. Claremont.

Grießhammer R. (2011): Prämien oder Green-Carrot-Preise? Optionen für nachfrage- und/oder angebotsorientierte ökonomische Anreize in Deutschland. Vortrag auf der Transpose Abschluss-Konferenz am 7. Juli 2011 in Berlin.

Helm D. (ed.) (1990): Economic Policy Towards the Environment. Cambridge: Cambridge University Press.

Hennessy M., Keane D. M. (1989): Lifeline Rates in California: Pricing Electricity to Attain Social Goals. In: Evaluation Review 13 (2): 123–140.

Herling J., Brohmann B. (2011): Finanzielle Kaufanreize bei Weißer Ware und TV: Instrumentenoptionen für Deutschland? TRANSPOSE Working Paper No. 12. Darmstadt. [http://www.uni-muenster.de/Transpose/en/publikationen/index.html; 02.07.2012]

Kaufmann-Hayoz R., Brohmann B., Defila R., Di Giulio A., Dunkelberg E., Erdmann L., Fuchs D., Gölz S., Homburg A., Matthies E., Nachreiner M., Tews K., Weiß J. (in this volume): Societal steering of consumption towards sustainability.

Mert W., Schreuer A. (2010): Fokusgruppen zur österreichischen Trennungsprämie. Unveröffentlichter TRANSPOSE Bericht.

Mert W., Schreuer A. (2011): Fokusgruppen zu progressiven Tarifen. Unveröffentlichter TRANSPOSE Bericht.

Noergaard J., Brange B., Guldbrandsen T., Karbo P. (2007): Turning the Appliance Market Around Towards A++. ECEEE 2007 Summer Study Saving Energy – Just do it. [http://www.eceee.org/conference_proceedings/eceee/2007/Panel_1/1.345/paper; 11.07.2012]

Öko-Institut e.V. (2009): Konzeption eines produktbezogenen Top-Runner-Impulsprogramms. Freiburg: Öko-Institut e.V.

Prontera A. (2010): Europeanization, Institutionalization and Policy Change in French and Italian Electricity Policy. In: Journal of Comparative Policy Analysis: Research and Practice 12 (5): 491–507.

Rambøll Management (2004): Evaluation of the Danish Electricity Saving Trust. Statement by the Board of the Danish Electricity Saving Trust regarding the evaluation of the Danish Electricity Saving Trust by Rambøll Management. Kopenhagen. [http://www.savingtrust.dk/publications/evaluations/evaluation-of-the-electricity-saving-trust/at_download/File; 11.07.2012]

Rausch L., Timpe C. (2010): Assessment of Energy Consumption Decisions in Private Households. Report im Rahmen des BMBF Vorhabens "Soziale, ökologische und ökonomische Dimensionen eines nachhaltigen Energiekonsums in Wohngebäuden (Seco@home)". Darmstadt.

Reiss P. C., White M. W. (2005): Household Electricity Demand Revisited. In: Review of Economic Studies 72 (3): 853–883.

Rüdenauer I., Seifried D., Gensch C.-O. (2007): Kosten- und Nutzen eines Prämienprogrammes für besonders effiziente Kühl- und Gefriergeräte. Studie im Auftrag des Zentralverbandes der Elektrotechnik- und Eletkroindustrie e.V., Freiburg: Öko-Institut e.V.

Ruhbaum C. (2010): Eine Netz AG für Deutschland? Die Debatte um die Neuordnung der Stromübertragungsnetze. Masterarbeit, Freie Universität Berlin, Fachbereich Politik- und Sozialwissenschaften, Masterstudiengang Öffentliches und betriebliches Umweltmanagement. Vorgelegt: September 2010. Berlin.

Tanzer Consulting GmbH (2010): Trennungsprämie. CO_2-Einsparung durch Austausch alter Kühl- und Gefriergeräte gegen energiesparende A++ Geräte. Studie im Auftrag des UFH (Umweltforum Haushalt). Wien: Tanzer Consulting.

Tews K. (2011): Stromeffizienztarife für Verbraucher in Deutschland? Vom Sinn, der Machbarkeit und den Alternativen einer progressiven Tarifsteuerung. FFU-Report 05-2011, Berlin, und TRANSPOSE Working Paper 11. [http://www.polsoz.fu-berlin.de/polwiss/forschung/systeme/ffu/publikationen/2011/tews_stromeffizienztarife/index.html; 11.07.2012]

Thomas S. (2007): Aktivitäten der Energiewirtschaft zur Förderung der Energieeffizienz auf der Nachfrageseite in liberalisierten Strom- und Gasmärkten europäischer Staaten. Kriteriengestützter Vergleich der politischen Rahmenbedingungen. Frankfurt am Main: Lan.

Appendix

Profiles of the project groups

In the focal topic "From Knowledge to Action – New Paths towards Sustainable Consumption", funded by the German Federal Ministry of Education and Research (BMBF) as part of the Social-ecological Research Programme (SÖF), ten thematic project groups and an accompanying research project have been funded:

"BINK"	The contribution of educational institutions to the promotion of sustainable consumption among teenagers and young adults
"Change"	Changing sustainability-relevant routines in organisations
"Consumer/Prosumer"	From Consumer to Prosumer – How changing consumer roles in the internet economy can hold potentials for sustainable consumption
"ENEF-Haus"	Activating and empowering homeowners of single-family and two-family dwellings to carry out energy-efficient refurbishments
"Intelliekon"	Sustainable residential energy consumption with intelligent metering systems, communication systems and tariff schemes
"LifeEvents"	Life events as windows of opportunity for change towards sustainable consumption patterns
"User Integration"	Fostering sustainable consumption by involving users in sustainability innovations
"Seco@home"	Social, ecological and economic dimensions of sustainable energy consumption in residential buildings
"Transpose"	Transfer of electricity-saving policies
"Heat Energy"	Consuming energy sustainably – consuming sustainable energy. Heat energy in the field of tension between social predictors, economic conditions and ecological consciousness
"SÖF-Konsum-BF"	Accompanying research project "Focusing knowledge – encouraging commitment – facilitating mastery"

Website of the focal topic
www.sozial-oekologische-forschung.org/en/947.php

BINK

The contribution of educational institutions to the promotion of sustainable consumption among teenagers and young adults

Project group management/coordination
Prof. Dr. Gerd Michelsen, Leuphana University of Lüneburg, Institute for Environmental & Sustainability Communication (INFU)

Sub-projects (SP)
SP 1: Inter- and transdisciplinarity
SP 2: Consumer culture: culture of consumption
SP 3: Young people as consumers
SP 4: Interventions
SP 5: Media intervention
SP 6: Evaluation

Research partners
- *Leuphana University of Lüneburg, Institute for Environmental & Sustainability Communication (INFU)*: Prof. Dr. Gerd Michelsen; Dr. habil. Maik Adomßent; Dr. Matthias Barth; Daniel Fischer, M.A.; Claudia Nemnich, 1. Staatsex. Lehramt GHS; Sonja Richter, Dipl.-Päd.; Marco Rieckmann, Dipl. Umweltwiss.; Dr. Horst Rode
- *German Youth Institute (DJI), München*: Prof. a.V. Dr. habil. Claus J. Tully; Wolfgang Krug, Dipl.-Soz.
- *Humboldt-Universität zu Berlin, Hans-Saurer-Professur für Metropolen- und Innovationsforschung*: Prof. Dr. Harald A. Mieg; Jana Werg, Dipl.-Psych.; Judith Bauer, Dipl.-Psych.
- *Fresenius University of Applied Sciences, Idstein*: Prof. Dr. Andreas Homburg; Malte Nachreiner, Dipl.-Psych.
- *e-fect*, Berlin and Trier: Dr. Christian Hoffmann; Dr. Dirk Scheffler

Field partners
- Gymnasium Grootmoor (Grootmoor grammar school)
- IES Bad Oldesloe (formerly IGS, Bad Oldesloe integrated comprehensive school)
- BBS Friedenstraße, Wilhelmshaven (Friedenstraße vocational schools)
- BBS Osnabrück-Haste (Osnabrück-Haste vocational schools)
- Bremen University of Applied Sciences
- Leuphana University of Lüneburg

Aims

Developmental aims
- Changing individual behaviour patterns towards sustainable consumption in the areas of energy, transport and food
- Supporting institutions in their endeavours towards an organisational "culture of sustainable consumption" (in formal and informal learning settings)

Research aims
- A common understanding of organisational "culture of sustainable consumption" across disciplines (compatibility with theory and practice)
- Analysis of the principles underlying the design of inter- und transdisciplinary research and developmental processes; synthesis development
- In-depth investigation into the conditions surrounding individual consumer behaviour of teenagers and young adults

- Extending theories within environmental psychology explaining individual behaviour, whilst taking into account conditions and processes in educational establishments (universities and schools)
- Identification of success factors for institutional changes towards greater sustainability, particularly where participation and identity formation is concerned

Questions

- How can formal and informal settings in educational establishments contribute to consumer awareness and self-reliance with regard to sustainable behaviour (conscious behaviour as well as everyday routines)?

Key results

In the course of the project an analytical framework was developed for the description of consumer culture within educational institutions; both formal and informal learning settings were taken into account. This framework formed the basis for a process of development for schools and universities (initiated by BINK) and for the resulting intervention measures.

The results of the accompanying empirical study showed that teenagers and young adults who actively participated in the BINK programme, or who were at least aware of the programme, could be distinguished – some significantly so – from their peers in two respects: first, in terms of their perceived self-efficacy as consumers, second, in terms of their self-reported sustainable consumer behaviour.

A qualitative study of external educational establishments identified those conditions that favoured and those that hindered the establishment of a sustainable consumer culture.

Complementary qualitative studies examined how teenagers and young adults related to consumption and to sustainability, and the results fed into the process of change.

As a separate accompanying measure, as part of the media intervention, video clips were shot at all the institutions involved in the programme. The accompanying research project was able to provide some comments as to the effectiveness of different media formats.

As a result of the formative evaluation of the whole process, the original intervention strategy was further refined and modified; it then served as a basis for designing the training programme and the guidelines.

Key publications

Michelsen G., Nemnich C. (eds.) (2011): Bildungsinstitutionen und nachhaltiger Konsum. Ein Leitfaden zur Förderung nachhaltigen Konsums. Bad Homburg: VAS.

Tully C. J., Krug W. (2011): Konsum im Jugendalter. Umweltfaktoren, Nachhaltigkeit, Kommerzialisierung. Schwalbach/Ts: Wochenschau-Verl.

Fischer D. (2011): Educational Organizations as "Cultures of Consumption". Cultural Contexts of Consumer Learning in Schools. In: European Educational Research Journal 10 (4): 595–610.

Nemnich C., Fischer D. (eds.) (2011): Bildung für nachhaltigen Konsum: ein Praxisbuch. Bad Homburg: VAS.

Fischer D. (2011): Monitoring Educational Organizations' "Culture of Sustainable Consumption". Towards a Participatory Initiation and Evaluation of Cultural Change in Schools and Universities. In: Journal of Social Science 7 (1): 66–78.

Barth M. (forthcoming): Many roads lead to sustainability: A process-oriented analysis of change in higher education. In: International Journal of Sustainability in Higher Education.

Barth M., Fischer D., Michelsen G., Nemnich C., Rode H. (2012, accepted): Tackling the knowledge-action gap in sustainable consumption: insights from a participatory school programme. In: Journal of Education for Sustainable Development 6 (2).

Project website

www.konsumkultur.de/index.php?id=2&L=1

Change

Changing sustainability-relevant routines in organisations

Project group management/coordination

Prof. Dr. Ellen Matthies, Otto von Guericke University of Magdeburg, Institut für Psychologie I

Sub-projects (SP)

SP 1: Development, evaluation and standardisation of intervention programmes for universities, from the perspective of habit formation

SP 2: Analysis of the barriers in universities, promotion and dissemination of the targeted instrument (see below), development of an effective format for information and advice

Research partners

- *University of Bochum, Workgroup of Environmental Psychology and Cognition:* Prof. Dr. Ellen Matthies; Max Scharwächter, B.Sc.-Psych.; Ingo Kastner, M.Sc.-Psych. Dipl.-Bw. (BA); Nadine Hansmeier, Dipl.-Psych. B.A. Geogr.; Jennifer Zielinski, M.Sc.-Psych.
- *University of Bochum, Energy Systems and Energy Economics:* Prof. Dr.-Ing. Hermann-Josef Wagner; Andreas Klesse, Dipl.-Ing.
- *HIS Hochschul-Informations-System GmbH, Hannover:* Joachim Müller, Dipl.-Geogr.; Ralf-Dieter Person, Dipl.-Ing.; Ruth Cordes, M.A.
- *in-summa GbR, Braunschweig:* Dr. Dirk Thomas; Torben Aberspach

Field partners

- Universities: Rheinische Friedrich-Wilhelms-Universität Bonn; Universität Bremen; Technische Universität Dortmund; Philipps-Universität Marburg; Westfälische Wilhelms-Universität Münster; Universität Rostock; Universität Siegen; Hochschule Zittau/Görlitz
- EnergyAgency NRW: Contact person: Elke Hollweg

Aims

- Identification of potential for energy savings by changing individual user behaviour of office workers
- Development of a targeted instrument for the promotion of energy-efficient user behaviour within universities – an instrument that can also be used by other actors in the future and that takes into account the habitual nature of energy use in the workplace
- Development of measures for disseminating the instrument in universities
- Analysis of how the instrument could be used in other organisations

Questions

- What are the energy saving potentials in typical public sector workplaces, and in particular in universities?
- To what extent is it possible to activate this potential by means of psychological interventions?
- How efficient are interventions that are also targeted at habitual behaviour (interventions supporting actions vs. interventions targeted at increasing motivation only)?
- How effective are the interventions in terms of increasing sustainability?
- Evaluation of a web-based advisory service (the "Change campaign portal")
- Is it possible to use the Change intervention in other organisations?

Key results

The analysis showed that by changing user behaviour, universities could potentially save up to 18% of electricity and 9% of heat energy.

During the heating season 2008–2009 the first intervention was carried out in eight buildings. Seven further buildings were used as a control group. Altogether four universities were involved. Two intervention variants were tested. The standard intervention focused on information strategies aimed at enhancing motivation, whilst the other intervention used additional measures aimed at breaking everyday user habits (habit intervention). The evaluation comprised measurements of energy usage, observations and staff surveys. In the short term, habit interventions proved particularly successful, and it is here that electricity savings were mainly observed. In the long term, electricity savings stabilised; in terms of heat consumption, however, the intervention effects declined over time. The standardised and web-based intervention programme (web intervention) was tested in 23 buildings (five universities) over the course of a second intervention period (heating season 2009/2010). It was evaluated by means of a longitudinal analysis, whereby energy consumption was measured. Energy savings were realised in just under half the universities involved. It can be assumed that the Change initiative is capable of achieving notable success only if it is implemented in its entirety.

An organisational-sociological study, conducted in public nonuniversity institutions as well as in private organisations, found that, in principle, interest and opportunities do exist to implement user-focused interventions for energy saving. However, in most cases, specific modifications tailored to the particular organisations are required.

Key publications

Matthies E., Klesse A., Kastner I., Wagner H.-J. (2011): Darstellung des Projekts Change. In: Matthies E., Wagner H.-J. (eds.): Change – Veränderung nachhaltigkeitsrelevanter Routinen in Organisationen. Münster: LIT Verlag. 25–182.

Kastner I., Matthies E. (2011): Chancen einer webbasierten Beratung zur Veränderung von nachhaltigkeitsrelevanten Routinen in Organisationen In: Matthies E., Wagner H.-J. (eds.): Change – Veränderung nachhaltigkeitsrelevanter Routinen in Organisationen. Münster: LIT Verlag. 257–284.

Matthies E., Kastner I., Klesse A., Wagner H.-J. (2011): High Reduction Potentials for Energy User Behavior in Public Buildings – How Much can Psychology Based Interventions Achieve? In: Environmental Studies and Sciences 1 (3): 241–255.

Klesse A., Hansmeier N., Zielinski J., Wagner H.-J., Matthies E. (2010): Energiesparen ohne Investitionen – ein Feldtest in öffentlichen Liegenschaften. In: Energiewirtschaftliche Tagesfragen 60 (4): 8–12.

Hansmeier N., Klesse A., Matthies E., Müller J., Person R.-D., Wagner H.-J., Zielinski J. (2009): Energieeinsparung durch Nutzerverhalten – Veränderung nachhaltigkeitsrelevanter Routinen in Organisationen. In: HIS Magazin 1: 9–10.

Project website

www.change-energie.de/english_version?lang=de

Consumer/Prosumer

From Consumer to Prosumer – How changing consumer roles in the internet economy can hold potentials for sustainable consumption

Project group management/coordination

Dr. Siegfried Behrendt, Institute for Futures Studies and Technology Assessment gGmbH (IZT), Berlin

Prof. Dr. Birgit Blättel-Mink, Goethe University Frankfurt/Main, Department for Social Sciences

Sub-projects (SP)

SP 1: Reference models and sustainability effects, IZT – Institut für Zukunftsstudien und Technologiebewertung gGmbH

SP 2: Indicators of the extent of sustainability in consumption patterns in online second-hand trading, Goethe University in Frankfurt/Main

SP 3: Future markets and business sectors, Borderstep Institut für Innovation und Nachhaltigkeit gGmbH

Research partners

- *IZT – Institut für Zukunftsstudien und Technologiebewertung gGmbH, Berlin*: Dr. Siegfried Behrendt; Lorenz Erdmann, Dipl.-Ing.; Christine Henseling, M.A.
- *Borderstep Institut für Innovation und Nachhaltigkeit gGmbH, Hannover*: Prof. Dr. Klaus Fichter; Dr. Jens Clausen; Wiebke Winter, Dipl.-Soz.Wiss.
- *Goethe University Frankfurt/Main*: Prof. Dr. Birgit Blättel-Mink; Dirk Dalichau, Dipl.- Soz.

Field partners

- eBay Deutschland GmbH
- Federal Association for Information Technology, Telecommunications and New Media (BITKOM)

Aims

- Assessment of existing sustainability effects as well as potentials of electronic trading platforms for second-hand goods, exemplified by eBay
- Optimisation of existing online trading of second-hand goods, from an ecological perspective – in cooperation with eBay (e.g. neutralising CO_2 emissions from vehicles etc.)
- Identification and development of new forms of trading (e.g. solar electricity produced by private households) and the concept of auctioning second-hand goods, with the aim of enhancing sustainability
- Comprehensive empirical study (backed by a solid theoretical grounding) of the changing roles of consumers in the internet economy (from Consumer to Prosumer)
- Overview of approaches and online platforms, in the context of online trading, that are capable of encouraging consumers towards a prosuming culture

Questions

- Process dynamics: how does eBay online trading change the consumer behaviour of eBay users? Are consumer goods treated more respectfully, if the intention is to sell them at auction at a later stage? Are there indications for "rotating possession" as a new life style choice? Is there a tendency to buy more robust and better quality consumer goods, with a view to selling them at auction at a later stage?
- Demographic analysis: who uses eBay – as buyers/sellers: gender, age, profession, family situation, social status, location? How do the user types map onto the various social groups?

- Sustainability effects: what ecological sustainability effects have been noted to date as a result of eBay trading? Which of its features (e.g. product group, value retention, transport, packaging) most influence its environmental impact? How could the net environmental impact be improved?
- Potentials: what is the extent of untapped potentials for reducing the environmental impact? How can one differentiate between product groups in this context? What factors favour or hinder the realisation of potentials for reducing the environmental impact?
- Design options and innovations for the promotion of sustainable consumption: how can existing online trading be ecologically optimised? In the context of promoting strategies for sustainable consumption, what is the role of these new auction practices that are aimed at increasing the longevity and value retention of products? What should communication strategies look like in this respect?

Key results

In terms of sustainability criteria, the trade in second-hand goods is a largely positive phenomenon. It tends to have a positive impact on the environment in those instances where products are concerned that are used by the second-hand buyer for a significant period of time and where the products either do not use any energy or are modest in their energy consumption. Furthermore, online trading creates jobs. Thus, it seems a politically sensible strategy to promote second-hand trading. Giving used consumer items a new lease of life by selling them holds potential as a strategy for sustainable consumption. The online survey clearly showed that sellers appreciated the fact that the products on offer – even though they had been used – retained a high value in the eyes of the prospective buyers.

So far, only a small number of consumers are aware of how second-hand trading can impact positively on the environment and can create employment. Combining different motives for second-hand buying is particularly characteristic of ecologically-minded people.

Essentially, the messages to take home with regard to second-hand goods should focus on their long service life, and their (often high) quality. Additionally, these arguments should be stressed by actors such as second-hand dealers and online platforms, as a means of encouraging a combination of motives in the mindsets of second-hand buyers and sellers.

Key publications

Henseling C., Blättel-Mink B., Clausen J. (2009): Wiederverkaufskultur im Internet: Chancen für nachhaltigen Konsum. In: Aus Politik und Zeitgeschichte 32–33: 32–38.

Dalichau D., Hattenhauer M., Blättel-Mink B., Bender S.-F. (2010): Wer nutzt den Online-Gebrauchtwarenmarkt (nachhaltig)? Umweltorientierte, Prosumenten und andere User auf eBay. In: Wissenschaftsmagazin "Forschung Frankfurt". Frankfurt/Main.

Clausen J., Blättel-Mink B., Erdmann L., Henseling C. (2010): Contribution of Online Trading of Used Goods to Resource Efficiency: An Empirical Study of eBay Users. In: Sustainability 2: 1810–1830.

Behrendt S., Blättel-Mink B., Clausen J. (eds.) (2011): Wiederverkaufskultur im Internet: Chancen für einen nachhaltigen Konsum. Berlin, Heidelberg, New York: Springer Verlag.

Project website

www.izt.de/projekte/abgeschlossene-projekte/projekt/prosumer

ENEF-Haus

Activating and empowering homeowners of single-family and two-family dwellings to carry out energy efficient refurbishments

Project group management/coordination
Prof. Dr. Stefan Zundel, Lausitz University of Applied Sciences, Senftenberg/Cottbus

Components (C)
C 1: Review and development of the decision model
C 2: Analysis of potential savings
C 3: Analysis of public acceptance of energy advice services
C 4: Analysis of policy instruments
C 5: Optimisation of communication and advice services
C 6: Synthesis
C 7: Dissemination into practice
C 8: Dissemination into research community
C 9: Project coordination

Research partners
- *Lausitz University of Applied Sciences, Senftenberg/Cottbus*: Prof. Dr. Stefan Zundel; Tanja Albrecht, Dipl.-Kffr. (FH); Prof. Dr. Winfried Schütz; Sebastian Tempel
- *Institute for Ecological Economy Research (IÖW), Berlin/Heidelberg*: Dr. Julika Weiß; Elisa Dunkelberg, Dipl.-Ing.; Dr. Bernd Hirschl; Thomas Vogelpohl, Dipl.-Pol.
- *ISOE – Institute for Social-Ecological Research, Frankfurt/Main*: Dr. Immanuel Stieß; Dr. Jutta Deffner; Victoria van der Land, M.A.; Barbara Birzle-Harder, M.A.

Field partners
- Bremer Energie-Konsens GmbH
- dena – German Energy Agency
- GIH – Bundesverband Gebäudeenergieberater Ingenieure Handwerker
- Haus und Grund – Bundesverband
- Verbraucherzentrale Nordrhein-Westfalen
- ZAB – Brandenburg Economic Development Board

Aims
The principal practical aim of the project is to increase the share of energy-efficient refurbishments of single-family and two-family dwellings. Additionally, the following sub-targets have been set:
- Identification of key obstacles to energy-efficient refurbishment projects and identification of approaches to overcome them
- Categorisation of home owners related to refurbishment activities according to socio-demographic and lifestyle criteria
- Prioritisation of alternative courses of action and refurbishment measures in view of the different target groups
- Identification of deficiencies in existing policy instruments for the promotion of sustainability; recommendations for further innovative development

- Development of an integrated communication and consultancy strategy tailored to individual target groups

Questions
- What does an integrated decision and action model for energy-efficient refurbishment look like?
- What are drivers for and barriers to energy-efficient refurbishments of private homes?
- What target groups for energy-efficient refurbishment can usefully be distinguished?
- In the light of the refurbishment potentials, what would a refurbishment plan look like?
- Do we have to adjust, further develop or complement the instruments, and if yes, how?
- What would a target group oriented communication and advice strategy entail?
- What would an integrated policy approach entail?

Key results

The project group was able to develop a model for making energy-efficient refurbishment decisions. This was validated by a qualitative and a quantitative survey and formed the basis for a typology of target groups that included "convinced energy savers", "open-minded sceptics", "dedicated home improvers", "non-reflective maintainers" and "uninterested resisters".

Additionally – on the back of the project group's own survey, and as a result of further data – the potentials for implementing different measures were explored, taking into account the various target groups and types of buildings. One of the findings was the recognition that, in order to attain the energy saving targets set by the German federal government, energy-efficient refurbishment measures need to be implemented not only by a few enthusiasts, but on a much broader front. In parallel, an in-depth inquiry was conducted on the level of acceptance of current communication offerings. It was found that communication services were most helpful when they were staggered (in terms of time), as well as tailored to the needs of potential users, especially when offered at times when individuals were making refurbishment plans or when they were buying a property.

As a next step in the project, an investigation into suitable policy instruments was conducted and recommendations as to their design were made. It was found that intelligent use of appropriate instruments as well as correct timing is crucial. In a final step – and with the target group typology in mind – the communication and guidance tools were optimised and a dialogue-based marketing approach was thereby developed.

Key publications

Albrecht T., Deffner J., Dunkelberg E., Hirschl B., Van der Land V., Stieß I., Vogelpohl T., Weiß J., Zundel S. (2010): Zum Sanieren motivieren. Eigenheimbesitzer zielgerichtet für eine energetische Sanierung gewinnen. Frankfurt/Main.

Stieß I., Van der Land V., Birzle-Harder B., Deffner J. (2010): Handlungsmotive, -hemmnisse und Zielgruppen für eine energetische Gebäudesanierung. Ergebnisse einer standardisierten Befragung von Eigenheimsanierern. Frankfurt/Main.

Weiß, J., Vogelpohl, T., Dunkelberg, E. (2012): Improving policy instruments to better tap into homeowner refurbishment potential: Lessons learned from a case study in Germany. Energy Policy 44: 406–415.

Zundel S., Stieß I. (2011): Beyond Profitability of Energy-Saving Measures – Attitudes towards Energy Saving. In: Journal of Consumer Policy 34 (1): 91–105.

Project website
www.enef-haus.de/index.php?id=13

Intelliekon

Sustainable residential energy consumption with intelligent metering systems, communication systems and tariff schemes

Project group management/coordination

Sebastian Gölz, Dipl.-Psych., Fraunhofer Institute for Solar Energy Systems ISE, Freiburg i. Br.

Sub-projects (SP)

SP 1: Exploration of the relationship between knowledge and practice
SP 2: Socio-technological optimisation of the feedback system
SP 3: Identification of synergies with other efficiency instruments
SP 4: New forms of environmental impact assessment

Research partners

- *Fraunhofer Institute for Solar Energy Systems ISE, Freiburg i. Br.*: Sebastian Gölz, Dipl.-Psych.; Dominik Noeren, Dipl.-Ing. M.Sc.; Heike Schiller, Dipl.-Psych.
- *Fraunhofer Institute for Systems and Innovation research ISI, Karlsruhe*: Dr. Marian Klobasa; Prof. Dr. Joachim Schleich
- *Institute for Social-Ecological Research (ISOE), Frankfurt/Main*: Dr. Konrad Götz; Georg Sunderer, Dipl.-Soz.; Dr. Jutta Deffner
- *EVB Energy Solutions*: Björn de Wever, Dipl-Ing. FH; Andreas Häferer, Dipl.-Inform.

Field partners

- Energieversorgung Oelde
- Stadtwerke Hassfurt
- Stadtwerke Münster
- Stadtwerke Schwerte
- Stadtwerke Ulm
- SVO Energie (Celle)
- swb (Bremen)
- SWK SETEC (Krefeld)
- Technische Werke Kaiserslautern
- Linz Strom GmbH

Aims

- The aim of the project is to identify the potential for electricity savings by using feedback instruments. To that end, a large field trial was conducted in ten cities and regions, involving approximately 5,000 electricity consumers. Compared to previous studies in this area, this trial involves a large number of participants, tests a range of different feedback instruments, and runs for a long period (at least one year). The aim is to develop and test interactive systems that raise individuals' awareness of their use of electricity. By supplying users with detailed feedback on their electricity usage, it is hoped that they will, of their own volition, curb their energy use.

Questions

- What are people's attitudes towards feedback systems and energy saving?
- How do participants use the web portal on offer? (How often? How long?)
- What contents do they favour?
- Is a time-dependent tariff an adequate incentive to shift the time of energy consumption?

- Are people willing to pay for feedback instruments?
- Can energy be saved by using feedback meters? (If yes, how much?)
- What is the ecological effect of the energy savings (CO_2 emissions) that have been achieved? If these savings were extrapolated to the federal German energy system, how much energy would be saved?

Key results

The participants found the feedback data informative, useful, beneficial, helpful and user friendly. It was generally welcomed as a motivating method of engaging with energy saving. Only a minority of people voiced reservations in terms of data protection, and even fewer people perceived the necessity to engage with the information as tiresome and time-consuming.

The internet portal was used particularly intensively during the first month, after which interest and engagement with the system declined. Approximately one third of users only engaged with the system once. One quarter was mainly interested in learning how to save energy. Two thirds actively engaged with the portal, but only five percent used it intensively and on an ongoing basis. Despite having been used to variable degrees, the entire range of information available on the Intelliekon feedback system appears to be useful. Linking the information to practical actions whenever possible is important – this can, for instance, be done by giving advice on different ways of saving energy.

The differences in electricity consumption found between the feedback group and the control group at the beginning of the trial can be attributed in part to householders' possessing different appliances. After taking the effect of that circumstance into account, an overall decrease in energy consumption of 3.7 percent was observed in the feedback group. The energy savings observed are somewhat smaller than expected.

Key publications

Birzle-Harder B., Deffner J., Götz K. (2008): Lust am Sparen oder totale Kontrolle? Akzeptanz von Stromverbrauchs-Feedback. Ergebnisse einer explorativen Studie in vier Pilotgebieten im Rahmen des Projektes Intelliekon. Intelliekon-Berichte. Frankfurt/Main.

Schleich J., Klobasa M., Brunner M., Gölz S., Götz K., Sunderer G. (2011): Smart metering in Germany and Austria – results of providing feedback information in a field trial. Working Paper Sustainability and Innovation. Fraunhofer ISI.

Schleich J., Klobasa M., Brunner M., Gölz S., Götz K., Sunderer G. (2011): Smart metering in Germany – results of providing feedback information in a field trial. In: Proceedings of the eceee 2011 Summer Study. Belambra Presqu'île de Giens, France.

Project website

www.intelliekon.de/intelliekon-in-english

LifeEvents

Life events as windows of opportunity for change towards sustainable consumption patterns

Project group management/coordination

Prof. Dr. Dr. Martina Schäfer, Center for Technology and Society (ZTG), Technische Universität Berlin

Sub-projects (SP)

SP 1: Sociological sub-project
SP 2: Psychological sub-project

Research partners

- *Technische Universität Berlin, Center for Technology and Society:* Prof. Dr. Dr. Martina Schäfer; Dr. Melanie Jaeger-Erben; Dr. Adina Herde; Sophie Scholz, Dipl.-Psych.
- *FH Bielefeld University of Applied Sciences; Abteilung Sozialwesen, Fachgebiet Sozialpsychologie und Methoden:* Prof. Dr. Sebastian Bamberg; Angelika Just, Dipl.-Psych.

Field partners

- Senatsverwaltung für Gesundheit, Umwelt und Verbraucherschutz (patron)
- Omniphon Gesellschaft für Dialogmarketing und Marktforschung
- Allgemeiner Deutscher Fahrradclub (ADFC)
- BUND und BUNDjugend Berlin
- ElektrizitätsWerke Schönau (EWS), Greenpeace Energy, NaturStromHandel, LichtBlick
- Fördergemeinschaft Ökologischer Landbau Berlin-Brandenburg (FÖL), pro agro e.V.
- Verbraucherzentrale Bundesverband – Energieteam
- Verkehrsverbund Berlin-Brandenburg (VBB)
- Businesses involved in the processing and trading of organic as well as local products

Aims

- Investigation into the potentials of life events (moving into a large city, birth of a child) for changing behaviour patterns towards greater sustainability; assessment of the effectiveness of different interventions targeted at specific groups.

Questions

- How do everyday consumption patterns change as a result of life events? Can such events represent 'windows of opportunity' for changing consumption behaviour towards greater sustainability?
- Are individuals who have experienced major changes to their lives – compared to people in 'stable' life situations – more amenable to interventions aimed at sustainable consumption, and are they more likely to change their everyday routines?
- What changes are likely to result from an interactive, theoretically grounded dialogue-based marketing strategy compared to a 'classic' intervention strategy (sending out information material only)?

Key results

Qualitative sub-project: The birth of the first child is often accompanied by changes in the everyday consumption routines of parents. Here, shifts towards greater sustainability (e.g. home cooking with fresh products, increased use of organic products) as well as changes towards less sustainable habits can be observed (e.g. using the car instead of public transport, more intensive use of energy at home).
It has been noted that people who move to a large city tend to swap car usage for public transport.

The provision of conditions and appropriate structures (e.g. infrastructures) that enable individuals to act sustainably is of key importance in attempts to change people's consumption patterns. People who have had a baby and people who have moved to a new place are coping with the new situation by adopting new or extending and adapting existing social practices. In the case of parents, the newly adopted practices are closely related to the culturally transmitted role as mother or father. For people who have moved to a new place, the changes tend to relate to other concurrent events (new job, retiring, moving in with partner). In both cases, the existence of infrastructures, and the regulations around accessing and using them, can act either as facilitators or as barriers to becoming more sustainable. The changes in people's lives after major life events tend to be preceded by longer or shorter periods of cognitive, organisational and spatial-material preparation.

Quantitative sub-project: A theoretically grounded dialogue-based marketing campaign is significantly more likely to bring about changes towards greater sustainability than a campaign based on information only. The hypothesis that people are more likely to shift their consumption patterns towards sustainability after a major life event could not be confirmed regarding the tested type of intervention.

An integrative consideration of the experiences and results of the project suggest that the group of project participants who experienced major life changes were addressed by the intervention in a period where they had already settled into new routines, which they were unwilling to change again. It is possible that the preparation phase for a major life change is a good starting point for sustainability interventions, as it is at that point that the course is set for later changes.

Key publications

Schäfer M., Jaeger-Erben M., Bamberg S. (2012): Life events as windows of opportunity for changing towards sustainable consumption patterns? Results from an intervention study. In: Journal of Consumer Policy 35 (1): 65–84.

Bamberg S. (2012): Processes of change. In: Steg L., van den Berg A. E., de Groot J. I. M. (eds.): Environmental psychology. An introduction. New York: Wiley-Blackwell. 267–280.

Jaeger-Erben M., Schäfer M., Bamberg S. (2011): Forschung zu nachhaltigem Konsum. Herausforderungen und Chancen der Methoden- und Perspektiventriangulation. In: Umweltpsychologie 15 (1): 7–29.

Jaeger-Erben M. (2010): Zwischen Routine, Reflektion und Transformation. Die Veränderung von alltäglichem Konsum durch Lebensereignisse und die Rolle von Nachhaltigkeit – eine empirische Untersuchung unter Berücksichtigung praxistheoretischer Konzepte. Dissertation. Berlin: Technische Universität Berlin. [available at: http://opus.kobv.de/tuberlin/volltexte/2010/2816/pdf/jaegererben_melanie.pdf]

Project website

www.sozial-oekologische-forschung.org/en/1297.php

User Integration

Fostering sustainable consumption by involving users in sustainability innovations

Project group management/coordination

Prof. Dr. Frank-Martin Belz, Technische Universität München, TUM School of Management

Dr. Marlen Arnold, Technische Universität München, TUM School of Management

Sub-projects (SP)

SP 1: Involving users in sustainable innovation in the context of passive houses
SP 2: Involving users in sustainable innovation in the context of bioplastics
SP 3: Involving users in sustainable innovation in the context of mobility/transport
SP 4: Motivational research
SP 5: Gender and diversity research
SP 6: Scenarios for diffusion
SP 7: Institutional frameworks
SP 8: Indicator-based sustainability assessment

Research partners

- *Technische Universität München, Professur für Betriebswirtschaftslehre:* Prof. Dr. Frank-Martin Belz; Dr. Marlen Arnold; Sunita Ramakrishnan, Dipl. Kulturwirtin; Marc Requardt, Dipl.-Oec.; Dr. Sandra Silvertant
- *Technische Universität München, Lecturers in psychology:* Prof. Dr. Hugo Kehr; Dr. Stefan Engeser; Susanne Steiner, Dipl.-Psych.
- *Technische Universität München, Professur für Gender Studies in den Ingenieurwissenschaften:* Prof. Dr. Susanne Ihsen; Sabrina Gebauer, Dipl.-Päd.
- *Wissenschaftszentrum Straubing, Lehrstuhl für Rohstoff- und Energietechnologie*: Prof. Dr. Martin Faulstich; PD Dr. Gabriele Weber-Blaschke; Henriette Cornet, Dipl.-Ing.
- *Technische Universität Berlin, Professur Arbeitslehre/Haushalt:* Prof. Dr. Ulf Schrader; Benjamin Diehl, Dipl.-Psych.
- *Universität Oldenburg, Institut für Betriebswirtschaftslehre und Wirtschaftspädagogik:* Prof. Dr. Bernd Siebenhüner; David Sichert, Dipl. Soz.-Wiss.
- *Münchner Projektgruppe für Sozialforschung e.V. (MPS):* Prof. Dr. Cordula Kropp; Gerald Beck, Dipl.-Soz.; Dennis Odukoya

Field partners

- 81fünf high-tech & holzbau AG
- Gundlach
- Andechser Molkerei
- Bernbacher
- Natura Packaging
- Rhein-Main-Verkehrsverbund (RMV)
- Münchner Verkehrs- und Tarifverbund GmbH (MVV)

Aims
- The overarching research aim is the continued development of concepts and methods for business innovation processes, with particular attention given to user integration and aspects of sustainability. Not only potentials but also the limits of user integration into business innovation processes are carefully considered; the resulting intentional and unintentional actions are taken into account, and new guidance information is generated.
- The overarching practical aim is to generate recommendations that businesses can use to productively involve users in the different stages of the innovation process of sustainable products and services.

Questions
- What methods are particularly suited to involving users in the various stages of the innovation process for sustainable products and services?
- Which users should be selected for which stages of the innovation process?
- How can users be motivated to take part in the innovation process?

Key results
Innovation workshops play an essential role in involving users in sustainability innovations.

In collaboration with businesses, the project group ran twelve innovation workshops. Overall, 150 participants were involved. Six workshops were tailored to forward-looking users (lead users) and the other six were tailored to average users (non-lead users). A comparison between the two groups shows the following:
- Whilst it is more difficult to identify lead users, they are more motivated and reliable in terms of attending the innovation workshops.
- Throughout the 1.5 days of the workshop, the lead users show consistently higher motivation
- Generally, lead users are more creative in their ideas and more pertinent in their suggestions than non-lead users.

Overall, the field partners rate the workshops and the results as positive. However, given the overall effort involved, it seems unlikely that such workshops will be organised in the future. This relates to the fact that, whereas it is easy to estimate the *costs* of user involvement in innovation processes, it is as yet unclear how much *benefit* it can bring. However, one important finding – namely that users seem to be able to identify with the sustainable products – is certainly one argument in favour of involving users in the innovation processes around sustainable products and services.

Key publications
Belz F.-M., Schrader U., Arnold M. (eds.) (2011): Nachhaltigkeitsinnovation durch Nutzerintegration. Marburg: Metropolis.

Project website
www.food.wi.tum.de/index.php?id=69&L=1

Seco@home

Social, ecological and economic dimensions of sustainable energy consumption in residential buildings

Project group management/coordination
Dr. Klaus Rennings, Centre for European Economic Research (ZEW), Mannheim

Sub-projects (SP)
SP 1: Analysing preferences in the context of energy consumption in residential buildings
SP 2: Criteria for purchasing energy-efficient household appliances
SP 3: Assessment of ecological and economic potentials of sustainable energy consumption
SP 4: Measures and strategies for the promotion of a sustainable energy supply for private households

Research partners
- *Centre for European Economic Research (ZEW), Mannheim:* Dr. Klaus Rennings
- *University of St.Gallen:* Prof. Dr. Rolf Wüstenhagen
- *Fraunhofer Institute for Systems and Innovation Research ISI, Karlsruhe:* Prof. Dr. Joachim Schleich
- *German Institute for Economic Research (DIW), Berlin:* Dr. Thure Traber
- *Öko-Institut, Darmstadt, Freiburg i. Br.:* Dr. Bettina Brohmann

Field partners
- European Commission, DG JRC, Institute for Energy: Dr. Arnulf Jäger-Waldau
- Wuppertal Institute: Dr. Ralf Schuele
- Swiss Federal Office of Energy: Lukas Gutzwiler
- EnBW AG, Karlsruhe: Dr. Wolfram Münch
- Vattenfall Europe Berlin AG & Co. KG, Berlin: Roland Hellmer
- Verband kommunaler Unternehmen e.V., VKU: Dr. Barbara Praetorius
- Solar-Fabrik AG, Freiburg: Christoph Paradeis (former chairman)
- Accera Venture Partners AG, Mannheim: Matthias Helfrich
- Bundesverband Gebäudeenergieberater, Ingenieure, Handwerker (GIH), Stuttgart: Fred Weigl
- Arbeitsgemeinschaft für sparsame Energie- und Wasserverwendung im VKU: Vera Litzka
- Germanwatch, Bonn/Berlin: Dr. Ulrich Denkhaus
- ifeu Institut, Heidelberg: Jan Maurice Bödeker
- Bund der Energieverbraucher e.V.: Dr. Aribert Peters
- Verbraucherzentrale NRW e.V.: Udo Sieverding
- hessenENERGIE GmbH: Dr. Horst Meixner
- Berliner Energieagentur GmbH: Michael Geißler

Aims
- As a basis for effective and efficient decision-making it is crucial to gain a better understanding of the determinants and motives behind purchasing behaviour at the micro-level. It is only with knowledge at that level that targeted and effective strategies for sustainable development can be defined and that sustainable energy products can emerge from their eco-niche onto the mass market. In this project, a conjoint analysis was used to identify consumer preferences relating to energy-efficient products.

* A further aim of this project is to simulate the effectiveness of those measures that are supposed to increase the share of environmentally friendly and efficient forms of energy in private households. The collaboration of different academic disciplines and by actively involving field partners, the insights gained will be used to formulate concrete and effective sustainability strategies at a political and economic level.

Questions

* (1) What are the factors behind the decision-making in private households as to which energy services are used for their electricity and heating requirements?
* (2) What political and entrepreneurial measures are capable of achieving an increase in sustainable energy consumption in residential buildings?
* (3) What are the social, ecological and economic effects of these measures?
* (4) What role does gender play in such decisions?

Key results

(Question 1) The factors determining people's willingness to pay for energy-efficient products were identified and their willingness to pay for them was quantified. The following findings were of interest:

* the significance of the new EU energy efficiency label, which seems to negatively affect sustainability considerations in a given purchase, and
* a high degree of willingness to pay for sustainable heating technologies (approx. 180 euro/t CO_2)

(Question 2) Most proposals put forward were designed to adapt to the habits of energy consumers, e.g. changing standards (defaults) in order to accommodate people's inertia.

(Question 3) It is mainly the ecological effects of changes in heating patterns that are considered most important in the context of sustainable consumption.

(Question 4) Different "gender scripts" were identified in relation to energy use; the feminine "home making" vs. the masculine "facility management".

Key publications

Rennings K., Brohmann B., Nentwich J., Schleich J., Traber T., Wüstenhagen R. (eds.) (2012): Sustainable Energy Consumption in Residential Buildings. ZEW Economic Studies, Heidelberg: Physica.

Achtnicht M. (2011): Do environmental benefits matter? Evidence from a choice experiment among house owners in Germany. In: Ecological Economics 70 (11): 2191–2200.

Heinzle S., Wüstenhagen R. (2010): Disimproving the European Energy Label's value for consumers? – Results of a consumer survey. Working Paper No. 5 within the project SECO@home.

Bradford M., Schleich J. (2010): What's driving energy-efficient appliance label awareness and purchase propensity? In: Energy Policy 38 (2): 814–825.

Offenberger U., Nentwich J. (2009): Are sustainable energy technologies gendered? Home heating and the co-construction of gender, technology and sustainability. In: Kvinder, Køn & Forskning (Women, Gender & Research) 3–4: 83–89.

Project website

www.zew.de/seco

Transpose

Transfer of electricity saving policies

Project group management/coordination

Prof. Doris Fuchs, Ph.D., University of Münster (WWU), Institute of Political Science (IfPol)

Dr. Kerstin Tews, Freie Universität Berlin (FU Berlin), Environmental Policy Research Centre (FFU)

Modules (M) and work packages (WP)

M 1: Framework analysis: identifying electricity saving potentials (WP 1), developing a portfolio of policy instruments (WP 2), and locating price elasticity (WP3)

M 2: Deduction and identification of effective policy instruments: developing an integrated psycho-sociological action model (WP 4), and conducting a quantitative policy-analysis by means of country comparison (WP 5)

M 3: Micro foundation: conducting qualitative case studies for reconstructing the effects of policy instruments (WP 6)

M 4: Transfer conditions and policy import: bringing policy innovations into German politics (WP 7 & 8)

Research partners

- *University of Münster (WWU), Institute of Political Science (IfPol)*: Prof. Doris Fuchs, Ph.D.; Dr. Ulrich Hamenstädt; Dr. Hildegard Pamme; Christian Dehmel, M.A.
- *Freie Universität Berlin (FU Berlin), Environmental Policy Research Centre*: Dr. Kerstin Tews
- *Öko-Institut e.V., Darmstadt, Freiburg i. Br.*: Dr. Bettina Brohmann; Veit Bürger, Dipl.-Phys. Dipl.-Energiewirt
- *Universität Kassel, Institut für Psychologie*: Jun.-Prof. Dr. Dörthe Krömker; Dr. Frank Eierdanz; Christian Dehmel, M.A.
- *University of Constance, Dept. of Politics and Public Administration:* Prof. Dr. Volker Schneider; Nadja Schorowsky, M.A.
- *Inter-University Research Centre for Technology, Work and Culture (IFZ) Graz:* Assoc. Prof. DI Mag. Dr. MSc Harald Rohracher; Anna Schreuer, Mag. MSc; Wilma Mert, Mag. MSc

Field partners

- Verbraucherzentrale NRW
- Northern Alliance for Sustainability (ANPED)
- Wittenberg Center for Global Ethics

Aims

- Identification of key potentials for savings; identification of range and types of existing instruments; assessment of price elasticities
- Identification of effective instruments, both at a micro level (users) and at a macro level (country comparison)
- Contextual reconstruction of the effects of selected relevant policy instruments
- Identification of transfer conditions and agenda setting in the German context
- Initiation of political and social discussions regarding the potential implementation of the measures

Questions

- Why are the potentials for domestic electricity savings so rarely exploited?
- Are the barriers only to be found with the users? What framework conditions for consumer behaviour act as barriers for the efficient use of electricity in private households? What policy instruments are capable of promoting a more efficient use of electricity in private households?
- What experiences from abroad can successfully be transferred to Germany?
- Under what conditions can selected instruments be transferred to the German context?

Key results

Cf. Transpose Working Paper Series:

http://www.uni-muenster.de/Transpose/en/publikationen/index.html

- Approx. 60 percent of the electricity consumption in German private households could be saved by investment measures (mainly replacing household appliances etc.); the savings potential by changing usage behaviour amounts to an estimated 20 percent.
- Large discrepancies between instruments in use (instrument typology); identification of barriers to electricity saving, as experienced by users themselves, and in their environment.
- Relative inelasticity (especially short-term) of electricity consumption at times of price fluctuations. Considerable price sensitivity with respect to cooling appliances and TVs. Identification of relevant instrument bundles combining economic incentives and information.
- When selecting suitable instruments, the following should be taken into account: i. different self-perception of individuals, ii. socio-demographic factors, iii. people's practical everyday concerns and iv. moral (ethical) attitudes; additionally, instruments should target specific groups and combine several approaches.
- Identification of criteria for successful implementation and effectiveness of: 1.bonus schemes for energy-efficient cooling appliances and TVs (A, NL, DK); 2. progressive electricity tariffs (IT, California); 3. promotion of replacement schemes for electric heating systems (DK, CH); 4. electricity saving quotas (GB, FR, IT).
- Identification of those implementation and design conditions that allow for successful incentive programmes and progressive electricity tariffs in Germany (cf. Brohmann/Dehmel et al. in this volume).

Key publications

Bürger V. (2010): Quantifizierung und Systematisierung der technischen und verhaltensbedingten Stromeinsparpotenziale der deutschen Privathaushalte. In: Zeitschrift für Energiewirtschaft 34 (1): 47–59.

Fuchs D. (ed.) (2011): Die politische Förderung des Stromsparens in Privathaushalten. Herausforderungen und Möglichkeiten. Berlin: Logos.

Hamenstädt U. (2012): Die Logik des politikwissenschaftlichen Experiments. Methodenentwicklung und Praxisbeispiel. Wiesbaden: VS Verlag.

Mayer I., Schneider V., Wagemann C. (2011): Energieeffizienz in privaten Haushalten im internationalen Vergleich. Eine Policy-Wirkungsanalyse mit QCA. In: Politische Vierteljahresschrift 52 (3).

Tews, K. (2011): Progressive Stromtarife für Verbraucher in Deutschland? In: Energiewirtschaftliche Tagesfragen 10.

Project website

www.uni-muenster.de/Transpose/en/index.html

Heat Energy

Consuming energy sustainably – consuming sustainable energy. Heat energy in the field of tension between social predictors, economic conditions and ecological consciousness

Project group management/coordination

Prof. Dr. Dr. Ortwin Renn, University of Stuttgart, ZIRN – Interdisciplinary Research Unit on Risk Governance and Sustainable Technology Development

Sandra Wassermann, M.A., University of Stuttgart, ZIRN Interdisciplinary Research Unit on Risk Governance and Sustainable Technology Development

Diana Gallego Carrera, M.A., University of Stuttgart, ZIRN Interdisciplinary Research Unit on Risk Governance and Sustainable Technology Development

Work packages (WP)

WP 1: Structural analysis
WP 2: Mesoanalysis
WP 3: Microanalysis
WP 4: Demography and gender
WP 5: Integration and assessment
WP 6: Recommendations for action
WP 7: Project coordination

Research partners

- *University of Stuttgart, ZIRN:* Prof. Dr. Dr. Ortwin Renn; Diana Gallego Carrera, M.A.; Dr. Marlen Schulz; Sandra Wassermann, M.A.; Dr. Wolfgang Weimer-Jehle
- *University of Stuttgart,* Institute for Energy Economics and the Rational Use of Energy (IER): Dr. Ludger Eltrop; Dr. Till Jenssen; Daniel Zech, Dipl.-Geogr.
- *European Institute for Energy Research (EIFER), Karlsruhe:* Andreas Koch, Dipl.-Ing.; Pia Laborgne, M.A.; Kerstin Fink, M.A.
- *Energy Institute Bremen (BEI):* Dr. Jürgen Gabriel; Katy Jahnke, Dipl. Volkswirtin; Marius Buchmann, M.A.
- *Goethe University Frankfurt/Main, Law School:* Prof. Dr. Marlen Schmidt

Field partners

- Verbraucherzentrale Baden-Württemberg e.V.
- Verbraucherzentrale Sachsen e.V.
- Umwelt- und Transferzentrum – Handwerkskammer zu Leipzig
- Baden-Württembergischer Handwerkstag
- Siedlungswerk GmbH
- Volkswohnung GmbH
- Wohnungsbau-Genossenschaft Kontakt eG
- Pro Potsdam GmbH
- Energieberatungszentrum Stuttgart e.V.
- Schornsteinfegerinnung Stuttgart

Aims

- The aim of this project was an integrated, interdisciplinary analysis of heat energy consumption of private households at a micro, meso and macro level, as well as an analysis of the respective interactions.
- A further, more practical, goal was the drawing up of a set of recommendations that would actively promote a sustainable use of heat energy of private households.

Questions

- The main focus of the research questions was on discovering existing obstacles to potential incentives and opportunities for a more sustainable use of heat energy of private households.

Key results

The results from the analysis of energy consumption and the structural analysis of the separate disciplines as well as working groups, together with the description of intermediary actors in the environment of consumers, were considered as parts of a single system. By conducting a cross-impact balance analysis (CIB), it was possible to represent the complex interactions in this system of 'heat consumption' at the micro, the meso and the macro level. Thus, an overall picture of the estimated heat consumption for the year 2040 was drafted. A quantitative methodology was used to analyse the interaction between user and building. The effects of user behaviour in buildings were calculated to establish an order of the most dominant influencing parameters for the residential heating needs.

The results of the interdisciplinary work were published and disseminated as recommendations for the concerned actors in a brochure. Under the three headings "information and advice", "promotion and quality assurance" and "legal and political directives" a total of nineteen recommendations for action were devised. Some target political decision makers, others were regarded more relevant for intermediary actors, such as consumer advice centres or energy consultants.

Key publications

Fink K., Koch A., Laborgne P., Wassermann S. (2011): Behavioural Changes through Consumer Empowerment. Evidence from German case studies. In: ECEEE 2011 Summer Study: Energy efficiency first: The foundation of a low carbon society. Proceedings: 1861–1866.

Gallego Carrera D., Wassermann S., Weimer-Jehle W., Renn O. (eds.) (2012): Nachhaltige Nutzung von Wärmeenergie. Eine soziale, ökonomische und technische Herausforderung. Wiesbaden: Vieweg+Teubner.

Jahnke K. (2010): Analyse der Mesoebene: Praxisakteure im Blickfeld nachhaltigen Wärmekonsums. Stuttgarter Beiträge zur Risiko- und Nachhaltigkeitsforschung 17. Stuttgart: Institut für Sozialwissenschaften, Abteilung für Technik- und Umweltsoziologie der Universität Stuttgart.

Koch A., Jenssen T. (eds.) (2010): Effiziente und konsistente Strukturen – Rahmenbedingungen für die Nutzung von Wärmeenergie in Privathaushalten. Stuttgarter Beiträge zur Risiko- und Nachhaltigkeitsforschung 16. Stuttgart: Institut für Sozialwissenschaften, Abteilung für Technik- und Umweltsoziologie der Universität Stuttgart.

Schmidt M. (2008): Energieeffizienz im Mietrecht: Der neue Energieausweis. In: Zeitschrift für Umweltrecht 10: 463–468.

Zech D., Jenssen T., Wassermann S., Eltrop L. (2010): Von Äpfeln und Birnen – Wärmetechnologien auf dem Prüfstand der Nachhaltigkeit. Jahrestagung 2010 des Arbeitskreises Geographische Energieforschung (Deutsche Gesellschaft für Geographie) "Energie als interdisziplinäres Forschungsfeld" am 23. und 24. April 2010 an der Universität Koblenz-Landau.

Zech D., Jenssen T., Eltrop L. (eds.) (2011): Informieren, Fördern, Fordern. Handlungsempfehlungen zur Unterstützung eines nachhaltigen Wärmekonsums. Stuttgart: Selbstverlag.

Project website

www.uni-stuttgart.de/nachhaltigerkonsum

SÖF-Konsum-BF

Accompanying research project "Focusing knowledge – encouraging commitment – facilitating mastery"

Project management
Atty. Rico Defila, Dr. Antonietta Di Giulio, Prof. em. Dr. Ruth Kaufmann-Hayoz, University of Bern (Switzerland), Interdisciplinary Centre for General Ecology (IKAÖ)

Staff
Rhea Belfanti; Thomas Brückmann, M.A.; Peter Kobel, B.A.; Dr. Arthur Mohr; Andrea Gian Mordasini, lic. phil. hist.; Lukas Oechslin, B.A., Sonja Schenkel, lic. phil. hist.; Markus Winkelmann, M.A.

Tasks
Purpose of the accompanying research project:
- It should generate knowledge across the project groups. In particular, it should make available guidance and practical knowledge for managing the transition towards sustainable consumption patterns (*praxisfähiges Orientierungs- und Handlungswissen*). It should also generate new knowledge relating to inter- and transdisciplinary research processes.
- It should support the thematic project groups in their tasks, particularly with regard to synergies and the maximisation of practical relevance of their results and products.
- It should accompany and support the dissemination to practice of the results achieved by the focal topic.

Aims
- *Synthesis:* The results of the project groups are to be integrated into a suitable synthesis framework and communicated to the academic community. Against this background, the practical relevance of the results, together with proposed solutions, are to be identified and put at the disposal of corresponding groups of actors, in a suitable form.
- *Linking:* The research activities of the project groups are to be linked with each other as far as possible. Potentials for synergies between the project groups are to be identified and exploited – particularly in terms of their products and the diffusion of these into wider society; the results of this linking process will also flow into the synthesis (as well as into the diffusion and transformation activities).
- *Management support:* By mutual exchange of their experiences the project coordinators may gain insights into how to effectively manage their project groups and how to improve the inter- and transdisciplinary processes. In addition, they are to receive support by the team of the accompanying research, if they wish so. The project coordinators are to be given the opportunity to discuss ideas and prepare the ground so that the exploitation stage can proceed in an optimal and innovative fashion.
- *Quality assurance:* In collaboration with the team of the accompanying research project, the project groups are to develop quality assurance criteria and apply them in order to improve their projects. Proposals for the design of an external evaluation of the entire focal topic are to be developed, if needed, for the use of those in charge of the evaluation.
- *Diffusion/transformation:* Relevant societal fields of action and policy areas as well as actors are to be identified; clear and realistic ideas about the desired impacts and outcomes of the research results are to be formulated; those diffusion measures are to be implemented that are deemed most likely to achieve the desired outcomes.

- *Observation:* Also to be investigated are questions of whether and how the work undertaken and the results achieved by the project groups can contribute to an emergence of sustainable consumption patterns in the relevant fields of action and policy areas; conclusions for improving future transdisciplinary research projects are to be formulated.

Working method

The activities necessary for achieving the aims are planned in discussion with the project groups, the project management agency and the BMBF. Exchange of information, collaboration and decision making mostly take place or are prepared at workshops assembling all project groups. It is the task of the accompanying research project team, in consultation with the project management agency, to prepare these workshops with regard to content and methodology, as well as to moderate and follow up the events.

Products

The cooperation between the accompanying research project and the project groups is expected to yield several synthesis products aimed at various academic and non-academic audiences. Some have been determined from the outset:

- The present book (peer reviewed): Defila R., Di Giulio A., Kaufmann-Hayoz R. (eds.) (2012): The Nature of Sustainable Consumption and How to Achieve it. Results from the Focal Topic "From Knowledge to Action – New Paths towards Sustainable Consumption". München: oekom verlag GmbH (original German version published in 2011).
- International academic conference: Sustainable Consumption – Towards Action and Impact. 6.–8. November 2011 in Hamburg.
- Book written by a team of authors from the focal topic, addressed to decision-makers (2012/13, in German).
- Conference organised for field partners and decision-makers (2012).

As a result of the synthesis and linking activities, further products are to be elaborated, particularly as a result of a range of synthesis activities, in which some, but not all project groups may be involved.

Project website

www.ikaoe.unibe.ch/forschung/soefkonsum/index.en.php

Authors

Sophia Alcántara
M.A. Sociologist. Academic researcher at ZIRN (University of Stuttgart) and at Dialogik gGmbH. Research interests: evaluation; participation procedures; energy topics. Member of project group "Heat Energy".

Sebastian Bamberg
Prof. Dr. phil. Psychologist. Professor of social psychology and quantitative research methods at FH Bielefeld University of Applied Sciences. Research interests: theoretical models for explaining behaviour and changes in behaviour; theory-led development of interventions. Sub-project leader within "LifeEvents".

Matthias Barth
Dr. Senior research fellow at RMIT University, Melbourne, Australia. Fellowship on the postdoctoral programme of the German Academic Exchange Service (DAAD). Until August 2010 senior lecturer at the Institute of Environmental and Sustainability Communication (INFU), Leuphana University of Lüneburg. Research interests: education for sustainable development, competence development and assessment. Member of project group "BINK".

Gerald Beck
Dipl.-Soz. Sociologist. Academic researcher at the Institute of Sociology, University of Munich. Board member of the "Munich project group for Social Research (MPS)". Research interests: science studies; sociology of risk; social-ecological transformation processes and science communication. Member of sub-project group "Diffusion scenarios" within "User Integration".

Frank-Martin Belz
Prof. Dr. Economist. Professor of business administration, Technische Universität München (TUM School of Management). Research interests: sustainability innovations; sustainability marketing; open innovation und lead user. Leader of project group "User Integration".

Birgit Blättel-Mink
Prof. Dr. phil. Sociologist. Professor of sociology, specialising in industrial and organisational sociology, at the Goethe University, Frankfurt/Main. Research interests: innovation as a social process, social science perspectives on sustainability; women in higher education. Joint leader of project group "Consumer/Prosumer".

Bettina Brohmann
Dr. phil. Social scientist. Head of energy and climate protection research at the Öko-Institut e.V. (Darmstadt). Research interests: consumer and motivation research; studies using "need areas" as an analytical concept; participation processes; social aspects of energy and climate policy; design and assessment of sustainability programmes; evaluation. Leader of sub-project "Assessing the ecological and economic potentials of sustainable energy consumption" within "Seco@home"; also member of project group "Transpose".

Veit Bürger
Dipl. Phys. Qualified energy economist. Deputy head of Energy and Climate Divison at the Öko-Institut e.V. (Freiburg i. Br.). Research interests: development and assessment of policy instruments in the field of energy policy. Sub-project leader within "Transpose".

Jens Clausen
Dr. rer. pol. Engineer, doctorate in economics. Senior researcher at the Borderstep Institute for Innovation and Sustainability gGmbH. Research interests: innovation, entrepreneurship research; sustainable future markets and corporate social responsibility. Borderstep senior researcher "Consumer/Prosumer". Leader of sub-project "Future markets and business sectors" within "Consumer/Prosumer".

Dirk Dalichau
Dipl.-Soz. Sociologist. Doctoral candidate and project member, Institute for the Analysis of Society and Policy, Department of Social Sciences, Goethe University, Frankfurt/Main. Research interests: (sustainable) consumption; consumption and experience; prosumption and user integration; electromobility. Member of project group "Consumer/Prosumer".

Rico Defila
Fürspr. Attorney at law. Head of the planning and operations department at the Interdisciplinary Centre for General Ecology (IKAÖ), University of Bern; deputy leader of the research group Inter-/Transdisciplinarity within the IKAÖ. Research interests: theory and methodology of interdisciplinary and transdisciplinary research and teaching; general propaedeutics of science (human and natural); organisational development of interdisciplinary institutions. Joint leader of the accompanying research project "SÖF-Konsum-BF".

Christian Dehmel
Mag. Art. Sociologist. Doctoral candidate in qualitative policy analysis, University of Münster and Kassel. Research interests: efficacy of policy instruments for the promotion of energy saving in private households; socio-psychological factors in individual electricity consumption; efficiency and sufficiency in the context of sustainable consumption. Member of project group "Transpose".

Antonietta Di Giulio
Dr. phil. Philosopher. University lecturer in general ecology, specialising in general propaedeutics of science (human and natural) and interdisciplinarity. Leader of the research group Inter-/Transdisciplinarity at the Interdisciplinary Centre for General Ecology (IKAÖ) at the University of Bern. Research interests: theory and methodology of interdisciplinary and transdisciplinary research and teaching; general propaedeutics of science (human and natural); sustainability as a concept; education and sustainable development. Joint leader of the accompanying research project "SÖF-Konsum-BF".

Elisa Dunkelberg
Dipl.-Ing. Environmental engineer. Researcher at the Institute for Ecological Economy Research (IÖW), Berlin. Research interests: energy-efficient refurbishment of buildings; bio energy; ecological assessment; life cycle assessment. Member of project group "ENEF-Haus".

Lorenz Erdmann
Dipl.-Ing. Environmental engineer. Researcher at the Institute of Futures Studies and Technology Assessment (IZT), Berlin; from October 2012 at the Fraunhofer Institute for Systems and Innovation Research (ISI), Karlsruhe. Research topics: sustainable ICT; foresight methods; industrial ecology; sustainability innovations; interdisciplinary and transdisciplinary method integration. Member of project group "Consumer/Prosumer".

Daniel Fischer
M.A. Educational scientist. Researcher at the Institute for Environmental and Sustainability Communication (INFU) at the Leuphana University of Lüneburg. Research interests: sustainable consumer education; school development; education for sustainable development. Member of project group "BINK".

Doris Fuchs
Prof., PhD. Political scientist and economist. Professor of international relations and development policy, University of Münster. Research interests: sustainable consumption; environmental, energy and food policy; economics and politics; power and legitimacy. Joint leader of project group "Transpose".

Wolfgang Glatzer
Prof. Dr. Emeritus. Sociologist. Institute for Social and Political Analysis, Goethe University, Frankfurt/Main. Research interests: social structure; social indicators; welfare state; household production and household mechanisation; former president of "International Society for Quality of Life Studies", leader of working group "Social Infrastructure". Social science consultancy within "Intelliekon".

Sebastian Gölz
Dipl.-Psych. Psychologist. Researcher at the Fraunhofer Institute for Solar Energy Systems in Freiburg i. Br.; team leader "User behaviour and field tests". Research interests: development and implementation of innovative energy supply concepts from a social science perspective, and the analysis of user behaviour and users' responses in the context of smart grids; impact of user feedback and tariff signals on consumer behaviour. Leader of project group "Intelliekon".

Konrad Götz
Dr. phil. Sociologist. Coordinator for target group policies and strategic consultancy at the Institute for Social-Ecological Research (ISOE). Research interests: empirical lifestyle research; development of target group models in the context of sustainability; transport-mobility research. Sub-project leader within "Intelliekon".

Ulrich Hamenstädt
Dr. phil. Political scientist. Academic researcher attached to the Chair of International Relations and Development Policy, University of Münster. Research interests: experiments in political science and international political economy. Member of project group "Transpose".

Andreas Homburg
Prof. Dr. phil. Psychologist. Professor and dean of studies, business psychology, Hochschule Fresenius, University of Applied Sciences, Idstein. Research interests: description, analysis and promotion of environmentally friendly and sustainable behaviours in various contexts (local authorities, companies, households, schools); evaluation research. Leader of the sub-project "Evaluation" within "BINK".

Melanie Jaeger-Erben
Dr. phil. Psychologist and sociologist. Academic researcher at the Center for Technology and Society at the Technische Universität Berlin. Research interests: sustainable consumption; daily routines; theories of social practices; social-scientific aspects in the context of introducing and using energy innovations; qualitative social research methods. Member of project group "LifeEvents".

Ruth Kaufmann-Hayoz
Prof. em. Dr. phil. Psychologist. 1992 to January 2011: professor of general ecology and director of the Interdisciplinary Centre for General Ecology (IKAÖ) at the University of Bern. Research interests: conditions for learning about and moving towards sustainable development – for individuals and for society as a whole; interdisciplinarity and transdisciplinarity. Joint leader of the accompanying research project "SÖF-Konsum-BF".

Andreas Klesse
Dipl. Ing. Academic researcher at Ruhr-Universität Bochum, in the fields of energy systems and energy economics. Research interests: resource-friendly energy supply for buildings and housing estates. Member of project group "Change".

Andreas Koch
Dipl. Ing. Architect, M.Sc. City Design and Social Science, M.Sc. Energy Management. Academic project leader at the European Institute for Energy Research (EIFER) within the research group Energy Planning and Geosimulation. Research interests: systemic modelling of cities; energy balancing and sustainability assessment in urban areas. Leader of WP1 (structural analysis of factors influencing heat consumption) within "Heat Energy".

Dörthe Krömker
Prof. Dr. Psychologist. Junior (assistant) professor of social and innovation psychology at the Institute for Psychology, University of Kassel. Research interests: social determinants of environmental behaviour; sustainable consumption; nutrition, environment and health; acceptance; identity and action; images of nature; participative intervention research; sub-project leader within "Transpose".

Cordula Kropp
Prof. Dr. Sociologist. Professor of social innovation and future studies at the Hochschule München, board member of the Munich project group for Social Research (MPS). Research interests: social-ecological transformation processes; participatory opinion-making, decision-making and implementation processes; sustainable urban and regional development. Leader of sub-project "Diffusion scenarios" within "User Integration".

Ellen Matthies
Prof. Dr. phil. Psychologist. Professor of environmental psychology at the Otto von Guericke Universität Magdeburg. Research interests: explanation of and changes in environmentally friendly behaviour / sustainable consumption, particularly transport modes, energy use and energy-related decisions in households; socio-psychologically based interventions; effects and steering mechanisms of participatory processes. Leader of project group "Change".

Wilma Mert
Mag., M.Sc. Psychologist. Academic researcher in energy and climate, IFZ – Inter-University Research Centre Graz; lecturer at Klagenfurt University. Research interests: energy efficiency; user acceptance of new energy technologies; sustainable communication, sustainable lifestyles. Member of project group "Transpose".

Gerd Michelsen
Prof. Dr. Economist. Institute of Environmental and Sustainability Communication, Leuphana University of Lüneburg; UNESCO Chair in Higher Education for Sustainable Development; member of the German National Committee "Education for Sustainable Development". Research interests: sustainability communication and education for sustainable development. Leader of project group "BINK".

Bradford F. Mills
Prof., PhD. Professor of agricultural and applied economics at Virginia Polytechnic Institute & State University, Blacksburg, Virginia, USA. Research interests: empirical analyses of
a) the adoption behaviour of private householders with regard to innovative energy technologies, b) the distribution effects of policy measures and c) the effects of energy pricing on the energy supply security of low income households. Contribution to "Seco@home".

Joachim Müller
Dipl.-Geogr. Natural scientist. Lecturer in environmental communication at the University of Hildesheim; Organisational consultant (project leader), deputy leader of the Higher Education Infrastructure (within Higher Education Development) working group at the Higher

Education Information System (HIS), Hannover. Research interests: sustainable development in the management of higher education institutions; health and safety in higher education institutions. Member of the sub-project "Barriers and potentials in higher education institutions" within "Change".

Malte Nachreiner
Dipl.-Psych. Academic researcher in the Faculty of Economics and Media at the Hochschule Fresenius, University of Applied Sciences, Idstein. Research interests: sustainable consumer behaviour, in particular among teenagers and young adults; evaluation research. Member of the sub-project "Evaluation" within "BINK".

Julia Nentwich
Prof. Dr. Psychologist. Assistant professor of organisational psychology and gender studies, University of St. Gallen. Head of the teaching programme "Gender und Diversity". Research interests: gender in management and organisations; gender and technology; integration and participation; change agency and resistance; discursive psychology. Sub-project leader within "Seco@home".

Ursula Offenberger
M.A. Sociologist. Doctoral candidate at the University of Tübingen. Research interests: science and technology studies; research on couples and gender; qualitative methods in social research. Member of project group "Seco@home".

Ralf-Dieter Person
Dipl-Ing. Electrical engineer. Organisational consultant (project leader), member of the Higher Education Infrastructure (within Higher Education Development) working group at the Higher Education Information System (HIS), Hannover. Research interests: consultancy in facility management in universities, including various energy-related projects (energy controlling, energy efficiency, CO_2 accounting). Leader of sub-project "Barriers and potentials in higher education institutions" within "Change".

Horst Rode
Dr. phil. Educational researcher. Academic researcher at the Institute for Environmental & Sustainability Communication, Leuphana University of Lüneburg. Research interests: education for sustainable development; evaluation; assessment of school quality; member of the sub-project "Consumer culture" within "BINK".

Martina Schäfer
Prof. Dr. Dr. Biologist and sociologist. Deputy Director of Center for Technology and Society at the Technische Universität Berlin. Research interests: sustainable consumption; sustainable land use; sustainable regional development; methods in interdisciplinary and transdisciplinary research. Leader of project group "LifeEvents".

Joachim Schleich
Prof. Dr. rer. pol. Economist. Professor of energy economics at the Grenoble École de Management (GEM); concurrently senior researcher at the Fraunhofer Institute for Systems and Innovation Research (ISI) in Karlsruhe. Research interests: economic analysis of energy and climate policy instruments, focusing on effects of innovation; diffusion of energy-efficient technologies in households and companies. Sub-project leader within "Seco@home".

Volker Schneider
Prof. Dr. phil. Political scientist. Professor of Empirical Theory of the State, specialising in political and administrative science, University of Constance. Research interests: state and democratic theory, complexity theory, policy research, social network analysis. Sub-project leader within "Transpose" (in charge of macro-quantitative policy impact analysis).

Ulf Schrader
Prof. Dr. Economist and political scientist. Head of the Division for Economic Education and Sustainable Consumption at the Technische Universität Berlin. Research interests: sustainable consumption and consumer policy; corporate social responsibility and sustainable marketing. Sub-project leader within "User Integration".

Anna Schreuer
Mag., M.Sc. Science and technology research. Academic researcher in the field of energy and climate, IFZ – Inter-University Research Center Graz, lecturer at the Alpen-Adria University of Klagenfurt. Research interests: social science research on innovation and transition, innovation networks, energy regions, the involvement of civil society in the energy sector. Member of project group "Transpose".

Marlen Schulz
Dr. rer. pol. Sociologist. Academic researcher at ZIRN (University of Stuttgart) and Head of "Science and Society" at Dialogik gGmbH. Research Interests: empirical social research methods; participation, evaluation, climate protection and adaptation, energy issues. Leader of WP 3 (microanalysis: the consumers of heat energy) within "Heat Energy".

Immanuel Stieß

Dr. rer. pol. Social scientist and planner. Head of research for energy and climate change in everyday life at the Institute for Social-Ecological Research (ISOE), Frankfurt/Main. Research interests: social-ecological research on everyday routines; CO_2-frugal lifestyles; climate protection and social justice; target-group-related sustainability communication. Leader of the sub-project "Analysis of public acceptance and target group oriented optimisation" within "ENEF-Haus".

Georg Sunderer

Dipl.-Soz. Sociologist. Academic researcher in "mobility and urban spaces" at the Institute for Social-Ecological Research (ISOE); PhD student at the University of Zurich. Research interests: environmental behaviour, acceptance of social-ecological innovations, ethical consumption. Member of project group "Intelliekon".

Kerstin Tews

Dr. rer. pol. Sociologist. Senior researcher and project leader at the Environmental Policy Research Centre at the Freie Universität Berlin. Research interests: energy efficiency policy; policy (transfer) analysis, effectiveness of policy instruments. Joint leader of project group "Transpose".

Dirk Thomas

Dr. phil. Sociologist. Managing Director of the Markt-, Meinungs- und Sozialforschungs-institut in-summa, Braunschweig. Research interests: urban sociology, environmental sociology, sustainable development. Implementation of an organisational-sociological study as part of "Change".

Sandra Wassermann

M.A. Social scientist. Academic researcher and project leader at ZIRN (University of Stuttgart). Research interests: innovations in energy technologies, energy policy, acceptance and communication of new energy technologies as well as energy-saving behavioural strategies. Coordinator within "Heat Energy" and leader of WP 4 (demographics and gender).

Julika Weiß

Dr. rer. nat., Dipl.-Ing. Environmental scientist. Senior researcher in the field of sustainable energy and climate change research at the Institute for Ecological Economy Research (IÖW), Berlin. Research interests: renewable energies; ecological assessment; energy efficiency; energy-efficient refurbishment of buildings. Sub-project leader within "ENEF-Haus".

Ines Weller
Prof. Dr. rer. nat. Professor at the Center for Gender Studies and at the Research Center for Sustainability Studies at the University of Bremen; post-doctoral thesis on sustainability and gender with a focus on sustainable product design at the Technical University of Berlin (TU); currently deputy spokesperson for the Research Center for Sustainability. Research interests: sustainable consumption and production patterns; sustainability and gender; technology and gender; social-ecological research. Involvement in the sub-synthesis "gender" within the focal topic.

Markus Winkelmann
Mag. Artium. Sociologist. Research Associate at the Interdisciplinary Centre for General Ecology (IKAÖ) at the University of Bern. Research interests: issues around sustainable consumption; sustainability and acceptance of renewable energies; diffusion of decentralized energy technologies, dietary behaviour and health consciousness. Member of the accompanying research project "SÖF-Konsum-BF".

Angelika Zahrnt
Prof. Dr. Economist and systems analyst. Fellow at the Institute for Ecological Economy Research. Publications on ecological tax reform, sustainability and post-growth society. Chairwoman of BUND from 1998–2007; since 2007 honorary chairwoman of BUND (Friends of the Earth). Winner (2009) of the German Environmental Award of the German Federal Foundation for the Environment. Member of the German Council for Sustainable Development and the Strategy Advisory Board within the SÖF.

Daniel Zech
Dipl. Geogr. Academic researcher at the University of Stuttgart. Research interests: comprehensive analysis and integrated assessment of renewable heat and energy technologies, in the context of residential buildings. Member of project group "Heat Energy".

Stefan Zundel
Prof. Dr. rer. pol. Environmental economist. Professor of economics, energy economics and environmental economics, University of Applied Sciences Lausitz. Research interests: environment and innovation; sustainable consumption; regional development. Leader of project group "ENEF-Haus".

UNDISCIPLINED !
SCIENCE BEYOND DISCIPLINES

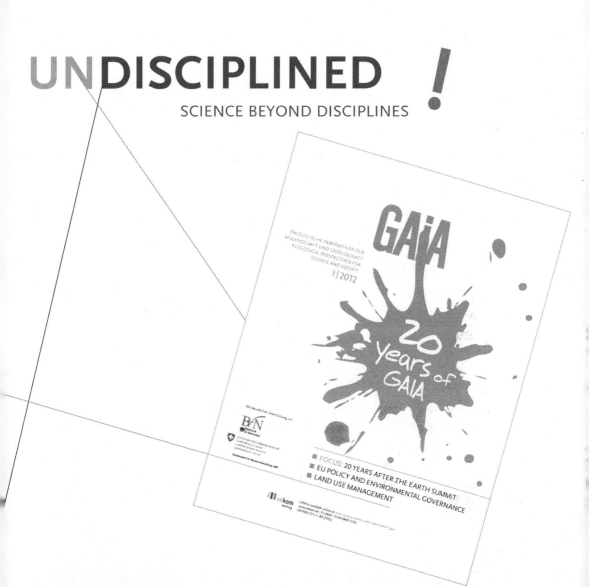

GAIA – ECOLOGICAL PERSPECTIVES FOR
SCIENCE AND SOCIETY

is a transdisciplinary journal for scientists and other interested
parties concerned with the causes and analyses of environmental
and sustainability problems and their solutions.

Get your TRIAL SUBSCRIPTION now!
More Information at www.gaia-online.net

Nachhaltigkeit von A-Z →

A wie Armutsbekämpfung

Die Erfolge von Muhammad Yunus' »Social Business« sind umstritten. Dieses Buch wirft einen Blick hinter die Kulissen, zeigt die Meinungen der Menschen vor Ort und analysiert Stärken wie Schwächen des Konzepts. Kerstin Maria Humberg gibt mit ihrer Dissertation die dringend ausstehende wissenschaftliche Antwort auf oberflächliche und skandalfixierte Medienberichte.

K. M. Humberg
Poverty Reduction through Social Business?
Lessons Learnt from Grameen Joint Ventures in Bangladesh
Hochschulschriften zur Nachhaltigkeit Band 53
348 Seiten, broschiert, in englischer Sprache, 34,95 Euro,
ISBN 978-3-86581-287-2

P wie Potentiale

Bergbauregionen überall in Mitteleuropa sind auf der Suche nach neuen Perspektiven. Sie setzen auf Tourismus, regenerative Energiegewinnung oder auf ihr reichhaltiges kulturelles Erbe. Trotz vielfältiger Herausforderungen ist die nachhaltige Entwicklung ehemaliger Bergbaugebiete ein ebenso lohnenswertes wie aussichtsreiches Unterfangen. Eine Vielzahl von Fallstudien bietet reichlich praktisches Anschauungsmaterial für regionale Akteure und Politik.

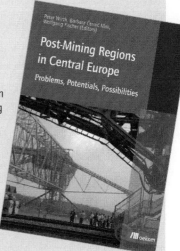

P. Wirth, B. Černič Mali, W. Fischer
Post-Mining Regions in Central Europe
Problems, Potentials, Possibilities
272 Seiten, broschiert, 29,95 Euro,
ISBN 978-3-86581-294-0

Erhältlich bei www.oekom.de, oekom@verlegerdienst.de

Die guten Seiten der Zukunft